# LIGHTNING, AURORAS, NOCTURNAL LIGHTS, AND RELATED LUMINOUS PHENOMENA

## *A CATALOG OF GEOPHYSICAL ANOMALIES*

Compiled by:

William R. Corliss

Published and Distributed by

The Sourcebook Project P.O. Box 107 Glen Arm, MD 21057

Library of Congress Catalog Number 82-099902

ISBN 0-915554-09-7



First printing: October 1982

Printed in the United States of America

# TABLE OF CONTENTS

# LIST OF PROJECT PUBLICATIONS

CATALOGS: Lightning, Auroras, Nocturnal Lights, and Related Luminous Phenomena

HANDBOOKS: The Unfathomed Mind: A Handbook of Unusual Mental Phenomena
Incredible Life: A Handbook of Biological Mysteries
Unknown Earth: A Handbook of Geological Enigmas
Mysterious Universe: A Handbook of Astronomical Anomalies
Ancient Man: A Handbook of Puzzling Artifacts
Handbook of Unusual Natural Phenomena

SOURCEBOOKS: Strange Phenomena, vols. G1 and G2
Strange Artifacts, vols. M1 and M2
Strange Universe, vols. A1 and A2
Strange Planet, vols. E1 and E2
Strange Life, vol. B1
Strange Minds, vol. P1

NEWSLETTERS: Science Frontiers (current anomaly reports)
Anomaly Register (Catalog addenda)

For information on the availability, prices, and ordering procedures, write:

SOURCEBOOK PROJECT
P. O. Box 107
Glen Arm, MD 21057

# PREFACE

After more than ten years of scouring the scientific and semiscientific literature for anomalies, my major conclusion is that this is an amazingly fruitful activity. In fact, organized science should have been doing the same searching for the past 200 years. It is simply astounding that a Catalog of Anomalies does not already exist to guide scientific thinking and research. It is at least as important to realize what is not known as it is to realize what is well-explained. Nevertheless, here begins the first publication of a Catalog of Anomalies; the product largely of one person's library research, carried forward without a single public or foundation dollar.

Under the aegis of the Sourcebook Project, I have already published 16 volumes, comprising well over 7,000 pages, of source material on scientific anomalies. (See page iv for a list of titles.) As of this moment these 16 volumes represent only about 20% of my collection of scientific anomalies. New material is being added at the rate of about 1,200 new articles and items per year, 300 of which are from the current literature. These rates could be easily multiplied several times over by spending more time in the libraries. After ten years only the scientific journals of the United States and England have received my serious attention. There remain the English-language journals of the rest of the world, those publications in other languages, university theses, government reports, bulletins of scientific research facilities, conference papers, and not the least, books and newspapers. The cataloging task is just beginning, for the anomalies in the world's scientific and semiscientific literature seem nearly infinite in number.

Given this rough assessment of the magnitude of the anomaly literature, one can understand why the planned Catalog of Anomalies will require at least 25 volumes of about the same size as the one you now hold. I visualize a shelf of 25 volumes, with master indexes, to be only the initial step in providing scientists with ready access to what is not, in my opinion, well-explained.

Will the Catalog of Anomalies revolutionize science? Probably not---at least not immediately. Quite often the initial reaction to the books of anomalies already published has been disbelief. The data must be in error; the data are mainly testimonial; the data are too old; the supposed anomaly was explained long ago. Germs of truth reside in all these complaints. But for every anomaly or example that can be legitimately demolished, ten more take its place. Nature is very anomalous or, equivalently, Nature is not yet well-understood by science. Much remains to be done.

William R. Corliss

P. O. Box 107
Glen Arm, MD 21057
August 1, 1982

# HOW THE CATALOG IS ORGANIZED

### Purpose of the Catalog

The Catalog of Anomalies is designed to collect and categorize all phenomena that cannot be explained readily by prevailing scientific theories. Following its definition, each recognized anomaly is rated in terms of: (1) its substantiating data; and (2) the challenge the anomaly poses to science. Next, all examples of the anomaly discovered so far are noted, some in detail. Finally, all examined references are listed. Thus, the Catalog is a descriptive guide as well as a compendium of examples and references. Scientific researchers therefore have a ready-made foundation for beginning further investigations into these intriguing phenomena. This is the basic purpose of the Catalog: the collection and consolidation of the unknown and poorly explained to facilitate future research and explanation.

### General Plan of the Catalog

It was tempting to organize this Catalog alphabetically, making it an "encyclopedia of anomalies." But many of the phenomena have obscure names or, even worse, no names at all. Under these circumstances, access to the data base would be difficult. Therefore, a system of classification was designed based upon readily recognized classes of phenomena and the means by which the observer detects them. Subject matter is first divided into nine general classes of scientific endeavor, as illustrated in the diagram. Few persons would have difficulty classifying a phenomenon as biological, astronomical, etc. The second, third, and fourth levels of classification are also based upon generally recognized attributes, such as luminosity, sound, etc. The similarity of this method of categorization to those employed in natural history field guides is purely intentional. Like bird identification, phenomenon classification soon becomes second nature. In fact, almost all of the phenomena described in the Catalog are accessible to everyone with five normal senses and perhaps binoculars and telescope.

Most catalogs boast numbering systems and this one is no exception. Rather than employ a purely numerical system, the first three levels of classification are designated with letters. The triplets of letters chosen have some mnemonic value. Thus, a GLA anomaly is easily recognized as being in the geophysics class (G), luminous (L) in nature, and resembling auroras (A). The number added to the triplet of letters marks the fourth level of classification, so that GLA4 would signify the category of low-level auroras. Every anomaly type has a unique alphanumeric code, like GLA4. All cross references and indexes are based upon this system. Catalog additions and revisions are made much easier with this scheme.

The Catalog codes may seem cumbersome at first but their mnemonic value to the compiler has been significant. Hopefully, they will help other users, too. The codes are simple, yet flexible enough to encompass the several thousand types of anomalies identified so far.

A glance through this volume of the Catalog will reveal that each example of a specific anomaly type bears an X-number, and each reference an R-number. GLA4-X12 therefore specifies the twelfth example of low-level auroras, and GLA4-R17 the seventeenth reference to low-level auroras. Indexes and cross references can consequently be made more precise than conventional page references.

### How Data and Anomalies Are Evaluated

Each anomaly type is rated twice on four-level scales for data "validity" and "anomalousness," as defined below. These evaluations represent only the opinion of the compiler and are really only rough guides.

DATA EVALUATION SCALE

1 Many high-quality observations. Almost certainly a real phenomenon.

2 Several good observations or one or two high quality observations. Probably real.

3 Only a few observations, some of doubtful quality. Pheonomena validity questionable.

4 Unacceptable, poor-quality data. Such phenomena are included only for purposes of comparison or amplification. Only one such item is registered in this volume.

| First-order classification | Second-order classification | Third-order classification | Fourth-order classification |
|---|---|---|---|
| A Astronomy | E Atmospheric optics | (A) Auroras | 1 Pillars |
| B Biology | F Falls | B Ball lightning | 2 Arches |
| C Chemistry & physics | H Hydrospheric phenomena | D Discharge phenomena | 3 Meteors |
| E Earth sciences | (L) Luminous phenomena | L Lightning | (4) Low-level |
| (G) Geophysics | M Magnetic phenomena | M Meteor-like phenomena | 5 Odors |
| L Logic & math | Q Earthquakes | N Nocturnal lights | ⋮ |
| M Archeology | S Sound | W Marine phosphorescence | 26 False dawns |
| P Psychology | V Volcanic phenomena | | |
| X Unclassified | W Weather | | |

CATALOG CODING SCHEME

## ANOMALY EVALUATION SCALE

1 Anomaly cannot be explained by modifications of present laws. Revolutionary.

2 Can probably be explained through relatively minor modifications of present laws.

3 Can probably be explained using current theories. Primarily of curiosity value.

4 Well-explained. Included only for purposes of comparison and amplification. Only one such item is registered in this volume.

Anomalies that rate "1" on both scales are very rare, with only a single entry registered in this volume of the Catalog. Such anomalies are, however, the most important because of their potential for forcing scientific revolutions. The Catalog code for this potentially revolutionary anomaly type is GLN2. As additional Catalog volumes are published, the relative proportion of double-1s will increase, especially in the fields of biology and psychology. The following matrix summarizes the ratings assigned in this volume.

| | | Anomaly Ratings | | | |
|---|---|---|---|---|---|
| | | 1 | 2 | 3 | 4 |
| Data Ratings | 1– | 1 | 16 | 12 | 1 |
| | 2– | 0 | 24 | 10 | 0 |
| | 3– | 0 | 26 | 12 | 0 |
| | 4– | 0 | 1 | 0 | 0 |

Total: 103 anomaly types

### Anomaly Examples

Examples of anomaly types are designated by the letter X in the body of the Catalog. All examples discovered so far are listed. If the example is of the event type, time and place are specified if available. Some of the more significant geophysical events, such as the remarkable auroral meteor of 1882, may appear in several references, all of which are provided in full and keyed to the X-number of the example. Ball lightning is represented by hundreds of examples (thousands of examples exist but have not yet been collected); but only a single example of multicolored marine phosphorescent rays is listed. When many examples of an anomaly type are available, several of the most interesting and instructive are reported in some detail. Direct quotations from the eye-witness accounts are frequently employed to convey accurately the characteristics of the phenomenon to the reader. These selected examples are printed full-width in the Catalog columns. Examples of less import are indented to separate them visually.

### The References and Sources

Each anomaly type and the examples of it are buttressed by all references that have been collected and examined. Since some references describe several examples, each reference includes the X-numbers of the examples mentioned. When a reference covers more than one type of anomaly, it is repeated in the bibliography following each anomaly type. Actually, there is little repetition of this sort in the Catalog.

Perusal of the Source Index will demonstrate that the great majority of references come from the scientific literature. Heavily represented in this volume of the Catalog are: Nature, Science, Monthly Weather Review, Meteorological Magazine, and the Quarterly Journal of the Royal Meteorological Society. Several less technical publications are also frequently mentioned: Marine Observer, Scientific American, and English Mechanic. The Marine Observer is a publication of the U. K. Meteorological Office. The English Mechanic was for many years an important English technical magazine, much like Scientific American. All of the serials mentioned so far are considered generally reliable, although one must always be wary. In addition to these often-referenced journals, there is a very wide selection of other publications. Some of these are more popular in character. In this context, it should be remembered that unusual phenomena do not seek out scientists, and the laymen who observe many anomalies have no knowledge of or access to scientific journals.

The time span covered by the sources ranges over 150 years. Many excellent reports of anomalies come from the latter half of the Nineteenth Century. Although many scientists frown on such old reports, the quality is often good, and they should not be discarded arbitrarily. Not only were the observers of a century ago competent, they were unbiased by knowledge of modern theories and lived in an environment unpolluted by modern vehicles, effluxes, and other contaminants. It should also be mentioned here that the modern meteorological literature tends heavily toward mathematical modelling and theory as opposed to eye-witness accounts of strange phenomena. Two journals that are exceptions to this rule are the Marine Observer and the U. K. Journal of Meteorology.

### The Indexes

Each Catalog volume concludes with five separate indexes. At first glance this may seem to be too much of a good thing. But in the context of a science-wide catalog of anomalous events and unusual natural phenomena, each idex has its special utility.

The subject index is essential in any work of this type. It is placed last for easy access. The time-of-event and place-of-event indexes are analytical tools for the anomaly researcher. They help identify phenomena that are reported separately (perhaps in widely different journals) but are really different aspects of the same event. This integrating feature will become more apparent as additional Catalog volumes appear. To illustrate, the 1908 Tunguska Event produced luminous and barometric phenomena as well as atmospheric optical effects. It is possible that composite time-of-event and place-of-event indexes covering all scientific fields will reveal cause-and-effect relationships that have not been recognized before. It is the intent of the Catalog effort to generate a composite set of indexes when all four Catalog volumes in geophysics are published as well as a final master index covering all of science.

The source index shows immediately the dependence of this Catalog upon scientific literature rather than newspapers and other popular publications. Its real purpose, though, is the rapid checking to determine if a specific reference has or has not been caught already in the fishing nets of this Catalog project. The source index is doubly valuable because many footnotes and bibliographies in the scientific literature display sources only; that is, titles and authors are omitted entirely. The researcher also comes across many vague references to such-and-such an article by so-and-so back in 1950 in Nature. The exhaustive and rather ponderous source and first-author indexes can help pin down many references lacking specifics.

All five indexes use the catalog codes described above rather than page numbers. The codes are permanent whereas the page numbers will change as addenda and revised volumes are produced. The mnenonic value of the catalog codes is useful here, too, because the approximate nature of each index entry is readily apparent, while page numbers give only location.

### Supporting Publications of the Sourcebook Project

The Catalog volumes are actually only distillations of huge quantities of source material. The Anomaly Data Research Center/Sourcebook Project has already published 16 volumes of source material, as detailed on page iv.

### Catalog Addenda and Revisions

Over 1,000 new reports of anomalies are collected from current and older scientific journals each year. New anomaly types and additional examples of types already cataloged are accumulating rapidly. When new material assignable to a Catalog volume already in print is acquired, it appears first in the serial ANOMALY REGISTER, published frequently by the Anomaly Data Research Center. (Ordering information on page iv.) When sufficient new material has been assembled in a Catalog area, that volume of the Catalog will be revised and published with all additions and corrections incorporated.

### Request for Additions and Corrections

The Anomaly Data Research Center welcomes reports of new anomalies and examples of unusual phenomena. If you discover a report of an anomaly not listed in the Catalog, send a xerox of the article to the Anomaly Data Research Center for evaluation. If the report is added to our data base, credit will be given to the submitter in the ANOMALY REGISTER and the revised edition of the Catalog. Reports from recognized scientific journals are preferred but everything is grist for the mill! The address of the Anomaly Data Research Center is: P.O. Box 107, Glen Arm, MD 21057.

# INTRODUCTION TO LUMINOUS PHENOMENA

The primary sensory channel for detecting unusual natural phenomena will always be the human eye. Nothing catches the attention faster than a mysterious light. Who hasn't jumped at a nearby flash of lightning or pondered a far-off light moving through the night sky? Was it just an airplane or some phenomenon unknown to science? Most strange lights of course turn out to have prosaic scientific explanations; but this catalog volume will demonstrate that some luminous phenomena still tax the ability of today's science to account for them.

The lights described here are self-luminous; like a light bulb they shine through the release of internal energy rather than the reflection of light from the sun, moon, or other external source. This category includes lightning, the auroras, meteors, marine bioluminescence, and other luminous displays that manufacture their own light. Halos, rainbows, mirages, and phenomena depending upon reflected light are covered in another volume of this catalog.

The earth's atmosphere absorbs sunlight and streams of subatomic particles spewed out by the sun. This influx of solar energy fuels many of our terrestrial luminous phenomena. Various chemical and electrical processes convert solar energy into the auroras, lightning, St. Elmo's Fire, the Andes Glow, and that host of curious and beautiful luminous phenomena that make up this volume. A meteor transforms its energy of motion into light as it plunges into the atmosphere. Closer to the earth's surface, the incredible geometrical displays of marine phosphorescence draw their energy from minute organisms in the seawater. Finally, the earth itself is a reservoir of energy. The ponderous motion of an earthquake somehow ignites the sky with electric flashes, glows, and moving balls of light. Gases escaping from the earth's crust may be converted into those soft lambent flames called will-o'-the-wisps. Nature is profligate with energy and has discovered many ways to turn it into visible light.

Genuine mysteries surround many of the luminous phenomena described in this volume. For example:

> The precise energy sources of some of the phenomena are unknown.
>
> Ball lightning, normal lightning, and the so-called ghost lights sometimes display prankish, uncanny behavior that seems to transcend objective science.
>
> Some auroras seem to descend close to the earth's surface where conditions supposedly preclude their formation.
>
> Some auroras, most marine phosphorescent displays, and a few electric discharge phenomena manifest themselves as curious geometrical patterns (rings, wheels, parallel bands, and other unlikely patterns) that are most difficult to explain.
>
> Ghost lights, ball lightning, marine light displays, and several other classes of phenomena involve the murky problems of human perception and the role of the mind. Indeed, we already know that some lights, such as those experienced by the astronauts when bombarded by cosmic rays, do not truly exist.

This rich lode of luminous phenomena provides a fertile ground for scientific exploration. Unfortunately, the controversial aspects of many of the phenomena and the highly subjective nature of many of the reports make this a risky field for scientists to tackle.

# GLA AURORA-LIKE PHENOMENA

## *Key to Phenomena*

# GLA0 Introduction

Auroras are luminous displays of the northern skies in the Northern Hemisphere and the southern skies in the opposite hemisphere. The repertoire of "normal" auroral displays is remarkable, ranging from an unimpressive, dull-greenish glow just tinting the northern horizon to kaleidoscopes of flickering red, yellow, and bluish flames and arches that fill half the heavens. The convoluted, eerie draperies, the silently flashing displays, and the cavorting flames we call the Merry Dancers grip the modern beholder and doubtless inspired awe and legend in primitive man.

Despite all our modern research with spacecraft and instrumented rockets, the auroras have not yielded up all their mysteries---even the "normal" auroras. Auroras are loosely associated with solar activity and geomagnetic storms, so we surmise that gusts of electrically charged particles emitted by the sun (the solar wind) help in some way to set fire to the polar heavens. Beyond these generalities, there is much we do not understand. In particular, the anomalous or abnormal auroras described below reveal even deeper layers of ignorance about one of Nature's most spectacular manifestations.

To begin, aurora-like phenomena of uncommon form, such as those that brush the starlit sky with misty parallel bands, pose serious problems of explanation. Why the geometrically arranged bands? Are there regularly spaced conduits in the upper atmosphere for solar electricity to flow or is there a regular sea of gravitational waves in the upper atmosphere? Next, how do some rare auroras seem to approach the earth's surface, even to the point of engulfing observers in luminous mists? Conventional wisdom has it that all auroras weave their luminous magic above 80 kilometers altitude. The abundant records of auroral "fogs", the crackling and hissing sounds, and the occasional acrid odor of ozone demonstrate that some auroras may actually record the slow discharge of terrestrial electricity to the upper atmosphere. It is almost as if a temporary hole had been punched in the planet's magnetic bottle, short-circuiting the earth to the solar-system dynamo, the sun. There are also puzzling observations linking auroras to daytime cloud formations, to thunderstorms, to earthquakes, to mountain-top glows. That the earth is electrically active cannot be denied; it is part of an enveloping cosmos seething with energy.

Auroras, though frequently strange and mystifying, are accepted denizens of the night skies, but there exist even stranger luminous displays that may or may not be related to auroras. What could cause an electric-blue, all-sky flash or expanding bubbles of ghostly light many times the size of the moon? What brightens the entire sky to the point where print can be read outdoors at midnight? For want of a better category, such phenomena are included with anomalous auroras.

# GLA1 Auroral Pillars: Natural Searchlight Beams

Description. Isolated, often breath-taking, white-to-greenish-white shafts of light 1-10° wide extending from the horizon to near and occasionally past zenith. Auroral pillars usually appear in the eastern or western horizon, well away from the zone where auroral activity normally occurs. The pillars or beams may remain visible from a few minutes to over an hour, drifting slowly north or south. Wavelets of light may seem to rush up and down the filmy but well-defined beams. Both straight-sided and fan-shaped pillars are known.

Background. Modern observers of the night sky often see beams of light rising from the horizon. Investigation almost always proves them to be searchlights at shopping centers or some other artificial illumination. In this environment, it is not surprising that there are few modern observations of auroral pillars and that most of the scientific discussions regarding their origin are confined to the older literature.

Data Evaluation. Many reports of auroral pillars reside in the older journals. A goodly fraction of these observations were made by scientists who provided direction, angular dimensions, lifetimes, and other pillar characteristics. Rating: 1.

Anomaly Evaluation. The long-lived, isolated auroral pillar with a bearing well away from magnetic north may have a different origin than the ever-changing beams and flickerings associated with the usual auroral display. While auroral pillars are doubtless caused by the channelling of charged particles, the exact mechanism is still controversial, as is the reason for the appearances of pillars so far removed from the scene of most auroral activity. Despite these unresolved problems, auroral pillars represent only minor scientific anomalies. Rating: 3.

Similar and Related Phenomena. The auroral arch (GLA2) is probably closely related to the auroral pillar, the latter being merely an abbreviated arch. Fan-shaped auroral pillars are sometimes confused with comets, but comets almost never appear and disappear in a matter of minutes. The nebulous pyramid of the zodiacal light (AZO) rises from the horizon much like a pillar but is triangular and slanted along the ecliptic. The position of the zodiacal light in the sky is also fixed. Sun pillars bear some resemblence to auroral pillars but can be distinguished because they always appear directly over the position of the setting or rising sun. A rare meteor track may be seen perpendicular to the horizon, but these glowing trails of ionized gases are usually quickly distorted by high altitude winds.

Examples of Auroral Pillars

X1. November 3, 1934. Liverpool, England. "...about 8 o'clock, a singular luminous appearance was seen in the heavens commencing near the western horizon and after extending through the meridian of the heavens, finally losing itself near the brilliant planet Jupiter...It presented the aspect of a beauteous transparent zone of light, of near equal width, from six to seven degrees...The stars were distinctly visible through its filmy structure and here and there a thin vapoury cloud crossed it at right angles." (R1)

X2. May 8, 1837. Bay of Toronto, Canada. R.H. Bonnycastle describes a straight-sided auroral pillar, "...my attention whilst regarding the heavens was forcibly attracted to the sudden appearance due east of a shining broad column of light...I was convinced that the meteor was an effluence of the sky, as I now saw it extend upwards from the eastern water horizon line to the zenith, as a well-defined, equal, broad column of white strong light, resembling in some degree that of the aurora, but of a steady brightness and unchanging body..." The pillar moved slowly northward disappearing in the northeast ten minutes after the first sighting. (R2)

X3. March 4, 1896. Great Britain. Of all the observations of auroral pillars, the extraordinary beam of light appearing in the evening sky on March 4, 1896, evoked the most discussion and scientific controversy. Seen from all over the British Isles, the beam arose bright and sharply defined from a bank of clouds on the western horizon in the early evening. Its initial location was almost coincident with the axis of the zodi-

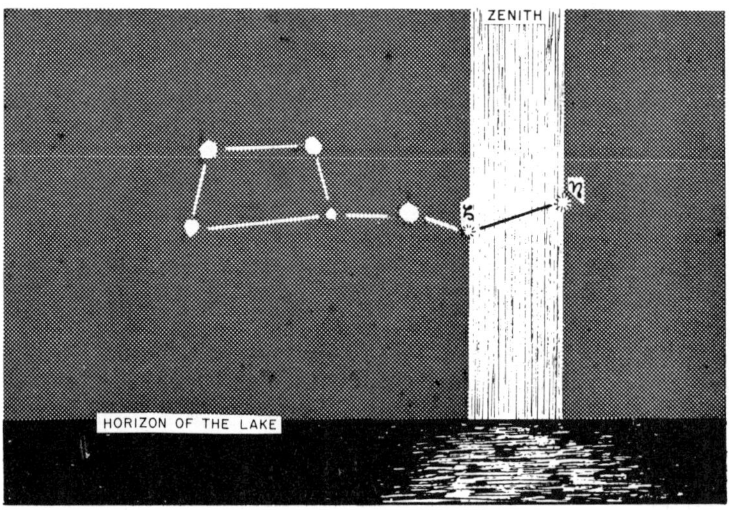


*A straight-sided auroral pillar rising from the Bay of Toronto*

acal light, and it pointed along the ecliptic. The March 4 beam, however, moved appreciably southward while in view. A comet was suggested when it became obvious that the phenomenon could not be the zodiacal light. However, the pillar's sudden appearance and disappearance upset that theory. Finally, the general scientific consensus settled on an auroral beam. Depending upon the location of the observer, the beam was straight-sided, fan-shaped, or pointed at both ends. Its duration varied likewise, from a few to many minutes. Some saw lesser beams nearby; others thought they saw transitory projections perpendicular to the beam. (R3-R8, R26)

X4. August 28, 1883. England. "I was just coming out of my observatory when, on the E.N.E. point of the horizon beneath the Pleiades, I saw a bright light. My first thought was that the moon was rising, but an instant's reflection sufficed to remind me that she would not be up for the next two hours. As I watched the light becoming brighter and brighter, I saw that it threw a kind of radial illumination upward, the effect of which I have tried to reproduce in the accompanying rough little sketch." (R9, R27)

*A fan-shaped auroral pillar over England*

X5. March 19, 1847. England. (R7)
X6. March 13, 1895. Kiel, Germany. (R7)
X7. March 17, 1843. Collingswood, England. The beam seemed to accompany the stars in their diurnal motion. (R10)
X8. April 22, 1852. New Haven, Connecticut. (R11)
X9. 1869. Braemar (Scotland?) (R12)
X10. April 1, 1880. Guildown, United Kingdom. (R13)
X11. August 4, 1882. Bloomington, Indiana. (R14)
X12. September 11, 1891. Ryde, Isle of Wight. (R15) Woolford Station, United Kingdom, (R16) Norway, latitude 62° north. (R17)
X13. November 8, 1897. Key West, Florida. (R18)
X14. March 29, 1902. Alta, Iowa. (R19)
X15. July 15, 1893. Alta, Iowa. (R19)
X16. May 3, 1899. Alta, Iowa. (R19)
X17. January 20, 1900. Alta, Iowa. (R19)
X18. October 16, 1903. London, England. A wriggling stream of light. (R20) This may have been rocket lightning (GLL1). (WRC)
X19. March 29, 1905. Cardiff, Wales. (R21)
X20. August 18, 1905. South Birmingham, England. (R22)
X21. October 16, 1916. Cardiff, Wales. Beam shifted north for 10 minutes and then returned to its initial position. (R23)
X22. April 28, 1916. Pittsburgh, Pennsylvania. (R24)
X23. April 27, 1937. Delaware, Ohio. Length 150°, almost a complete auroral arch (GLA2). Streaks of light ran perpendicular to the ray. The observer noted the similarity between the scintillations and flickerings of this auroral beam and those seen in the tails of some comets. (ACO) (R25)
X24. August 21, 1903. Seen from many locations from Maine to Iowa. (R25) See GLA2-X2.
X25. September 9, 1898. Hazelmere, England. Shaft of light rises in southwest to near zenith. It appeared and disappeared at intervals of a few seconds. (R28)
X26. March 5, 1764. Oxford, England. A bright tapering column of light seen to the south, wider at its base, 30° high. After a few minutes several rays darted out horizontally to the west, giving it a ladder-like appearance. (R29)

References

R1. "Celestial Phenomenon Seen at Liverpool," Nature, 134:709, 1934. (X1)
R2. Bonnycastle, R.H.; "Auroral Appearance," American Journal of Science, 1:32:393, 1837. (X2)
R3. Turner, H.H., et al; "Note on a Curi-

ous Light (The Zodiacal Light?) As Seen at Oxford, 1896, March 4," Royal Astronomical Society, Monthly Notices, 56: 332, 1896. (X3)
R4. Cope, E. J.; "The Aurora of March 4, 1896," British Astronomical Association, Journal, 6:295, 1896. (X3)
R5. Monck, W. H. S., et al; "Zodiacal Light (?)" Knowledge, 19:112, 1896. (X3)
R6. Wood, W. H.; "Zodiacal Streamers," English Mechanic, 63:170, 1896. (X3)
R7. Ellis, William; "On the Curious Light Seen at Oxford and Other Places, 1896 March 4," Observatory, 19:228, 1896. (X3, X5, X6)
R8. Hadden, David E.; "Auroral Phenomena or Zodiacal Light?" Popular Astronomy, 10:388, 1902. (X3)
R9. Noble, William; "A Curious Phenomenon," Knowledge, 4:173, 1883. (X4)
R10. Herschel, John F. W.; "Notice of an Extraordinary Luminous Appearance..." Royal Society, Proceedings, 4:450, 1842. (X7)
R11. Herrick, E. C.; "Observations of an Auroral Beam," Report of the British Association, part 2, 130, 1852. (X8)
R12. "Strange Phenomenon," Eclectic Magazine, 9:637, 1869. (X9)
R13. Capron, J. Rand, "Zodiacal Light, or Aurora?" English Mechanic, 31:111, 1880. (X10)
R14. Wylie, T. A.; "Auroral Phenomenon," Scientific American, 47:117, 1882. (X11)
R15. Levander, F. C.; "A Rare Phenomenon," Nature, 44:519, 1891. (X12)
R16. Rix, Herbert; "A Rare Phenomenon," Nature, 44:541, 1891. (X12)
R17. Duppa-Crotch, W.; "A Rare Phenomena," Nature, 44:614, 1891. (X12)
R18. "Anomalous and Sporadic Auroras," Monthly Weather Review, 26:260, 1898. (X13)
R19. Hadden, David E.; Auroral Phenomena at Alta Iowa," Popular Astronomy, 10:249, 1902. (X14-X17)
R20. Nature, 68:627, 1903. (X18)
R21. Mee, Arthur; "Strange Light in the Sky," English Mechanic, 81:220, 1905. (X19)
R22. Packer, David E.; "The Luminous Beam of August 18," English Mechanic, 82:88, 1905. (X20)
R23. Nature, 98:153, 1916. (X21)
R24. Jordan, Frank C.; "A Peculiar Aurora," Popular Astronomy, 24:401, 1916. (X22)
R25. Bobrovnikoff, N. T.; "An Unusual Auroral Display," Popular Astronomy, 45:299, 1937. (X23)
R26. Browne, W. Lyon, Jun.; "Zodiacal Light," Knowledge, 19:84, 1896. (X3)
R27. Bradgate, W. K.; "A Curious Phenomenon," Knowledge, 4:207, 1883. (X4)
R28. Russell, Rollo; "The Aurora of September 9," Nature, 58:496, 1898. (X25)
R29. Swinton, John; "An Account of a Remarkable Meteor Seen at Oxford...," Philosophical Transactions, 54:326, 1764. (X26)

# GLA2 Sky-Spanning Auroral Arches

Description. A narrow, bright arch of light stretching from the eastern to the western horizon passing near zenith. The color of the arch is almost always white or greenish white, although reddish examples are known. The arch may be as narrow as $1^{o}$ or as wide as $15^{o}$. Rapidly moving wavelets or masses of light seem to rush along the beam. Almost always perpendicular to the magnetic meridian, the arch often drifts slowly north or south. Duration: typically 30 minutes or so; rarely, several hours.

Background. Very broad arches of various hues are rather common in the far north. The temperate zone arch is rare, much narrower, and seems a different phenomenon---at least scientist-observers usually separate it from normal polar arches.

Data Evaluation. Dozens of fine observations in the scientific literature. Rating: 1.

Anomaly Evaluation. No specific explanations have been found so far. The low latitudes and narrowness may imply an origin different from polar arches. The drifting, splitting, persistence, and internal motion make the auroral arch a fascinating phenomenon for detailed study, although it presents no special challenge to auroral theory. Rating: 3.

Possible Explanation. Ring current of charged particles.

Similar and Related Phenomena. Auroral arches are difficult to confuse with the zodiacal light because the latter phenomenon does not usually span the heavens. (Note: some tropical observations of the zodiacal light have it visible well beyond zenith. AZO) Lunar rainbows are always associated with the moon. Glowing meteor trails may span the sky but usually break up quickly and would only coincidently be perpendicular to the magnetic meridian. The banded sky phenomenon (GLA23) consists of several filmy, widely separated bands, with no preferred orientation. Finally, curious bands and arches of light are associated with the setting and rising sun; viz., the crepuscular and anticrepuscular rays (GES); but they are always seen emanating from the direction of the sun or the antisolar point.

Examples of Auroral Arches

X1. August 2, 1937. Perkins Observatory, Ohio. "At 1:25 a brilliant white bow was observed projected like the beam of a powerful searchlight across the zenith from west by north to east by south. It extended almost from horizon to horizon. At this instant the western half of the bow was very bright. Gradually this portion split into two fainter bows, while the eastern half increased in intensity until its light predominated. A slow, swirling movement was detected in the portion of the bow overhead as it drifted lazily southward. Fully fifteen minutes passed before the bow disappeared." (R1)

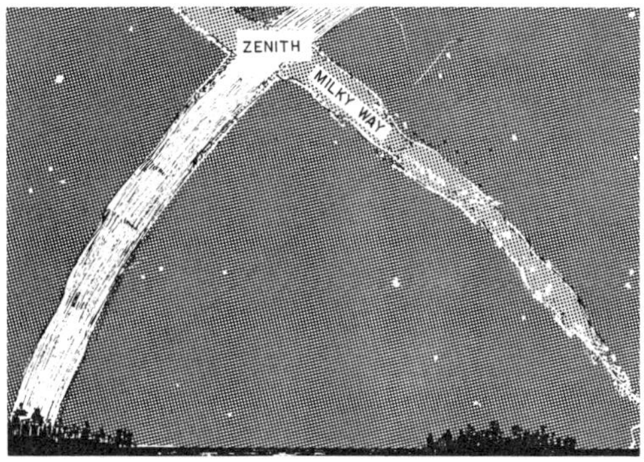

*An auroral arch seen from the Adirondacks*

X2. August 21, 1903. Cranberry Lake, 75 miles east of Watertown, New York. "At about half past nine, however, I was suddenly summoned out of doors to witness a spectacle such as I had never looked upon before nor had ever heard or read of. The heavens were spanned by two great bows of light, crossing each other near the zenith by a wide angle. The one was the familiar Milky Way or Galaxy, then in its glory, no Moon being present. The other was a remarkable archway of light, comparable with the Galaxy in width, but much brighter, and seemingly, stretching from horizon to horizon, though in one direction its light, like that of the stars, faded near the horizon, and in the other it was obscured by a small cloudbank. I estimated the width of the arch to be three degrees or that of six full moons placed side by side. This bow stretched from a little north of west, to a little south of east... The arch remained in place for a long time, possibly half an hour or more. It changed very little in appearance or brilliancy. In places it had a fluted, wavy aspect, being completely broken only here and there, and then only by a very narrow interval. The waves appeared as if blowing in the direction of the arch, toward the northwest, like curtains waving in a breeze, or a little like steam being blown along." (R2) Also seen at Cape Porpoise, Maine. (R3)

X3. January 12, 1890. Texas. "The luminous mist was first observed by the engineer, when it was still several hundred yards ahead of the train, and thinking it a prairie fire, he slowed up, thus arousing the passengers, who, with the crew, crowded to the windows and on to the platforms to look at the vast, hueless rainbow spanning the neavens. As the arch was more closely approached its dim, white radiance was seen to be clearly defined against the sky as though painted there by the sweep of a brush dipped in white fire. The stars could be seen shining close against the rim of it, and all around and under the arch... When the train passed directly under the bridge of light, the surrounding country spanned by it became plainly visible, appearing to be bathed in pale moonlight." (R4) It is more likely that the motion of the arch itself rather than that of the train produced the sensation of passing under the "bridge of light."(WRC)

X4. March 11, 1929. North Pacific. "From behind Fracto Nimbus cloud stretching from horizon to a height of 11° and bearing S. 75° E. True, a shaft of light appeared, making an angle of 45° with the horizon and following a straight line 45' wide for a distance of 60°; in the sky the light broke off into two curved sections passing through the zenith and finished $3^{o}$ above the horizon bearing N. $75^{o}$ W. True. The straight portion of the phenomenon had the appearance of a gigantic searchlight. The whole streak of light was motionless over its arc of $166^{o}$ and commenced fading away at 10.20 p.m. and finally disappeared entirely ten minutes later." Total duration 30 minutes. (R5) See GLA2-X32 for simultaneous phenomenon in Oregon.

X5. July 16, 1892. Knox College Observatory, Galesburg, Illinois. A peculiar auroral display was seen. "This was a streamer of nearly white light, that, starting in a sharp point almost on the horizon, in the north-west, shot with great velocity north of Arcturus, passed over Corona Borealis, which constellation it equalled in diameter, crossed Hercules and Cerberus, and, passing over Altair, descended almost to Mars in the south-east, terminating also in a fine point. This majestic sword moved bodily $10^{o}$ to the south, and, after shivering and pulsating throughout its length three times, vanished, after existing fourteen minutes." (R6) Mount Pleasant, Iowa. (R7) Dundas, Canada. (R8) See also GLA15-X1 for a simultaneous phenomenon in Indiana.

X6. June 1, 1961. Indian Ocean. "At 1800 GMT the upper of the two arcs illustrated appeared in the sky; 10 min. later, the lower narrower arc became visible. The upper one was very bright, while the lower was of slightly less intensity; both arcs attained maximum brilliance at 1820 and disappeared at 1832. They extended from $090^{o}$, through s, to $240^{o}$." Since there was little magnetic activity at the time, an auroral origin for these arcs was considered unlikely. No other explanations were offerred. (R9)

X7. November 10, 1813. Sunderland, England. (R10)

X8. September 29, 1828. Chelmsford, England. Shortened to an auroral pillar. (R11)

X9. March 21, 1833. Athboy, Ireland. Stream-like motion observed in the beam. Also seen in England. (R12)

X10. November 3, 1834. Near Manchester, England. Beam $4$-$5^{o}$ wide. (R13)

X11. May 29, 1840. New Haven, Connecticut. Initially $85^{o}$ above the southern horizon. Drifted south at about $1^{o}$ per minute. (R14)

X12. April 29, 1859. New Haven, Connecticut. Lasted 46 minutes. Western leg drifted southward. (R15)

X13. April 9, 1863. Newburyport, Connecticut. (R16)

X14. April 15, 1869. Toronto, Ontario. Torrent of luminous material appeared to move "with prodigious velocity." (R17)

X15. 1869. Near Glasgow, Scotland. (R18)

X16. October 24 and 25, 1871. Many observations of this remarkable phenomenon from the northeastern United States. The beam was $15^{o}$ wide, about three times normal. It was observed in the same position for two nights in a row. Composed of blood-red masses, with waves that moved at about $112^{o}$ per second. The color and width of this arch suggest that it may have been a normal polar arch located abnormally far to the south. (R19)

X17. February 4, 1872. New Haven, Connecticut. Three parallel arches that persisted over four hours. Northward band was white; rosy red band in between near zenith; white streak south of zenith. (R20)

X18. Fall 1872. Autun, Scotland. (R21)

X19. May 20, 1888. Buffalo, New York. Top located a few degrees south of zenith. Bow became "bent." (R22, R23)

X20. September 9-10, 1891. Lyons, New York. (R24) Nashua, New Hampshire. (R25) Dunecht, Scotland. Beam was only $1^{o}$ wide. (R26)

X21. September 11, 1891. Cape Breton, Nova Scotia. (R27)

X22. July 15, 1894. Toronto and Bala, Ontario. 166 miles high. (R28)

X23. August 1903. Lincoln, Illinois. Wavy motion progressed east to west. (R29)

X24. July 3, 1907. Bath, Maine. Bow drifted south. (R30)

X25. 1910. Sheffield, England. (R31)

X26. March 18, 1915. Garden Plain, Alberta. Arch lasted 1 hour. (R32)

X27. April 27, 1916. Chippewa Falls, Wisconsin and Northfield, Minnesota. (R33)

X28. May 2, 1919. West Reading, Pennsylvania. Double arch separated by a dark rift. One arch faded away, then the remaining arch split. (R34)

X29. September 18, 1919. Fargo, North Dakota. $15^{o}$ south of east to $15^{o}$ north of west. Like a "powerful searchlight." (R35)

X30. May 13, 1921. Wellington, England. A perfect arch, $2$-$3^{o}$ wide, passing through zenith. Incomplete arches on either side. (R36) This tendency to split may infer some unrecognized property of the terrestrial magnetic field. (WRC)

X31. April 14, 1926. Missoula, Montana. Formed and disappeared three times.

(R37)

X32. March 11, 1929. Portland, Oregon. (R38) See GLA2-X4.

X33. August 21, 1930. Littleton, New Hampshire. (R39) Greensboro, Vermont. (R40)

X34. March 22, 1841. England. Isolated auroral arch stretching WSW to ENE. (R41)

X35. November 19, 1841. England. Same as X34. (R41)

X36. March 29, 1843. England. Same as X34. (R41)

X37. August 29, 1845. England. Same as X34. (R41)

X38. February 6, 1847. England. Same as X34. (R41)

X39. 1890s. Durham, England. Uniform broad band WSW to ENE. It was motionless and looked like a searchlight. After half an hour it broke up into auroral streamers. (R42)

References

R1. Cherrington, Ernest, Jr.; "Observations of a Brilliant Aurora," Science, 86:265, 1937. (X1)

R2. Campbell, Frederick; "A Remarkable Phenomenon," Popular Astronomy," 11:484, 1903. (X2)

R3. Coit, Judson B.; "The Aurora of August 21, 1903," Popular Astronomy, 11:534, 1903. (X2)

R4. "Luminous Arch," American Meteorological Journal, 8:35, 1891. (X3)

R5. Groves, C. A.; "Light Rays at Night," Marine Observer, 7:65, 1930. (X4)

R6. Larkin, Edgar L.; "Magnetic Storm, Aurora, and Sun-Spots," Science, 20:65, 1892. (X5)

R7. Bereman, T. A.; "Auroral Display," Science, 20:65, 1892. (X5)

R8. McLellan, R. J.; "Aurora," English Mechanic, 55:553, 1892. (X5)

R9. Williams, J. O., and Craig, W. B.; "Unidentified Phenomena," Marine Observer, 32:64, 1962. (X6)

R10. Renney, Robert; "Account of a Luminous Meteor Seen at Sunderland," Annals of Philosophy, 2:456, 1813. (X7)

R11. Forster, T.; "On the Zodiacal Light," Philosophical Magazine, 2:4:389, 1828. (X8)

R12. Children, John George; "An Account of an Extraordinary Luminous Appearance in the Heavens...," Royal Society, Proceedings, 3:186, 1833. (X9)

R13. "Aurora Borealis of November 3, 1834," Nature, 134:709, 1934. (X10)

R14. Herrick, E. C.; "Auroral Belt," American Journal of Science, 1:39:194, 1840. (X11)

R15. "Auroral Arch," American Journal of Science, 2:28:154, 1859. (X12)

R16. "Auroral Arch of April 9th, 1863," American Journal of Science, 2:35:461, 1863. (X13)

R17. Kingston, G. T.; "Aurora at Toronto, Canada," American Journal of Science, 2:48:65, 1869. (X14)

R18. S., J. G.; "A Wonderful Arch in the Heavens," English Mechanic, 10:189, 1869. (X15)

R19. T., A. C.; "Auroral Belt of October 24th-25th, 1870," American Journal of Science, 3:1:126, 1871. (X16)

R20. Twining, Alex. C.; "The Aurora of February 4th, 1872," American Journal of Science, 3:3:273, 1872. (X17)

R21. "The Aurora Borealis," English Mechanic, 14:564, 1872. (X18)

R22. Kellicott, D. S.; "An Unusual Auroral Bow," Science, 11:266, 1888. (X19)

R23. Clayton, H. Helm; "An Unusual Auroral Bow," Science, 11:289, 1888. (X19)

R24. Veeder, M. A.; "Auroral Phenomena," Science, 18:261, 1891. (X20)

R25. Fergusson, S. P.; "Auroral Phenomena," Science, 18:305, 1891. (X20)

R26. Copeland, Ralph; "A Rare Phenomena," Nature, 44:494, 1891. (X20)

R27. Bell, Alexander G.; "A Rare Phenomena," Nature, 45:79, 1891. (X21)

R28. Wood, W. H.; "The Unnatural in 'Nature'," English Mechanic, 59:249, 1894. (X22)

R29. Allen, O. A.; "Auroral Band, Lincoln, Ill.," Popular Astronomy, 12:65, 1904. (X23)

R30. Hervey, A. B.; "'Northern Lights'," Popular Astronomy, 15:446, 1907. (X24)

R31. Alcor; "Weather Phenomena," English Mechanic, 90:612, 1910. (X25)

R32. Wimbush, B. E.; "Extraordinary Phenomenon," English Mechanic, 101:385, 1915. (X26)

R33. Jordan, Frank C., et al; "A Peculiar Aurora," Popular Astronomy," 24:401, 1916. (X27)

R34. Wagner, William H.; "A Strange Aurora?" Popular Astronomy, 27:405, 1919. (X28)

R35. Waldron, C. B.; "The Auroral Display of September 18," Science, 50:347, 1919. (X29)

R36. Price, W. S.; "Aurora," English Mechanic, 113:200, 1921. (X30)

R37. Merrill, A. S.; "Auroral Display April 14," Popular Astronomy, 34:344, 1926. (X31)

R38. Millard, Robert E.; "Auroral Display," Popular Astronomy, 37:305,

1929. (X32)
R39. Brooks, Charles F.; "A Celestial Searchlight," Science, 72:244, 1930. (X33)
R40. "Possible New Northern Light Like Celestial Searchlight," Science News-letter, 19:45, 1931. (X33)
R41. Stevenson, William; "Abstract of Observations on the Aurora....," Philosophical Magazine, 4:6:20, 1853. (X34-X38)
R42. Ellison, Wm. F.A.; "Mysterious Phenomenon," English Mechanic, 111:154, 1920. (X39)

# GLA3 Auroral Meteors: Moving Luminous Patches

Description. Large, well-defined patches of light that proceed steadily and rather deliberately across the sky. Generally elongate, several degrees wide and several times as long, the patches are white to greenish white, rarely reddish. Some auroral meteors, like true meteors, cross the sky in a few seconds, while others, particularly luminous bands and ripples, drift slowly like clouds.

Background. The absence of clear-cut form and structure, as in the auroral pillars and arches, opens the door to many natural agents capable of creating luminosity; viz., auroras, air glow mechanisms, meteors, electrical storms, etc. The wide range of patch velocities emphasizes the possibility of different causes.

Data Evaluation. Considerable high quality testimony supports the existence of fast-moving auroral meteors. Slow-moving bright patches, ripples, and other shapes have been photographed in modern times in both visible and infrared light. The subject of "nebulous meteors," however, is still controversial. It will be touched upon here and treated more fully later (AYO). Rating: 1.

Anomaly Evaluation. There are several reasonable, unmysterious mechanisms for producing luminous patches. The scientific task remaining seems primarily that of working out specific details. Rating: 3.

Possible Explanations. Rapidly moving patches, such as the reknowned "auroral meteor" of November 17, 1882, likely have auroral origins---perhaps some sort of burst of charged particles. The ascending luminosities associated with electrical storms may be electrical discharges of some sort. (GLA9) Slowly moving bright regions, bands and ripples in particular, are probably air-glow events initiated by gravity waves or other atmospheric disturbances.

Similar Phenomena. Normal auroras generate a great variety of rapidly moving beams, arches, rays, draperies, etc., but these cold flames and flickerings are qualitatively different from the isolated, more deliberate auroral meteors. Glowing meteor trails drifting in the upper atmospheric winds may emulate auroral meteors.

Examples of Auroral Meteors

X1. November 17, 1882. Much of Europe. "About 6 P.M., while the aurora was fitfully blazing in the north, north-east, and north-western sky, in the east there rose from the horizon a long beam of detached bright light, which, apparently lengthening as it advanced, crossed rapidly the southern horizon in front of or near the moon, and then sank in the west, shortening in length as it did so. The light emitted from it was described by one observer as of a glowing pearly white; and the general effect of this huge shining mass sailing majestically across the sky, even upon those accustomed to kindred phenomena, was at least one of wonder and surprise, while in the less experienced in such matters it created a feeling of absolute awe." (R1-R11, R20)

X2. July 15, 1893. Adrian, Michigan. "The peculiar feature of this aurora was the movement of a series or succession of whitish flecks across the sky from east to west, resembling somewhat the waves of a body of water... The white flecks or

*The remarkable auroral meteor of 1882*

streaks were about $10^{o}$ in length, strictly parallel north and south, and quite uniform in distance apart. They grew brighter and more distinct as they approached and passed the meridian. Their motion was very regular and quite rapid,---comparable to the swiftest apparent motion of light clouds." (R12) Note the strong resemblence to the wavelets observed in some auroral arches (GLA2). (WRC)

X3. April 4, 1978. Haleakala, Hawaii. Two spectacular naked-eye events were observed from the University of Hawaii's zodiacal light observatory. These ripples travelled at about 91 meters per second, with crest-to-crest wavelengths of 16 kilometers. Altitude: about 90 kilometers. (R13)

*Two bright ripples observed from Hawaii in 1978*

X4. December 30, 1737. Chiloe Island, Chile. A fiery cloud from the north traversed the archipelago. (R14)
X5. September 1828. Lynn Regis, England. White, smoke-like mass. (R15)
X6. July 16, 1850. Manchester, England. Luminous ball-shaped masses moved through the sky. Associated with an electrical storm. (R16)
X7. June 25, 1875. Norfolk, England. Aurora-like blue streaks darted from the earth to the sky. (R17)
X8. January 4, 1891. St. Charles, Missouri. Elliptical, stationary. (R18)
X9. May 1894. Dublin, Ireland. Elongated mass of white light shot west to east and then turned north. Phenomenon repeated three times. Followed by a mass of concentric rings. (R19)
X10. August 21, 1894. England. (R20)
X11. July 24, 1900. England. Dull nebulous meteors. (R20) See AYO for more on the so-called "nebulous meteors."
X12. August 18, 1905. England. Comet-like beam, $1^{o}$ x $80^{o}$, moved south by east. (R21)
X13. May 15, 1909. Blue Hill Observatory, Massachusetts. Comet-like glowing mass moved across sky. In view about two hours. (R22)
X14. July 19, 1916. Huntington, West Virginia. Pulsating, dirigible-like mass, $1/2^{o}$ x $2^{o}$. (R23)
X15. October 16, 1916. Glamorgan, Great Britain. Moving, angry-red, cigar-shaped mass. (R24, R25)
X16. March 24, 1920. Limerick, Ireland. Pulsating, bluish-white, elongated patch. Changed shape, split. Highly variable. (R26)
X17. June 27, 1927. Location unknown. (R27)

References

R1. Capron, J. Rand, et al; "The Magnetic Storm and Aurora," Nature, 27:82, 1882. (X1)
R2. McLeod, Herbert, et al; "The Aurora," Nature, 27:99, 1882. (X1)
R3. Groneman, H. J. H.; "Remarks and Observations of the Meteoric Auroral Phenomenon of November 17, 1882," Nature, 27:296, 1883. (X1)
R4. Taylor, H. Dennis; "Meteor of November 17," Nature, 27:365, 1882. (X1)
R5. Batson, Alfred; "The Auroral 'Meteoric Phenomena' of November 17, 1882," Nature, 27:412, 1883. (X1)
R6. Groneman, H. J. H.; "The True Orbit of the Auroral Meteoroid of November 17, 1882," Nature, 28:105, 1883. (X1)
R7. Capron, J. Rand; "The Auroral Beam of November 17, 1882," Philosophical Magazine, 5:15:318, 1883. (X1)
R8. Corder, H.; "Aurorae and Meteors,"

Observatory, 5:372, 1882. (X1)
R9. M., E. W.; "The Auroral Beam of 1882, November 17," Observatory, 6: 192, 1883. (X1)
R10. "The Aurora," Knowledge, 2:419, 1882. (X1)
R11. Maunder, E. Walter; "A Strange Celestial Visitor," Observatory, 39:213, 1916. (X1)
R12. Howard, W. H.; "An Unusual Aurora," Science, 22:39, 1893. (X2)
R13. Peterson, Alan W.; "Airglow Events Visible to the Naked Eye," Applied Optics, 18:3390, 1979. (X3)
R14. Botley, C. M.; "Unusual Auroral Periods," British Astronomical Association, Journal, 77:328, 1967. (X4)
R15. "Aurora Borealis?" Philosophical Magazine, 2:4:293, 1828. (X5)
R16. Clare, Peter; "On Some Extraordinary Electrical Appearances...," Report of the British Association, 1850, part 2, 31. (X6)
R17. Gape, Charles; "A Singular Phenomena," English Mechanic, 21:488, 1875. (X7)
R18. "Interesting Phenomena," Sidereal Messenger, 10:103, 1891. (X8)
R19. O'Neill, A.; "A Curious Phenomenon," English Mechanic, 59:267, 1894. (X9)
R20. Packer, D. E.; "Nebulous Meteors---Relation between Meteors and Aurorae," English Mechanic, 74:178, 1901. (X1, X10, X11)
R21. Packer, David E.; "The Luminous Beam of August 18," English Mechcanic, 82:88, 1905. (X12)
R22. Palmer, Andrew H.; "A Remarkable Aurora Borealis," Science, 30:57, 1909. (X13)
R23. Eager, Walter H.; "An Unusual Aurora," Scientific American, 115:241, 1916. (X14)
R24. Mee, Arthur; "Curious Phenomenon in Glamorgan," English Mechanic, 104: 271, 1916. (X15)
R25. Dixon, Lewis W., and Jenkins, T. K.; "Curious Phenomenon in Glamorgan, English Mechanic, 104:310, 1916. (X15)
R26. Ferguson, James; "Mysterious Phenomenon," English Mechanic, 111:120, 1920. (X16)
R27. Hollis, H. P.; "A Curious Phenomenon," English Mechanics, 4:96, 1928. (X17)

# GLA4 Low-Level Auroras

Description. Bright streamers, beams, curtains, and other auroral manifestations that seem to descend below the generally accepted lower limit of about 80 kilometers for auroral displays.

Background. Much controversy surrounds the alleged appearance of auroral phenomena below about 80 kilometers. Skeptics cite at least three objections: (1) The physical conditions conducive to the formation of auroras (pressure, etc.) do not exist at very low levels in the atmosphere; (2) Several long series of optical and photographic studies of auroras provide no evidence of low-level auroras; (3) The data supporting the reality of low-level auroras are mostly anecdotal and probably contaminated by illusions (viz., reflections from snow and fog) and nonauroral luminous phenomena (viz., lunar halos).

Data Evaluation. Many experienced observers have been certain they have seen auroras appearing between themselves and nearby terrestrial objects. A few observations with optical instruments have put auroras as low as 1-4 miles above the surface. Coupled with such data are numerous instances of other phenomena suggestive of low-level auroras, such as luminous fogs (GLA21), sounds (GSH), and even odor (GLA5). All in all, the evidence for low-level auroras seems quite good, although it is likely contaminated with some misinterpretations and illusions. Rating: 2.

Anomaly Evaluation. Low-level auroras, if real, directly challenge our concepts of auroral formation, which make auroras wholly a high-altitude phenomenon. If auroras do approach the earth's surface, these concepts must be wrong in some way or, possibly, aurora-like phenomena exist at low altitudes with a different mode of origin. Rating: 2.

Possible Explanation. Low-level auroras may represent the slow, large-scale discharge of terrestrial electricity to space, perhaps induced by or coincident with normal, high-altitude auroral activity.

Similar and Related Phenomena. Similar phenomena: auroral fogs (GLA21); artificial auroras (GLA6); mountain-top glows (GLD). Related phenomena: auroral sounds (GSH); the possible association of auroras with thunderstorms (GLA9), earthquakes (GLA10), lunar halos (GLA19), surface electrical effects (GLA20).

## Examples of Low Auroras

X1. Winter 1901. Eagle, Alaska. On the Yukon River, near a 1200 foot bluff. "One cold night, with the temperature about 45 degrees below zero, (F.), I witnessed a singular aurora---an array of dancing streamers, having prismatic colors---which aurora was in line between me and the bluff. The streamers were about one-quarter of the way down below the summit of the bluff, and no part thereof reached to that summit. There was a swishing sound at times as they moved." (R1)

X2. February 19, 1852. Olean, New York. A low cloud of auroral vapor came from the east. "When it came against us, apparently not over thirty rods from us, it was about half the length of several pine trees which stood in the field, from the ground. There were several of these trees, some were on this side of it, and others were just beyond it, which we could distinctly see, and upon which we remarked at the time. There were no streamers proceeding from it, but a curtain of auroral light slowly passing across the valley, at the distance above stated from the earth.... It must have been ten minutes from the time we first noticed it, before it seemed to strike the hill on the west side of the valley. From the base of this sheet, or curtain of light, the most beautiful, brilliant appearance was seen. Masses of auroral light, like off-shoots from a sky-rocket, would fall towards earth, scattering itself in the air, the largest particles of which would not disappear till they were within a few feet, say eight or ten, of the earth. It was the most splendid sight I ever beheld." (R2)

*An aurora seems to descend below mountain-top levels.*

X3. September 2, 1906. On the Yukon River. "I was on the south bank of the Yukon river, which at this point is about half a mile wide. The north shore is high wooded hills. The streamers came between me and the north shore and came right to the water's edge. Of this there was not the slightest doubt, as the trees could be seen through the openings in the auroral streamers." (R3)

X4. October 1925. Fort Smith, Northwest Territory. "Q. M. S. Griswold, engineer at this radio station, states that at about eleven p. m. one night.... he personally saw a curtain of the aurora approach to within four feet of the surface of the earth. The weather at the time was calm, clear and frosty, but not very cold, and the display was particularly brilliant and coloured, though lasting for only about an hour between eleven and twelve. The curtain, which approached the ground, was pale green, and a two-story building could be seen beyond and through the curtain of light. No sensation could be felt on walking right into the curtain, which disappeared from the view of anyone approaching closely; however; other people about a hundred yards from the curtain could see the observer enter and pass through the curtain." (R3)

X5. November 16, 1929. Abisko, Sweden. "I observed a rather intensive auroral ray of about $10^{o}$ apparent length and about $1/2^{o}$ apparent breadth in the west-south-west below a completely cloudy sky." This ray subsequently disappeared and reappeared several times. (R4-R6)

X6. Early 19th. Century. Scotland. Two-station triangulation on aurora indicated it was 4,000 feet above the earth. (R7)

X7. February 13, 1821. The Arctic. The aurora appeared below a dense layer of cloud. (R8)

X8. February 24, 1842. England. Aurora seen below the clouds. (R9)

X9. 1868. Sweden. Aurora seen below a mountain peak. (R7)

X10. Winter 1878-9. Norway. (R10)

X11. December 8, 1881. Point Barrow, Alaska. Aurora seen below the clouds. (R11)

X12. 1887. Godthaab, Greenland. Simultaneous observations from two stations put the altitude at 2,000 feet. (R7)

X13. 1892. Godthaad (sic), Greenland. Theodolites 4 miles apart produced altitudes between 1 and 4 miles. (R12)

X14. May 23, May 31, June 21, no year given but observations were made in the Antarctic on the First Scott Expedition, 1902-1904. Three observations of auroral streamers descending below Mt. Erebus. (R13)

X15. May 21, June 22, July 19, 1911. Antarctica. Three observations of auroras between observers and Mt. Erebus. Some observers considered these to be illusions, mock moon halos. (R14)

X16. October 31, 1903. Calgary, Alberta. Tips of auroral shafts touched the ground, and objects could be seen through them. (R15)

X17. Winter 1917-8. No location given. Kaleidoscope of colors seen between the hand and body. (R16)

X18. August 8, 1924. Valparaiso, Saskatchewan. Auroral beams seen among trees. (R3)

X19. August 10, 1928. Fort Smith, Canada. Aurora seen against a river bank less than a mile away. (R17)

X20. Winter 1929. Norwood, Ontario. Aurora extended to ground level. Radio receiver went dead. (R18) See GLA20-X2.

X21. March 8, 1932. Tromso, Norway. Photographic measurements of aurora at 65 kilometers. (R19)

X22. September 18, 1941. Cambridge, Massachusetts. Aurora only 20-30 feet overhead moving in waves. Seemed to consist of snow-like particles moving in a severe wind. (R20)

X23. No date. Canada. Engineer sees men walk into aurora. (R21) (Same as X4?)

**X24. April 22, 1915. Viking, Alberta. Numerous bright, milky streamers observed against the clouds. (R22)**

References

R1. Johnson, James Halvor; "On the Altitude of the Aurora," Astronomical Society of the Pacific, Publications, 39:347, 1927. (X1)

R2. Olmsted, D.; "Great Aurora Borealis of February 19th, 1852," American Journal of Science, 2:13:426, 1852. (X2)

R3. Beals, C. S.; "The Audibility of the Aurora and Its Appearance at Low Atmospheric Levels," Royal Astronomical Society of Canada, Journal, 27:184, 1933. (X3, X4, X18)

R4. Corlin, Axel; "Observations of a Low Altitude Aurora and Simultaneous Phenomena," Nature, 127:553, 1931. (X5)

R5. Corlin, Axel; "The Low Altitude Aurora of Nov. 16, 1929," Nature, 127:928, 1931. (X5)

R6. Simpson, G. C.; "Low Altitude Aurora," Nature, 127:663, 1931. (X5)

R7. "How High Is the Aurora?" Scientific American, 117:378, 1917. (X6, X9, X12)

R8. Espy, James P.; "Remarks on the

Height of the Aurora Borealis...," Franklin Institute, Journal, 17:363, 1834. (X7)
R9. Farquharson, James; "Report of a Remarkable Appearance of the Aurora Borealis below the Clouds," Royal Society, Proceedings, 4:382, 1842. (X8)
R10. English Mechanic, 33:184, 1881. (X10)
R11. "The Aurora as Seen in Alaska," Scientific American, 57:65, 1887. (X11)
R12. "Height of Auroras," Scientific American, 67:49, 1892. (X13)
R13. Chree, C.; "Auroral Observations in the Antarctic," Nature, 101:114, 1918. (X14)
R14. Simpson, G. C.; "Auroral Observations in the Antarctic," Nature, 102:24, 1918. (X15)
R15. Nature, 69:158, 1903. (X16)
R16. Chapman, S.; "The Audibility and Lowermost Altitude of the Aurora Polaris," Nature, 127:341, 1931. (X17)
R17. Gates, H.; "The Audibility and Lowermost Altitude of the Aurora Polaris," Nature, 127:486, 1931. (X19)
R18. "A Low Aurora and Its Effect on a Radio Receiver," Nature, 127:108, 1931. (X20)
R19. Chapman, S.; "Low Altitude Aurorae," Nature, 130:764, 1932. (X21)
R20. Brown, F. C.; "One Unusual Observation in the Auroral Display of September 18," Science, 94:562, 1941. (X22)
R21. Eve, A. S.; "Northern Lights," Smithsonian Institution Annual Report, 1936, p. 149. (X23)
R22. "An Aurora Seen in the Daytime," Scientific American, 113:90, 1915. (X24)

# GLA5 The Odor of the Aurora

Description. The odors of ozone, sulfur, and/or "electricity" detected in conjunction with low-level and high active auroras.

Background. The odors accompanying geophysical phenomena are rarely considered seriously, perhaps because instruments and theories are lacking. Nevertheless, a few students of unusual auroral attributes (Beals, Silverman) have recorded some of the sparse though persistent accounts of auroral odors.

Data Evaluation. Reports of auroral odors generally come from the older literature and without much in the way of helpful details. Unquestionably, the data are weak, but they constitute just the kind of supplementary information needed by those who believe that terrestrial electricity may sometimes be discharged during auroral displays. Rating: 2.

Anomaly Evaluation. Since the odors of sulfur and ozone are characteristic of electrical discharges, verification that they occur during auroras, especially low-level auroras, would add support to the hypothesis of large-area (brush) discharges of electricity during some auroras. The current view of the aurora makes it a high-altitude (over 80 kilometer) phenomenon with little room for accompanying terrestrial electrical activity. Rating: 2.

Possible Explanation. Low-level, active auroras may sometimes be associated with large-area discharges of terrestrial electricity.

Similar and Related Phenomena. Similar phenomena: the odors detected during lightning and brush discharge phenomena, such as ball lightning and corona discharge. Related phenomena: odor detected in mountain-top glows (GLD1).

Examples of Auroral Odors

X1. No date. Great Britain. "Mr. Trevelyan observed, that the Aurora borealis in Faroe and Shetland was often seen very low, not more than 40 or 50 feet above the level of the sea; and he learned that it both countries it is frequently heard. In Faroe Mr. T. met one person who stated, that when the colour of the Aurora borealis is dark red, and extends from west to east with a violent motion, he had experienced a smell similar to that which is perceived when an electric machine is in action. (R1, R3)

X2. September 18, 1941. Radburn, New Jersey. "The odor of ozone was reported by several persons in the vicinity of Radburn, N.J., during the great auroral display of Sept. 18..... The sensation was strongest when the auroral display was at its height, entering the zenith with coronal light and spreading toward the horizon in sheets. Odors have been reported occasionally from earlier displays." (R2)

X3. 1870. Norway. "In 1870, Rollier, the balloonist, descended on a mountain in Norway 1300 meters high, saw auroral rays across a thin mist, and heard a muttering. When the sound ceased he perceived a very strong smell of sulfur." (R3, R4)

X4. October 19, 1726. London, England. (R3)

X5. No date. Sweden. Aurora accompanied by the smell of smoke or burnt salt. (R5)

X6. No date. Canada. (R6)

References

R1. "Aurora Borealis in Faroe and Shetland," American Journal of Science, 1:8:392, 1824. (X1)

R2. "Ozone Odor Detected during Auroral Display," Science News Letter, 40:295, 1941. (X2)

R3. Silverman, S. M., and Tuan, T. F.; "Auroral Audibility," Advances in Geophysics, 16:193, 1973. (X1, X3, X4)

R4. Rufus, W. Carl; "The Beauty and Mystery of the Northern Lights," Sky and Telescope, 5:3, September 1946. (X3)

R5. "On the Noises That Sometimes Accompany the Aurora Borealis," Edinburgh New Philosophical Journal, 1:156-1826. (X5)

R6. Beals, C. S.; "Low Auroras and Terrestrial Discharges," Nature, 132:245, 1933. (X6)

# GLA6 Artificial Low-Level Auroras

Description. Terrestrial pillars or streamers of aurora-like light stimulated by the presence of man-made terrestrial electrodes, acting in conjunction with active natural displays of auroras.

Background. The 1881-1882 experiments of Selim Lemstrom in Lapland are the only attempts found so far at inducing artificial low-level auroras. There have been, however, a number of artificial auroras created at normal auroral heights by the high-altitude detonation of nuclear explosives, the release of chemicals (barium), and electron accelerators.

Data Evaluation. Lemstrom's experiments were widely reported but apparently never replicated. Rating: 2.

Anomaly Evaluation. Artificial low-level auroras basically have the same significance as natural low-level auroras (GLA5); that is, they suggest the possibility that very active auroras may be accompanied by large-area terrestrial discharges. Brush discharge and St. Elmo's Fire are related types of discharge and, like corona discharges, do not constitute anomalies. It is the appearance of aurora-like luminous phenomena in concert with normal auroral activity that signifies an unknown dimension to auroral processes. Rating: 2.

Possible Explanation. Large-scale discharges of terrestrial electricity to space.

Similar and Related Phenomena. "Pseudo-auroras" occasionally rise over powerful artificial lights, but these are not self-luminous and are instead akin to sun pillars.

Example of an Artificial Aurora

X1. 1881-1882. Several mountain tops in Lapland. Selim Lemstrom mounted arrays of wires studded with sharp points atop insulated posts. His most successful experiment was on Mount Pietarintunturi, about 40 miles west of Sodankyla. "Here experiments were possible on only two days, but on the first of them, December 29th, 1882, an auroral streamer 400 feet long was formed, which observations from several azimuths proved to be exactly over the apparatus." (R1-R3)

References

R1. Lemstrom, Prof.; "An Artificial Aurora," Symons's Monthly Meteorological Magazine, 18:51, 1883. (X1)

R2. "Artificial Aurorae," Symons's Monthly Meteorological Magazine, 18:33, 1883. (X1)
R3. Science, 4:465, 1884. (X1)

*Lemstrom induces an artificial aurora in Lapland.*

# GLA7 Geographically Displaced Auroras

Description. Auroras seen exclusively to the south in the North Temperate Zone and vice versa in the South Temperate Zone, with little or no activity in the direction that auroras usually appear.

Background. Auroras concentrate in annular regions centered on the earth's magnetic poles. With exceptionally active auroras, this region may expand toward the Equator, thus pushing luminous phenomena in the same direction.

Data Evaluation. Southward auroras are very rare in the latitudes of the United States. Only a few cases have been discovered in the literature. Rating: 1.

Anomaly Evaluation. Our understanding of the aurora is not challenged by geographically displaced auroras; they are really extreme instances of auroral activity rather then anything fundamentally different. Rating: 3.

Possible Explanation. Exceptionally powerful magnetic storms push the visible auroras Equator-ward.

Similar and Related Phenomena. Auroral pillars (GLA1) and arches (GLA2) frequently drift south of zenith---usually with significant "normal" auroral activity remaining in the polar direction.

Examples of Geographically Displaced Auroras

X1. March 30, 1894. Lyons, New York. "The aurora of Friday evening, March 30, was first seen at Lyons, N. Y., at 7.40 P. M. It then consisted of flickering, irregular patches which filled the entire heavens from the zenith down to the southern horizon, there being nothing whatever to be seen in the usual location of auroras toward the north... From the reports thus far received, it appears that at the hour above indicated, the aurora was seen toward the south exclusively at stations directly north of Lyons, while southward in Pennsylvania the entire sky was more or less covered with flickering patches and masses of light, and still further south the aurora was seen toward the north exclusively, at Charleston, S. C., rising about 25° above the northern horizon." (R1)

X2. May 29, 1932. Ottawa, Ontario. "The remarkable thing about it was that the northern part of the sky seemed entirely clear while the rest of the sky was in great turmoil." (R2)

X3. May 15, 1909. Blue Hills Observatory, Massachusetts. Entirely to the south. (R3)

X4. September 18, 1941. Mount Washington, New Hampshire. Aurora to the south; accompanied by mountain-top glows. (R4)

References.

R1. Veeder, M. A.; "The Aurora of March 30, 1894," American Meteorological Journal, 11:77, 1894. (X1)
R2. McFarlane, G.; "The Aurora of May 29, 1932," Popular Astronomy, 40:590, 1932. (X2)

R3. Palmer, Andrew H.; "A Remarkable Aurora Borealis," Science, 30:57, 1909. (X3)

R4. Botley, Cicely M., and Howell, W.E.; "Mystery on Mount Adams," Mount Washington Observatory News Bulletin, 8:9, 1967. (X4)

# GLA8 Auroras with Unusual Geometries

Description. Auroras with geometries that are difficult-to-explain in terms of present-day theories about the origin of auroras

Background. Aurora geometries are usually characterized as arches, rays, draperies, etc. Beyond these common shapes, which are not necessarily well-understood, are many curious and unusual configurations. Most are unique; some are bizarre; all seem to challenge students of auroras.

Data Evaluation. The observations reported here are mostly one-of-a-kind; that is, without confirming testimony. Rating: 3.

Anomaly Evaluation. A superficial judgment would be that all of the phenomena cited below might force aurora theorists to do some rethinking. Rating: 3.

Possible Explanations. None.

Similar and Related Phenomena. Speaking in generalities, most complex natural phenomena are characterized by rare "wild points" and observations that do not fit the mold. Ball lightning (GLB), halos (GEH), and marine light wheels (GLW), for example, offer just as many unique and unexpected varieties.

Examples of Auroras with Unusual Shapes

X1. November 16, 1871. Finland. "Fig. 3 gives an idea of the variety of forms that the phenomenon may effect. It represents an aurora that was observed at the presbytery of Enare on the 16th of November, 1871. The aurora this time took on the form of a glowing red band, curved as shown in the figure. The two extremities bordered on yellow and green." (R1)

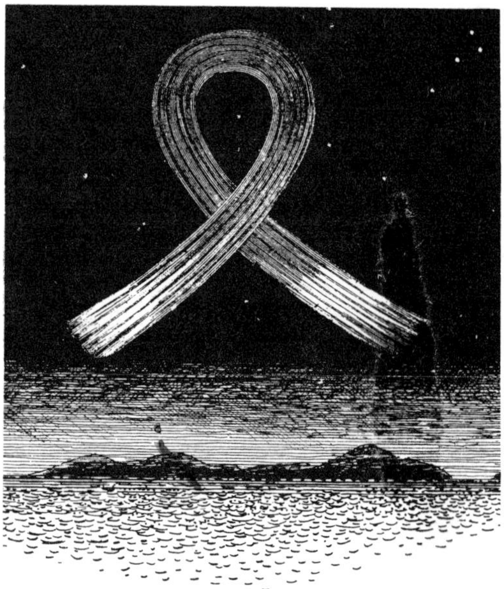

*An auroral "loop"*

X2. August 21, 1903. York Harbor, Maine. "Beginning a little east of the zenith and continuing almost to the western horizon, there appeared what might easily be likened to a string of tremendous comets. These pennants of light, however---unlike comets ---were more brilliant at their apices. Their bases were directed upwards, their points down. They were constantly changing, appearing and disappearing, but not very rapidly. Some would remain a minute or more without much variation. The number varied from ten to fifteen. They were shorter toward the zenith, longer toward the horizon. At the western end of the arch, one long half-luminous streak shot upward obliquely (as shown in the figure) and re-

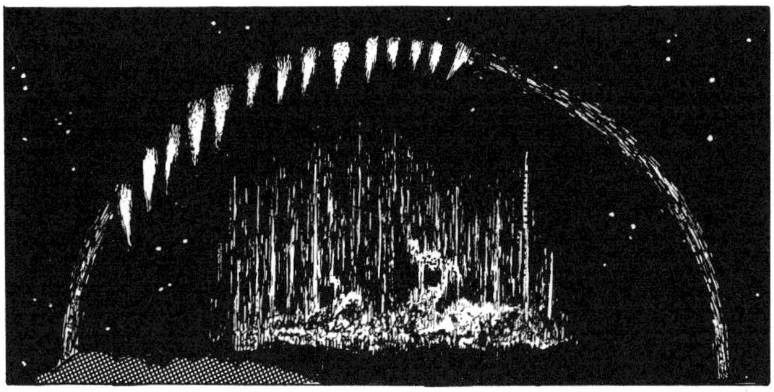

*Pennants of light suspended from an auroral arch over the Maine coast*

mained for some minutes after the arch itself had disappeared. The arch lasted from 9.30 to nearly 10 P. M. In size the comet-like pendants appeared about as wide at their bases as the diameter of a full moon, and four or five times such a diameter in length." (R2)

X3. September 12, 1881. Off Newfoundland. "A peculiar and interesting auroral phenomenon witnessed from the steamship Atlantic off the Newfoundland coast on September 12 last year, has been described by Mr. Engler to the St. Louis Academy. While an aurora of normal type was clearly seen in the northern sky, there appeared in the southeast, about 30 to 35 deg. above the horizon two horizontal streaks of light, about 5 deg. apart, and 15 or 20 deg. in length. Their pale hazy light resembled moonlight. From the upper streak were suspended, by small cords of light, a number of balls, brighter than either of the streaks, which were continually jumping up and down in vertical lines much like pith-balls when charged with electricity." (R3)

X4. Circa 1930. Colorado. "Shortly after midnight, away to the south, there arose from the crest of a long high mountain, the trend of which was east and west, a curtain of fire several miles long and from the ground straight upwards to a very great height. It seemed to be made up of threads of fire rising rapidly upward, and at the extreme top of this curtain of fire all of these threads turned abruptly at right angles and flowed rapidly westward. Immediately the

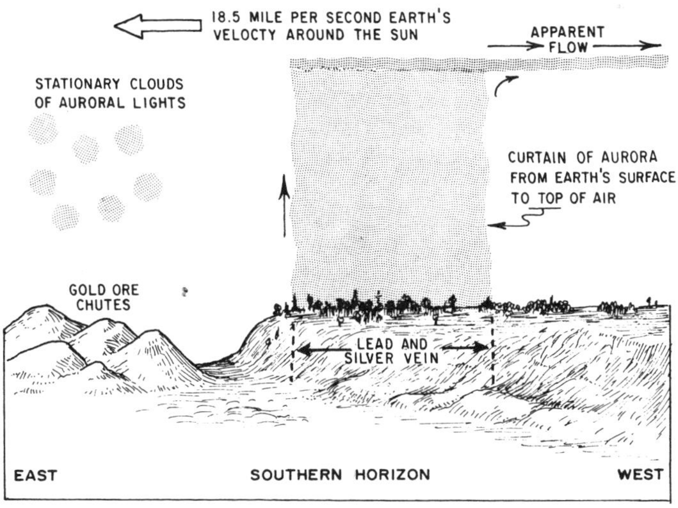


*A fiery curtain rises from a Colorado mountain.*

conclusion was that this curtain had reached the top of the air and that its rapid movement westward was a visual demonstration that the earth had a tail like a comet, and further that the outer so-called void was not a void at all but contained sufficient of a gaseous filling to cause this drag on our outer air as a result of the great speed of the earth on its way around the sun" (R4) This phenomenon might have been a mountain-top glow (GLD1) or a low-level aurora. (WRC)

X5. 1849. Cincinnati, Ohio. On the western horizon a streak of light shot up to 45° and burst, spreading across the heavens. (R5)

X6. June 26, 1873. North America. Observers in several places saw bars of light moving rapidly from south to north. (R6)

X7. June 25, 1875. Norfolk, England. Very narrow blue streaks darted up from the ESE horizon. (R7)

X8. August 12, 1892. Dewsbury, England. Waves of phosphorescent light, like billows of smoke, swept through zenith and were lost in the south. (R8)

X9. September 25, 1926. North Atlantic. A ball of bluish light, altitude 5°, bearing 345°, emitted a fan of rays across the sky. The rays were bright and illuminated the cloud layers. Four minutes later, the rays disappeared and an auroral arch was observed. (R9)

X10. General feature of auroras. Satellite photos have revealed a sawtooth structure to the aurora. The teeth have amplitudes of about 250 miles and probably cannot be distinguished on the ground. (R10)

References

R1. "The Northern Lights," Scientific American, 56:135, 1887. (X1)

R2. King, A. F. A.; "An Unusual Aurora Borealis," Popular Science Monthly, 63:563, 1903. (X2)

R3. Nature, 26:160, 1882. (X3)

R4. Finn, Oliver B.; "A Very Unusual Auroral Display," Popular Astronomy, 41:287, 1933. (X4)

R5. "Singular Phenomena," Scientific American, 4:193, 1849. (X5)

R6. "The Auroral Phenomenon of June 26," Scientific American, 29:49, 1873. (X6)

R7. Gape, Charles; "A Singular Phenomenon," English Mechanic, 21:488, 1875. (X7)

R8. Blakeley, E. Reginald; "Peculiar Auroral Phenomenon," English Mechanic, 56:15, 1892. (X8)

R9. Quirk, W. E.; Marine Observer, 27: 147, 1957. (X9)

R10. Sehlstedt, Albert, Jr.; "Study of Satellite Pictures Reveals Saw-Tooth Pattern in Aurora Borealis," Baltimore Sun, May 31, 1981, p. B2. (X10)

# GLA9 Auroras Correlated with Thunderstorms

Description. Auroral activity associated with thunderstorms in any of three ways: (1) The direct visual observation of auroral displays over active thunderstorms; (2) The temporal and geographic coincidence of auroras and thunderstorms; and (3) Statistical observations that thunderstorms are frequent when auroras are rare and vice versa---a reciprocal relationship.

Background. Prevailing theory has auroral phenomena originating in the thin reaches of the outer atmosphere and magnetosphere beyond heights of 50 miles. Thunderstorms, on the other hand, are denizens of the lower atmosphere. Furthermore, the energies involved in kindling auroras are many times less than those involved in active thunderstorms. (R11) The issue here is whether such disparate phenomena can exert coupling forces on one another.

Data Evaluation. (1) Only one account has been found of apparent direct association between an active thunderstorm and auroral phenomena; (2) A few not-too-convincing circumstantial accounts of coincident auroras and thunderstorms exist; and (3) A very few observers of the two phenomena have suggested, with no published backup data, that a reciprocal relation prevails between aurora and thunderstorm frequencies. In other words, the relationship between auroras and thunderstorms is rather tenuous. Rating: 2.

Anomaly Evaluation. As discussed under "background" any verified connection between au-

Anomaly Evaluation. Meteors and auroras occupy the same region of the atmosphere and magnetosphere. In addition, ionized meteor trails constitute a reasonable physical mechanism for stimulating auroral activity. For these reasons, the meteoric enhancement of auroral activity is primarily a curiosity. If, however, meteors really do originate in auroral structures, a different explanation is required. Rating: 3.

Possible Explanation. The eruption of meteors from already existing auroral structures could be due either to: (1) The mere coincidence of an auroral structure with a meteor radiant; or (2) The actual creation of the auroral structure by incoming meteoric material from a radiant point, followed by additional meteors from the same radiant point.

Similar and Related Phenomena. Nebulous meteors (AYO) leave diffuse luminous streaks that may resemble auroral phenomena. Very large meteors, such as the 1908 Tunguska Meteor, produce a wide variety of luminous effects, some of which may be taken for auroras. At times, auroras and lunar halos seem to mutually enhance one another. (GLA19) A further connection between meteors and auroras is their apparent mutual capability of generating anomalous sounds (GSH) and electrical effects (GLA20) at the earth's surface.

## Examples of Auroras Correlated with Meteor Activity

X1. February 22, 1909. Great Britain. A remarkable meteor left a bright streak that was seen from many locations. "Several observers noted flashes like very faint lightning during the early period of the projection of the streak. Others allude to the fact that it exhibited bright pulsations, as though the lingering embers were fanned into brilliancy by the breeze. A few of the most careful spectators state that they noticed scintillations of the beam similar to the temporary light-waves which affect the streamers of Aurorae." (R1)

X2. September 4, 1908. Worcestershire, England. A bright, aurora-like cloud appeared. "The bright cloud was in a constant state of fluctuation, flashing up and fading with almost the same rapidity as the well-known scintillation effect observed in bright stars. The colour was pale white, and as far as could be judged, the cloud was stationary over the magnetic meridian during the whole period of observation. At 8.45 a meteor from Perseus approached it, and on entering the cloud at its eastern extremity burst out with extraordinary splendor, far surpassing Venus in lustre, and exhibiting a string of many smaller nuclei on each side of the main nucleus. At 8.50 the cloud at this point had thrown out a streamer, which was traceable between Auriga and Perseus into Aries." (R2)

X3. February 19, 1852. Wyoming, New York. A spectacular aurora with an arch south of zenith. "On turning to the south again, they saw a number of meteors shoot off from the circumference of the corona, some west, some north, and in all directions. They appeared in succession, a minute or two apart. He is familiar with the flashes of light seen at times during the aurora, and said the meteors had no resemblence to it. They appeared as bodies or masses shot off from the corona, and had trains after them, though they were not very bright, the auroral light itself being too intense to permit them to appear as bright as in the dark. The number seen in fifteen or twenty minutes was ten or twelve." (R3)

X4. A general observation. "Hoffmeister has collected various instances of auroral-like glows connected with meteoric showers, Perseids, Leonids, 1872 Bielids. Such was also noted in Germany and England during the famous Leonid shower of November 13-14. It was indeed remarked it might be possible that 'brilliant meteor streams electrify the higher regions of the atmosphere and produce phenomena analogous to the aurora borealis.' The detection of H and the red line of NI is interesting in view of the observation by von Wrangel in Siberian waters 1820-1824 that dark areas of aurora traversed by meteors lighted up and continued to glow. (R4)

X5. November 13, 1833. Eastern North America. One of the greatest meteor showers in historical times. Auroras observed in many places, but totally absent elsewhere. (R5)

X6. November 17, 1848. France. (R6)

X7. February 20, 1849. Edinburgh, Scotland. Meteor-like objects in an auroral arch. (R7)

X8. 1852. Galveston, Texas. A pale blue light traversed a meteor trail from the terminus back to the starting point. (R8)

X9. March 30, 1922. Wick, England. (R9)

X10. A general observation. Great auroras tend to coincide with the larger meteor showers, especially the Leonids, during the Nineteenth Century. (R10)

X11. 1941. Near Ithaca, New York. Bright meteor followed by aurora. (R11)

References

R1. Denning, W. F.; "The Meteoric Streak of February 22," Nature, 80:42, 1909. (X1)

R2. Packer, D. E.; "Aurora and Meteor Display of September 14," English Mechanic, 88:211, 1908. (X2)

R3. Olmstead, D.; "Great Aurora Borealis of February 19th, 1852," American Journal of Science, 2:13:426, 1852. (X3)

R4. Botley, Cicely M.; "Spectra of Meteor Wakes," Nature, 200:1194, 1963. (X4)

R5. Olmsted, Denison; "Observations on the Meteors of November 13th, 1833," American Journal of Science, 1:25:363, 1833. (X5)

R6. Comptes Rendus, 49:401, 1848. (X6)

R7. Pringle, W.; "Peculiarity Observed in a Luminous Arch," Philosophical Magazine, 4:5:308, 1853. (X7)

R8. "Meteoric Phenomenon at Texas," Scientific American, 8:18, 1852. (X8)

R9. English Mechanic, 115:134, 1922. (X9)

R10. Hollis, H. P.; "Auroras and Meteors," English Mechanic, 82:494, 1906. (X10)

R11. "Bright Meteor Followed by Northern Lights," Science News Letter, 40:72, 1941. (X11)

# GLA12 The Close Relationship between Auroral Displays and Clouds

Description. The apparent physical association between some auroral displays and high-altitude clouds and haze, as in the following types of observations: (1) The seeming conversion of auroral displays into similarly configured clouds at daybreak and vice versa at nightfall; (2) The apparent origin of radiating auroral streamers in luminous clouds; (3) The formation of high hazes after auroral displays; (4) The increased incidence of halo phenomena during and after auroral displays (GLA19); and (5) The increased twinkling of stars during auroral displays.

Background. Several Nineteenth Century observers, particularly some arctic explorers, were quite certain that auroral displays turned into cloud formations during the day. Some insisted that the aurora was "material" in nature. (R1)

Data Evaluation. Almost all the relevant data are pre-1900. Furthermore, this phenomenon is not clear-cut due to the large number of similar phenomena. (See: Similar and Related Phenomena below.) Rating: 3

Anomaly Evaluation. Since conventional scientific wisdom locates auroral displays above 80 kilometers, any intimate connection between auroras and clouds, which are much lower, would be sharply anomalous. Rating: 2.

Possible Explanations. The precipitating particles that apparently initiate auroras may penetrate far enough into the denser atmosphere to act as nuclei for water vapor condensation, producing clouds with a superficial resemblence to higher altitude auroras. Both the auroras and clouds might be stimulated by meteoric debris. (GLA11)

Similar and Related Phenomena. Magnetically aligned cirrus clouds (GWC); banded skies (GLA23); steady night sky glows (GLA13); auroral meteors (GLA3); low-level auroras (GLA4); auroras and meteor activity GLA11); auroras interacting with lunar halos (GLA19); black auroras (GLA22); and just possibly the strange sky-spanning bands seen rarely at sunset and sunrise (GES).

Examples of Auroral Displays Associated with Clouds

X1. General observations. "The records of the great nineteenth century Arctic expeditions contain many accounts of 'clouds' becoming aurorae after dark, and aurorae becoming 'clouds' at dawn. Indeed A. Paulsen regarded the presence of such clouds as an indicator of daylight aurora; the opinion has also been expressed that the aurora is a material object: the clouds luminous by night and white by day are an aerosol, formed in the ionosphere. Indeed the formation was actually seen during the second Franklin expedition 1826-1827, with active aurorae, haziness developing around the coruscations.....As regards the classic observation of Usher (1788) a strong proof of its reality is the accompanying strong scintillation of the stars the association of which with aurora had recently been discovered by W. Herschel, and was confirmed in the following century by Montigny. It would be interesting to confirm the opinion of Forbes that at Edinburgh the stars never twinkle save during aurora." (R1)

X2. General observations. "Cirrostrati. These clouds resemble cirri in their tendency to an arrangement in long parallel bands; they differ, however, from them in not being composed of ice-crystals. They appear very frequently as concomitants of the aurora, so much so as to leave no doubt of a connecting influence between them and that phaenomenon.....Cirrostrati as well as cirri exhibit very decided luminosity at night. I have often observed the luminosity between the dark cirrostratus bands nearly equal to that of a faint aurora." (R2)

X3. January 28, 1841. England. "From the N.N.W. point, dark continuous cirrous (sic) bands extended across the sky to S.S.E. Between the bands throughout their whole extent a very decided luminosity prevailed, of a fine deep yellow tint, like that of the aurora. The luminous bands did not, however, show any tendency to converge to the coronal point, which the streamers of the aurora always do, but kept distantly separate to near the S.S.E. point. The light was strongest between the dark rays in N.N.W., but was very distinct throughout the whole extent of the bands. The absence of moonlight rendered the phaenomenon much more striking and unequivocal."(R2)

X4. March 14, 1841. England. "A splendid display of luminous cirri or cirrostrati? From the N.N.W. point proceeded dark lines or bands, which extended across the sky and appeared to converge to the S.S.E. point. These were apparently at a much inferior altitude than that of true cirri. They sometimes appeared to shoot suddenly over a large area of the sky. Between them a very distinct luminosity was apparent, which from the N.N.W. point, up to nearly 45°, equalled in brightness that of an ordinary aurora....The light emitted by this phenomenon was very considerable, and exactly resembled the pale greenish light of phosphorescent substances." (R2)

X5. General observations. England. "The allusions to self-luminous night haze remind me of some phenomena which I witnessed on the Durham coast between the years 1895 and 1900. On several occasions I noticed that slight displays of aurora seemed to associate themselves with a certain form of cirrostratus cloud. On one night I watched from a back yard in Sunderland curved auroral streamers shooting up the Northern heavens in pairs, each pair enclosing a lens-shaped space of clear sky. But within a few seconds a cloud would form in the enclosed space and persist long after the enclosing streamers had faded. This occurred several times in the course of two hours. There could be no doubt about the association between the luminous bands and the cloud." (R3)

X6. January 25, 1881. Folkestone, England. "It is perhaps worth a note that my daughter saw at Folkestone a very unusual phenomenon on the evening of January 25, a little before 6.30. Some distance to the left of Orion (for the night was clear and starry) she observed a small cloud of a bright golden hue, from which streamers of great brilliancy darted in various directions, the cloud alternately paling and brightening. She describes the streamers as like small meteors, leaving trails of light behind them." (R4)

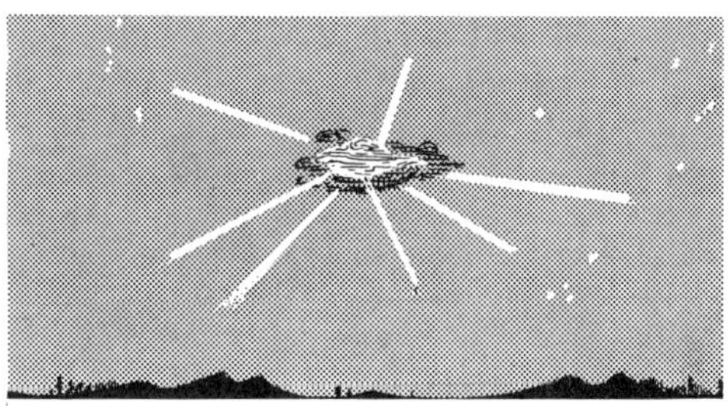

*Bright streamers dart from a cloud over England.*

X7. September 9, 1827. Canonmills, Great Britain. "The morning of Sunday the 9th September was rainy, with a light gale from the north-east. Before midday the wind began to veer to the west, and the clouds in the north-western horizon cleared away: the blue sky in that quarter assumed the form of a segment of a very large circle, with a well defined line, the clouds above continuing dense, and covering the rest of the heavens. The centre of the azure arch gradually inclined more to the north, and reached an elevation of nearly 20°. In a short time very thin fleecy clouds began to rise from the horizon, within the blue arch; and, through these, very faint perpendicular streaks of a sort of milky light could be perceived shooting: the eye being thus guided, could likewise detect the same pale streaks passing over the intense azure arch; but they were extremely slight and evanescent. Between 9 and 10 in the evening of the same day, the aurora borealis was very brilliant: so that there is no reason to doubt that the azure arch in the morning, and the pale light seen shooting across it, were connect-with the same phenomenon." (R11)

X8. April 8, 1880. Great Britain. White, self-luminous haze covered the entire sky, brightening and fading. Radiating beams shot up from the northeast. (R5)

X9. February 19, 1882. Germany. Peculiar cloud on southwest horizon emitting bright, aurora-like beams. (R6)

X10. September 9, 1898. New Zealand. Luminous cloud near zenith shot forth streamers of light. (R7)

X11. May 30, 1901. Braidentown, Florida. Thin streamers of cloud pulsing with auroral light. (R8)

X12. April 22, 1915. Viking, Alberta, Canada. Numbers of bright, hazy, milk-like streamers darting from a source just below a cloud. (R9)

X13. February 2, 1978. Minnesota. Beams of light emanating from the center of a cirrus cloud. (R10)

X14. General observations. Siberia, by von Wrangel. "When the streamers extend to the zenith or nearly so, they sometimes resolve themselves into small luminous and cloud-like patches of milk white colour, which not infrequently continue to be visible on the following day in the shape of <u>white, wave-like clouds</u>." (R12)

X15. General observation. Several comments by European observers to the effect that the stars twinkle only or mainly during auroras, possibly indicating the presence of refracting matter. (R13)

References

R1. Botley, Ciceley M.; "Daylight Aurorae," British Astronomical Association, Journal, 73:234, 1963. (X1)

R2. Stevenson, William; "Abstract of Observations on the Aurora....," Philosophical Magazine, 4:6:20, 1853. (X2, X3, X4)

R3. Ellison, Wm. F.A.; "Mysterious Phenomenon," English Mechanic, 111:154, 1920. (X5)

R4. Ingleby, C.M.; "Auroral Phenomenon," Nature, 23:363, 1881. (X6)

R5. Procter, Henry R.; "Peculiar Aurora," English Mechanic, 31:157, 1880. (X8)

R6. Abbe, Cleveland; Smithsonian Annual Report, 1883, p. 547. (X9)

R7. Purnell, Chas. W.; "The Aurora of September 9, 1898," Nature, 59:320, 1899. (X10)

R8. "Electrical Phenomena: Incandescent Cloud," Monthly Weather Review, 29: 466, 1901. (X11)

R9. Botley, C.M.; "Aurorae Seen In Daylight," British Astronomical Association, Journal, 69:222, 1959. (X12)

R10. Brandt, Curtis; Personal Communication, February 14, 1978. (X13)

R11. "Aurora Seen in the Daytime at Cannonmills," Edinburgh New Philosophical Journal, 3:378, 1827. (X7)

R12. Botley, Cicely M.; "Some Neglected Aspects of the Aurora," Weather, 18: 217, 1963. (X14)

R13. "Twinkling during Auroras," Franklin Institute, Journal, 115:468, 1883. (X15)

# GLA13 Glowing Night Skies

Description. Strongly enhanced general illumination of all or most of the night sky, exclusive of starlight, auroral light, and the normal airglow due to the bombardment of the solar wind and micrometeoroids. These rare milky-white enhancements may be strong enough to permit the reading of newspapers outdoors at midnight. The Milky Way and Zodiacal Light are usually obliterated. The increased levels of brightness may last several hours or persist for several nights.

Background. Country dwellers have always been aware that some moonless nights are much brighter than others. The observations recorded here, however, are so striking that they found their way into many scientific journals, as in the case of the long series of remarkably bright nights in England from 1916 through 1919.

Data Evaluation. Many naked-eye observations by scientists and amateur astronomers plus a few photometric determinations. Rating: 1.

Anomaly Evaluation. Bursts of solar wind or micrometeoroids provide a reasonable physical explanation of night sky enhancements. Some bright nights are probably auroral in nature. The real challenge here is most likely in correlating the bright nights with some astronomical event. The English bright nights of 1916-1919 must have had some specific geoastronomical cause. Rating: 3.

Possible Explanations. Sudden (but unexplained) increases in the solar wind and micrometeoroid flux probably suffice in most instances. The long 1916-1919 episode might conceivably be connected with fine particulate matter raised by the war, but this is speculation. Bright nights and high winds might be linked by solar activity changes. See, for example, weather lights, GLA16.

Similar and Related Phenomena. Transient sky brightenings (GLA14); restricted sky glows (GLA15 and 16); banded skies (GLA23); and episodes of milky sky (GWC).

Examples of Bright Night Skies

X1. November 8, 1929. Essex, England. "The moon had set, and the sky was clear, which it had not been on several previous nights. On looking out from a north window it was noticed at once without instrumental aid that the sky was exceptionally bright, and on going out into the open the exceptional brightness was seen all over the visible hemisphere, no direction being obviously favored. Although the sky was apparently quite clear, and the brighter stars gave the impression of being very bright, the luminous background was strong enough to make it difficult to distinguish the Milky Way, which, it was judged, was invisible for the same reason that it invisible when the moon is up---namely, that there is too much "false light" superposed upon it." Lord Rayleigh made photometric measurements that indicated the sky was four times its normal brightness. The auroral line was not conspicuous and no magnetic disturbances were recorded. (R1, R2)

X2. October 9, 1931. No location given, but probably England. "A misty, white light filled the northern horizon, obliterating the Zodiacal Band limit on the northern side, and so intense that it almost obliterated the Galaxy itself, where it arched up on the northern side. The sky itself was cloudless and filled with stars, which shone through this strange nebulosity. But though cloudless this white mistiness entirely destroyed the normal bluish-black contrast between the Galaxy and the night sky." (R3)

X3. About 1886. Florida. "I awoke a little before 3:00 A.M. and noticed that it was very light, and as there was no moon, got up and went out on the deck, and to my surprise saw everything illuminated with a pale greenish light so intense that we could read by it.....The atmosphere was hazy, and we could not make out the stars, and the light seemed to be general and from no particular direction." (R4)

X4. Late 1916 through 1919. England. Many reports of very bright nights during this extended period. December 23, 1916; bright sky, especially in the northeast. (R5) December 23, 1916. Much lighter than a summer solstice night. (R6) Year ending September 30, 1918. Unusual quantity of light in night skies. Astronomical photography adversely affected. (R7) September 30, 1918. Bright night. (R8) Winter nights of 1918, especially December 14, 1917 and January 13, 1918. Luminous skies. (R9) December 5, 7, and 14, 1918. Entire sky glowing with a soft white light. (R10) December 14, 1919. A remarkably light night. "On several nights the earlier hours were intensely dark, but toward midnight a soft twilight suffused the sky, which, as there was no moon, was very remarkable." (R11) November 25, 1919. Remarkable sky luminosity. Brightest moonless night ever recorded. (R12)

X5. Fall 1453. Constantinople. A dazzling light covered the town all night. (R13)
X6. June 18, 1783. (R14)
X7. January 7, 1831. Europe. Print could be read outdoors. (R14, R16)
X8. August 3, 1831. (R14)
X9. October 1855. (R14)
X10. November 18, 1859. (R14)
X11. November 26, 1859. (R14)
X12. June 30, 1861. Earth supposed to have passed through tail of Comet I 1861. (R14, R15)
X13. October 24, 1862. (X14)
X14. November 14, 1871. (X14)
X15. 1788. Mount Blanc. Whole horizon was bright, decreasing toward zenith. (R15)
X16. November 13 and 14, 1866. Attributed to Leonids. (R17)
X17. 1872. Attributed to Bielids. (R17)
X18. August 1880. Sky unusually bright the entire month. (R18)
X19. July 31, 1908. England. Sky bright, especially toward the north. (R19)
X20. August 4, 1908. England. Bright night, clouds illumined by a pink light. (R19)
X21. October 17, 1949, through the rest of the month. England. On the 17th., the whole sky was as bright as the Milky Way from 2340-0055. (R20)
X22. May 20, 1958. Indian Ocean. Whole sky bright, particularly toward the north. (R21)
X23. General observation. "When the sky is clear during a very high wind, a faint diffused luminosity has been often observed covering the whole sky." (R22, R23)
X24. General observation. Spring nights are the darkest of the year. (R24)

X25. September 29, 1910. Yerkes Observatory, Williams Bay, Wisconsin. "This was one of the brightest of the luminous nights that I have even seen. The matter seemed to be only ordinary haze but luminous for some reason. There was no trace of aurora. The sky on which the luminous haze was seen was, at this time, brightened with a pale uniform illumination covering the entire heavens and nearly blotting out the milky way." (R25)

X26. January 26, 1857. Quito, Ecuador, Sky was mostly clouded over, but it was bright enough outside to read print easily. The atmosphere itself seemed luminous. (R26, R27)
X27. July 1, 1859. St. Louis, Missouri. Everything outdoors seemed luminous. The air seemed made up of luminous particles. (R26)
X28. December 26, 1737. Kilkenny, Ireland. Much of Southern Europe was covered by a reddish, luminous haze. At Kilkenny it appeared like a globe of fire in the air, which lasted about an hour and then burst into pieces. (R28)

References


R1. Lord Rayleigh; "On a Night Sky of Exceptional Brightness....." Royal Society, Proceedings, A131:376, 1931. (X1)
R2. "An Exceptional Night Sky," Nature, 127:871, 1931. (X1)
R3. Housman, W. B.; "Luminous Sky," British Astronomical Association, Journal, 43:48, 1932. (X2)
R4. Fleming, David; "Aurora in 1886," Popular Astronomy, 29:448, 1921. (X3)
R5. Jenner, R. J.; "The Night of December 23, 1916," English Mechanic, 104:508, 1917. (X4)
R6. Burnerd, F.; "The Night of December 23," English Mechanic, 105:8, 1917. (X4)
R7. "Bright Night Skies in England," Scientific American, 120:49, 1919. (X4)
R8. Wilson, Fiammetta; "Bright Sky," English Mechanic, 108:158, 1918. (X4)
R9. "Luminous Night Skies and Zodiacal Light," British Astronomical Association, Journal, 28:167, 1918. (X4)
R10. Cook, A. Grace; "Luminous Night Skies," English Mechanic, 107:5, 1918. (X4)
R11. Grover, C.; "Luminous Night Sky," English Mechanic, 106:273, 1918. (X4)
R12. Ferguson, James; "Remarkable Sky Luminosity on November 25," English Mechanic, 110:257, 1919. (X4)
R13. Antoniadi, E. M.; "The Light Which Spread over Constantinople at Night....," British Astronomical Association, Journal, 16:239, 1906. (X5)
R14. Burns, Gavin J.; "Earthlight," Observatory, 33:169, 1910. (X6-X14)
R15. Burns, Gavin J.; "Earthlight," British Astronomical Association, Journal, 20:372, 1910. (X12, X15)
R16. von Humboldt; "On the Luminousness of the Earth," Edinburgh New Philosophical Journal, 39:339, 1845. (X7)
R17. "Meteoritic Night-Glow," Sky and Telescope, 35:485, 1968. (X16, X17)
R18. Burns, G. J.; "The Light of the Sky," British Astronomical Association, Journal, 16:308, 1906. (X18)
R19. "Abnormal Twilight," English Mechanic, 88:67, 1908. (X19, X20)

R20. "Luminous Night Skies," British Astronomical Association, Journal, 60: 247, 1950. (X21)
R21. Bradfield, F.; "Unidentified Phenomenon," Marine Observer, 29:60, 1959. (X22)
R22. Stevenson, William; "Abstract of Observations on the Aurora.....;" Philosophical Magazine, 4:6:20, 1853. (X23)
R23. "The Atmosphere Never Dark on a Windy Night," Scientific American, 3: 226, 1848. (X23)
R24. "Less Glow on Spring Nights," Science News Letter, 57:309, 1950. (X24)
R25. Barnard, E. E.; "Self-Luminous Night Haze," American Philosophical Society, Proceedings, 50:246, 1911. (X25)
R26. Jones, George; "On the Occasional Luminousness of the Atmosphere at Night on the Andes," American Association for the Advancement of Science, Proceedings, 13:291, 1859. (X26, X27)
R27. "Luminousness of the Atmosphere," Scientific American, 1:118, 1859. (X26)
R28. Guillemin, Amedee; "Length of Time during Which Aerolites Are Visible," Scientific American Supplement, 17: 6948, 1884. (X28)

# GLA14 Transient Sky Brightenings

Description. Sudden, bright flashes of light covering all or most of the sky, lasting from one to several seconds. The landscape may be lit briefly as bright as day by these generally bluish flashes.

Data Evaluation. The observations of this phenomenon are rare but of rather good quality, many being reported by scientists and amateur astronomers familiar with meteors and other luminous phenomena. The data, however, are often complicated by the presence of clouds which might conceal bright fireballs. Rating: 2.

Anomaly Evaluation. If meteors and lightning can be truly eliminated, sudden all-sky flashes constitute an important geophysical enigma. Rating: 2.

Possible Explanations. The available data are best explained by hypothesizing large, unseen meteors, possibly hidden by clouds. In most reports, however, the observers can exclude the possibility of meteors. If meteors are not the answer, sharp bursts of solar wind or micrometeoroid flux might trigger pulses of auroral or air-glow light.

Similar and Related Phenomena. Steady glows in the night sky (GLA13); superbolts of lightning (GLL); and possibly unannounced tests of nuclear weapons, such as that of September 22, 1979. (R9)

## Examples of Transient Sky Brightenings

X1. May 24, 1931. North Atlantic. "Sea and sky were suddenly quite brilliantly lit for about three seconds with a flickering purplish light, which did not appear to emanate from any particular point." The sky was overcast, so an unseen meteor is a possibility. (R1)

X2. December 5, 1956. South Pacific. "At 0643 G. M. T. a harsh, brilliant flash, bluish white in appearance, covering the whole of the night sky, appeared, lasting for approximately 1 1/2 sec. The sky was completely covered by a thick layer of Sc. The possibility of a meteor was ruled out as the cloud would not have permitted it to be visible, and the character and intensity of the flash did not conform to lightning or other electric phenomena. The forecastle head was plainly visible from the bridge and the horizon clearly defined. The source of the light appeared to issue from the NE. sky at an altitude of approximately 60°." (R2)

X3. December 1, 1959. North Atlantic. "About 2000 G. M. T. when low cloud was passing across, a bright blue diffused light suddenly grew, as from the clouds overhead.....No actual flash was seen---it was as though a diffused blue light had been gradually unshaded behind the clouds and then gradually covered again." (R3)

X4. June 9, 1970. Uttar Pradesh, India. "On a moonless and cloudless night, two Indian astronomers at a Smithsonian affiliated observatory noticed a sudden and extraordinary sky brightening. 'For about one second,' one scientist said, 'the total local sky suddenly brightened up to an intensity comparable to that of the Milky Way during these summer months.' According to the report, the event took place in two stages: first, the sky became bright; then, immediately thereafter, the brightening intensified momentarily. No explanation of the phenomenon was given." (R4)

X5. May 15, 1972. Western Alaska. The weather: overcast. "The sky suddenly brightened to what eyewitnesses call daylight intensity, remaining illuminated for approximately two seconds. Reports have come in to us from Saint Lawrence Island, Tin City (at the tip of Seward Peninsula), Galena, Bethel, and villages in between. Although the entire sky was affected, some witnesses say that the light was most intense in the west. It appeared suddenly, 'like a flashbulb,' changing color from bluish through green to white, fading to orange or reddish. Estimates of duration range from one to seven seconds. Two observers claim to have seen two closely spaced flashes..... Could this phenomenon have been a great fireball, so brilliant as to light up the clouds that hid it from view?" (R5)

X6. December 2, 1814. Peckham, England. "The night was cloudy and dark, the lower part of the atmosphere clear and calm, a very slight wind blowing from the E. Suddenly I was surrounded by a great light. I remember that at the instant I shrunk downward and stooped forward; as I was apprehensive of some danger behind me, I instantly ran a few paces......I saw nothing to cause this light. It did not give me the idea of the force and intensity of lightning; its brilliancy was not so instantaneous and fierce; but it was a softer and paler kind of light, and lasted perhaps three seconds. I could discover no noise, though immediately I expected an explosion. The strength of the light was nearly equal to that of common day-light; all near objects were distinctly visible." (R6)

X7. February 3, 1873. Huddersfield, England. "Shortly after ten o'clock the heavens were suddenly suffused with a brilliant bluish-purple colour; the greatest degree of intensity was in the zenith, gradually fading till, at the horizon, the colour was scarcely perceptible. The phenomenon lasted only a few seconds; but in a few moments a sound great resembling thunder, was heard, and this was followed in about three minutes afterwards, by the wind 'rising' and blowing with considerable force for about ten minutes." (R7) The thunder sound suggests a large unseen meteor, but the accompanying burst of wind is harder to explain. (WRC)

X8. 1884. England. "Whilst walking in the country last Monday evening, at about 10.20, I saw the following strange phenomenon. The night was dark and calm, and a drizzling misty rain was falling, when suddenly the whole heavens became illuminated for about two seconds, and then all was darkness again. The light was not vivid, but soft, and bright enough to show clearly the surrounding objects." (R8)

X9. April 24, 1894. Glasgow, Scotland. "The night of the 24th inst. was, to begin with, intensely dark. The sky was very clear, and the air cold and damp. Suddenly, however, the whole sky began to soften in aspect (if I may use the phrase), and a delicate glimmering light overspread the whole, or nearly the whole, of the heavens. This brightness increased perceptibly, and the stars paled visibly in lustre. In some ten minutes the light was gone, and the sky as black, and the stars as vivid as before. The illumination seemed to come from the N. W. I have often witnessed phenomena of this kind, the whole heavens appearing to become suffused with light which afterwards faded." (R10)

X10. December 28, 1980. South Atlantic Ocean. "At approximately 2245 GMT on a moonless night the entire ship and immediate surrounding area were illuminated by what can be best described as a great camera flash. The flash was bluish white and a small bolt of lightning appeared to be centered just above the vessel's samson posts. No noise was heard and the flash lasted only a second. The sky was clear at the time and stars of all magnitudes were clearly visible. The only clouds that could be seen were two or three small cumulus clouds; one of these was above the vessel and the others were moving towards us from the south, our course being 142°(T) and the wind being S'E, force 3. The cloud above the vessel was at a height of about 600 feet." (R11)

References

R1. Furneaux, S. J.; "Purple Light at Night," Marine Observer, 9:93, 1932. (X1)
R2. Hopkins, L. J.; "Unidentified Phenomena," Marine Observer, 27:209, 1957. (X2)
R3. Williams, J.; "Unidentified Phenomenon," Marine Observer, 30:194, 1960. (X3)
R4. Cornell, James, and Surowiecki, John; "Naini Tal Sudden Sky Brightening," The Pulse of the Planet, New York, 1972, p. 73. (X4)
R5. Busch, Thomas A.; Letter, Sky and Telescope, 44:19, 1972. (X5)
R6. Wallis, John; "Notice of a Remarkable Meteor Which Appeared on the 2d of December, 1814," Annals of Philosophy, 5:235, 1815. (X6) Note that in 1815, the word "meteor" meant any kind of atmospheric phenomenon.
R7. "Singular Atmospheric Phenomenon," English Mechanic, 16:530, 1873. (X7)
R8. Rogers, J.; "Strange Phenomenon," Knowledge, 5:171, 1884. (X8)
R9. Marshall, Eliot; "Scientists Fail to Solve Vela Mystery," Science, 205:504, 1980.
R10. Gemmill, S. Maitland Baird; "Comparative Lightness of Nights," English Mechanic, 59:244. 1894. (X9)
R11. Rutherford, N. W. C.; "Unidentified Phenomena," Marine Observer, 51:186, 1981. (X10)

# GLA15 Bright, Luminous Patches on the Horizon

Description. Bright, luminous regions or patches on or close to the horizon. These bright areas may wax and wane in intensity as well as move along the horizon. Some patches rise above the horizon as expanding discs of light; others form arches. The duration of these phenomena ranges from several minutes to several days. When sharp weather changes follow, these apparitions are classified (rather arbitrarily) as Weather Lights (GLA16).

Background. This type of phenomenon is complex with several reasonable physical causes. Luminous horizon phenomena were much discussed in the older scientific literature under such headings as "weather lights," feux d'horizon (fires on the horizon), etc. The most spectacular observation in this category occurred June 30-July 2, 1908, in conjunction with the famous Tunguska Event.

Data Evaluation. The older data are primarily testimonial with limited scientific content. The 1908 series, however, and most modern shipboard observations contain substantial detail. Rating: 2.

Anomaly Evaluation. So many likely physical explanations exist for this phenomenon (see below) that it must be rated low on the anomaly scale. Some features of the observations remain enigmatic; e.g., the possible precursors of the 1908 Tunguska Event (sky glows and halos) and the curious arches that are sometimes formed. Rating: 3.

Possible Explanations. (1) Unseen meteors, comets, and space debris; (2) Auroral phenomena (usually eliminated by spectroscopic and magnetic tests); (3) Spacecraft and rocket launches (a common cause after 1950); (4) Volcanos; (5) Distant conflagrations; (6) Steady sheet lightning beyond the horizon; (7) Refraction phenomena, especially for bright areas close to the horizon.

Similar and Related Phenomena. Weather Lights (GLA16), steady night sky glows (GLA13), transient sky brightenings (GLA14), earthquake lights (GLD8), false dawns (GLA25), the post-Krakatoa sunsets (GES), halos presumably due to the introduction of nuclei by meteors and comets.

Examples of Luminous Patches

X1. July 16, 1892. Indiana. "As we were ascending the hill on the east side of the creek and near its summit, we saw in our front the reflection of a great light from behind us. It was so noticable as to cause us both to turn about on our buggy seat and look backward. There, at a bearing S. 60° W. (that is the bearing of the road, with which the light was in alinement), we saw a great white light radiating from a point at the horizon where it was brightest, right, left, and upward to a height of 10° to 15°, weakening in brilliancy as it radiated and terminated in a dark band or segment of rainbow shape, some 10° wide. The light seemed to radiate from a point a half-radius above the centre of the circle which the black segment would indicate. Above the dark segment another segment or band of light, not so bright as the one at the horizon, formed a rainbow, or arch, some 10° to 15° wide. Above that second band of light was a light haze, or mist, through which the stars could be easily distinguished. Some 10° up in that mist, and directly over the centre of the light at the horizon, was a light about as large as a man would appear to be if suspended from a balloon a thousand feet distant. It was about four times as long vertically as wide horizontally.... When I first caught sight of it, it had the appearance of the head of a comet, only it was long vertically. When young McIntosh first saw it, it seemed to be a blaze such as large meteor appears to carry at its front. We halted and watched it about ten minutes, during which time it (the small light) slowly faded till only its locality could barely be noticed, then suddenly loomed bright almost to a white blaze, then slowly faded as before. It would loom up in five seconds, and consume five minutes in fading away. It kept the same position all the time, for we watched its position with relation to the stars to see if it moved..... This little light loomed up and faded four times when the big light under it faded also and made it dark there." (R1) See also GLA2-X5 for a simultaneous phenomenon.

X2. June 30, July 1-3, 1908. The British Isles and northern Europe. Glowing night skies to the north, unrecognized at the time as due to the Tunguska Meteor fall. "On the night of June 30, and to some extent on the nights succeeding, brilliant skyglows were observed in various parts of the country, and lasted throughout the night. At 9h 30m P. M. at Greenwich on June 30, the sky along the northwest and north horizon was of brilliant red, in fact there was what is usually termed a 'brilliant sunset,' the only peculiarity being that the brightness stretched more to the north than is usual, and endured, so that at one o'clock in the morning it extended well across the north of the horizon, and the northern sky above was of a brightness approaching that of the southern sky at the time of Full Moon. The light, indeed, was sufficient to take photographs of terrestrial objects." There was no sign of the auroral spectrum. "A long-lasting solar halo was seen in the forenoon and afternoon of June 30, and another on July 1. (R2) "In common with many, I observed the brilliant skies, June 30-July 1 and July 1-2. The colour was nearly or quite white, except close down to the northern horizon, where it was rosy. Also, an occasional high cirrus reflected reddish light. The dark segment and the radial lights of the aurora were absent. The point of greatest brilliance followed the sun, whilst aurora radiates about the magnetic North. The spectroscope gave negative results..... The magnets at the time were in a state of restfulness---not disturbed, as in the case of aurora..... All through June and early in July the upper atmosphere has contained a very considerable amount of ice crystal, which made itself visible by producing solar and lunar halos." (R3) "..... June 29 already giving signs of something unusual at night, but nothing like the night of June 30, or even that of July 1." (R4) (Could the muted display of June 29 and the earlier halos been precursors? (WRC) Peculiar clouds were also seen in the United States. (R5) Many other observations of the nocturnal glows are on record. (R6-R21, R39)

X3. July 31, 1916. Ireland. "On the night of July 31 a wonderful light appeared in the sky just above the horizon. I saw it a little after 11 p. m., and it looked like an immense fire. At 11.15 p. m. the light moved along the sky to the North-east, and then appeared as a great luminous star. At 12 p. m. it moved back again in a Northerly direction, and showed as a great blaze of fire in the same place as I first saw it, and there it shone until 4 a. m. the following morning, when it disappeared." (R22)

X4. September 1922. Sussex, England. "While standing about twenty yards from the seashore and looking south out to sea, the horizon and a region slightly above it (elevation only about 1° or 2°) were lit up by a faint white light which extended laterally over a segment subtending an angle of about 30°..... The appearances of the light were not the same to my wife as to myself. Her impression of it was that it was a light which she saw only if her eyes followed it, yet it consisted of a long streak of light parallel to the horizon with a break in it and then another small streak. My impression was that of a light which appeared to flash up over the horizon, subtending the angles already noted, the flashes not succeeding each other regularly." (R23)

X5. March 1, 1941. North Carolina. "A sky phenomenon, which attracted considerable attention, was observed here and at other places in the vicinity of Raleigh on the evening of March 1. It was a striking light effect visible in the western sky from 7:00 to 9:00 P. M. As I observed it about 8:00 o'clock its position and form were almost identical with that of the Zodiacal Light. Its appearance as to brightness and color was similar to that of a solar corona at the time of an eclipse. It was brightest just above the horizon. It was triangular shaped, as the Zodiacal Light, and the new moon was almost in the center of the apex towards the top. Numbers of our students here saw it and they report that it reached its maximum brightness about 7:30. Some of them reported coloring effects associated with it, and some observed a luminous patch in the East." The author also noted communications problems due to magnetic storms at the time. (R24)

X6. October 13, 1951. South Atlantic. "Between sunset at 1954 G. M. T. and 2040, a reddish glare spread from the western horizon until it covered the sky half-way to the zenith over an arc of 90°. The glare was reflected on the sea surface and combined with the silvery light from the moon gave an eerie effect which was most unusual." (R25)

X7. March 20, 1969. Caribbean Sea and Western North Atlantic. "At 2315 GMT, almost an hour after sunset, a semicircle of milky-white light became visible in the western sky and rapidly expanded upward and outward during the next 10 min. When first seen it was quite bright (e. g., as clearly defined as a high cloud across the face of a full moon) but, as it expanded, it became more diffuse. It was possible to follow the expansion to an altitude of 50° when the circumference of the arc of the light cut the horizon at points bearing 280° and 350° respectively." (R26) This and the following example may be spacecraft launches. (WRC)

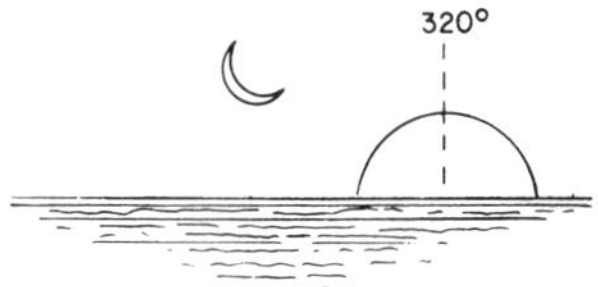


*Semicircle of milky-white light expands in the western sky.*

X8. June 22, 1976. North Atlantic. "At 2113 GMT a pale orange glow was seen to be coming from behind a bank of towering cumulus to the west. At 2115 a ghostly white disc, see sketches, was observed at an approximate altitude of 10° and bearing 290°. The glow from behind the cloud persisted." The glowing region developed as indicated in the figure. Stars could be seen through the disc at all times. By 2140 the disc had disappeared. (R27)

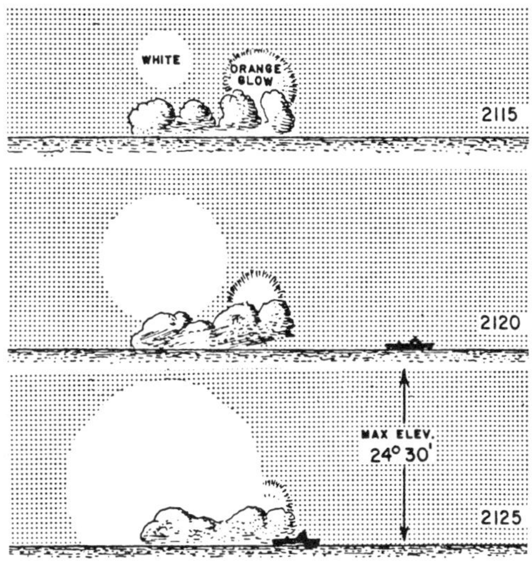


*Ghostly white disc seen in the North Atlantic.*

X9. March 24, 1977. North Atlantic. "At 0855 the look-out observed what appeared to be a searchlight shining downwards for about 10 seconds on a bearing of 300°T and 20° above the horizon. This light was extinguished and was replaced by a luminescent patch of approximately one degree in diameter. A semi-circular area of overall moderate luminosity formed about the luminescent patch. This took about three minutes to form and the dimensions are shown in the sketch. When this had formed another luminescent patch was also observed above the semi-circular area and after a total period of seven minutes the phenomenon dispersed completely." (R28)

X10. General observation. "Mr. C. Atwater has communicated an account of a spot or spots, near the horizon, appearing as if lighted, and giving rise to a belief that there was a great fire in that direction. He remarks that he has often noticed these light spots in Ohio, but not on the east of the Alleghanies. I would only remark that I have observed similar phenomena in New-England." (R29)

X11. September 11, 1898. West Virginia. Oval luminous "cloud" 5-6° long appeared and disappeared 10-20 times. No auroral activity noted. (R30)

X12. December 1915. England. A glow in the northern sky, but no magnetic activity to suggest an aurora. (R31)

X13. February 11, 1930. South Pacific. A bright light on the southern horizon after 30 seconds formed into an arc, which moved about 10° southward along the horizon. (R32)

X14. October 7, 1957. Aden. Pale green glow in sky over Indian Ocean, 25° high. Lasted 2.25 hours. (R33)

X15. August 22, 1969. West Indies. Luminous area bearing 310° grew in size and rose in altitude, then turned into an arch or crescent. (R34)

X16. June 30, 1970. Six locations in the Caribbean and Atlantic. Semicircle of milk-white light appeared on horizon and grew rapidly in size. Correlated with a spacecraft launch from Florida. (R35)

X17. September 25, 1972. North Atlantic. Expanding luminous area on horizon, bearing 310°. (R36)

X18. September 20, 1977. Northwestern Soviet Union. Huge flare of light in the sky, spreading like a "jellyfish." Thin light rays emanated from the central mass. Correlated with Soviet satellite launch. (R37, R38)

X19. November 27, 1963. North Atlantic Ocean. "At 1926 GMT an illuminated body was observed bearing 270°, altitude 30°. It appeared at first to have a suffused glow about it, but as the object moved parallel with the ship's course, the glow assumed the definite form of a tight spiral of blue-white light. The spiral expanded to a maximum radius of about 5° with about 12 turns visible at one time....The whole phenomenon gave the impression of looking into a conically formed spring and was indeed a most sensational sight." This phenomenon was correlated with the Florida launch of the _Centaur 2_ rocket. (R40)

## References

R1. Campbell, John T.; "Auroral Display," _Science_, 20:66, 1892. (X1)

R2. "A Protracted Twilight," _Observatory_, 31:325, 1908. (X2)

R3. Dennett, Frank C.; "The Glorious Midnight Sky," _English Mechanic_, 87:534, 1908. (X2)

R4. Buss, Albert Alfred; "Nocturnal Northern Glows," _English Mechanic_, 87:581, 1908. (X2)

R5. "Sky Glows," _Scientific American_, 99:139, 1908. (X2)

R6. de Roy, Felix; "Nocturnal Glows,"

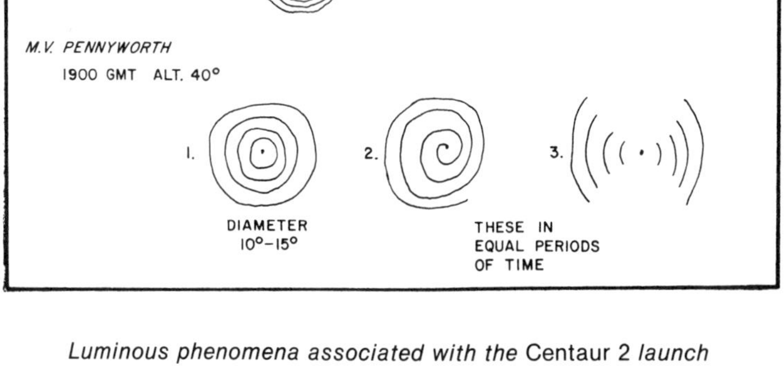


*Luminous phenomena associated with the* Centaur 2 *launch*

English Mechanic, 87:559, 1908. (X2)
R7. Hollis, H. P.; "The 'Aurora' of June 30," English Mechanic, 87:553, 1908. (X2)
R8. Gemmill, S. Maitland Baird; "The Glorious Midnight Sky," English Mechanic, 87:558, 1908. (X2)
R9. McHarg, John; "The Abnormal Twilight," English Mechanic, 87:581, 1908. (X2)
R10. Brauner, Bohuslav; "The Recent Nocturnal Glows," Nature, 78:221, 1908. (X2)
R11. Phillips, J. W., and Brauner, Boruslav; "Brilliant Sky Glows, June 30 and July 1, 1908," Royal Meteorological Society, Quarterly Journal, 34:202, 1908. (X2)
R12. Dennett, Frank C.; "The Brilliant Midnight Heavens, July 1 and 2," Knowledge, 5:187, 1908. (X2)
R13. "Brilliant Sky Glows," Nature, 125: 994, 1930. (X2)
R14. Crowther, J. G.; "More about the Great Siberian Meteorite," Scientific American, 144:314, 1931. (X2)
R15. "The Siberian Meteor of June 30, 1908," Nature, 127:719, 1931. (X2)
R16. Burns, Gavin J.; "The Unusual Luminosity of the Night Sky on the 30th of June, 1908," British Astronomical Association, Journal, 42:223, 1932. (X2)
R17. Burns, Gavin J.; "The Great Siberian Meteor of 1908," Popular Astronomy, 41:477, 1933. (X2)
R18. "The Siberian Meteor of June 30, 1908," Nature, 134:816, 1934. (X2)
R19. Cowan, Clyde, et al; "Possible Anti-Matter Content of the Tunguska Meteor of 1908," Nature, 206:861, 1965. (X2)
R20. Vasil'yev, N. V., and Fast, N. P.; "New Material on the 'Light Nights' of Summer 1908," Physics of Mesospheric (Noctilucent) Clouds, Jerusalem, 1973, pp. 80-85, (X2)
R21. Rich, Vera; "The 70-Year-Old Mystery of Siberia's Big Bang," Nature, 274:207, 1978. (X2)
R22. Rigel; "A Wonderful Light in the Sky," English Mechanic, 104:71, 1916. (X3)
R23. R., S.; "A Curious Luminous Phenomenon," Nature, 110:481, 1922. (X4)
R24. Speas, W. E.; "Sky Illumination Observed," Popular Astronomy, 49:226, 1941. (X5)
R25. Lougheed, F.; "Unusual Sky Phenomenon," Marine Observer, 22:196, 1952. (X6)
R26. Angus, F. S., and Carling, G.; "Optical Phenomenon," Marine Observer, 40:17, 1970. (X7)
R27. Moore, R.; "Unidentified Phenomenon," Marine Observer, 47:66, 1977. (X8)
R28. Brackenbridge, M.; "Unidentified Phenomenon," Marine Observer, 48:21, 1978. (X9)
R29. Webster, N.; "Luminous Appearance in the Atmosphere," American Journal of Science, 1:12:380, 1827. (X10)
R30. Newcomb, Simon; "An Unusual Aurora," Science, 8:410, 1898. (X11)
R31. Legh-Powell, R. T.; "Luminosity in Northern Sky," English Mechanic, 102: 464, 1915. (X12)
R32. Higgs, W. G.; "Arc of Light in Sky," Marine Observer, 8:39, 1931. (X13)
R33. Hirst, N. F.; "Observation of Optical Phenomena at Sirah Island, Aden, 7 October 1957," Meteorological Magazine, 87:278, 1958. (X14)
R34. Holmes, R. A.; "Unidentified Phenomenon," Marine Observer, 40:107, 1970. (X15)
R35. Kissane, D. M., et al; "Unidentified Phenomenon," Marine Observer, 41: 63, 1971. (X16)
R36. Dyer, H. K.; "Unidentified Phenomenon," Marine Observer, 43:114, 1973. (X17)
R37. "Tass Reports 'Jellyfish' of Light in Sky," Baltimore Sun, September 23, 1977, p. A3. AP item. (X18)
R38. "Soviet UFO Due to Secret Launch," Science News, 112:230, 1977. (X18)
R39. Whipple, F. J. W.; "On Phenomena Related to the Great Siberian Meteor," Royal Meteorological Society, Quarterly Journal, 60:505, 1934. (X2)
R40. Smith, Captain; "Unidentified Phenomenon," Marine Observer, 34:181, 1964. (X19)

# GLA16 Weather Lights or Storm Lights

Description. Luminous whitish or reddish patches on the horizon, often mobile, and located at any azimuth. The appearance of weather lights is by definition followed by a sharp weather change, generally a storm with high winds.

Background. European folklore and local belief has long held that the aurora presages stormy weather, as elaborated upon in the examples below. The Nineteenth Century American concept of weather lights or storm lights is closely related, perhaps identical in terms of geophysical phenomena. It should be noted that the artificial illumination of modern civilization almost precludes the detection and study of weather lights, assuming they exist.

Data Evaluation. No satisfactory scientific studies of the aurora-weather relationship and weather lights have been discovered. Only general, qualitative, rather old observations support the existence of this class of phenomena. Rating: 3.

Anomaly Evaluation. If storms or sharp weather changes are truly announced by luminous appearances on the horizon, this is an unrecognized phenomenon in meteorology. Rating: 2.

Possible Explanations. Solar activity is strongly correlated with auroral activity and rather weakly related to terrestrial weather. There is, in consequence, a recognized link between general auroral activity and weather, though not necessarily weather lights. If weather lights are not auroral in nature, they could be associated with distant sheet lightning or other electrical phenomena. Another possibility is that advancing weather changes modify the optical conditions of the atmosphere, refracting light, mirage-like, over the horizon. Finally, some nocturnal lights (GLN1) seem to presage storms. See especially GLN1-X42.

Similar and Related Phenomena. Auroras correlated with thunderstorms. (GLA9) The post-auroral appearance of cirrus clouds and high hazes. (GLA12) The purported brightness of windy nights. (GLA13) See also silent lightning (GLL4).

Examples of Weather Lights

X1. General observation. "Before a storm here, I have often noticed, in an evening of the latter part of autumn, and sometimes in the winter, a phenomenon not recollected by me to have been seen on the east side of the Alleghanies: Some one spot or spots near the horizon, in a cloudy night, appeared so lighted up, that the common people believed there was some great fire in the direction from which the light came. I have seen at once, two or three of these luminous spots, not far from each other: generally there is but one, and a storm invariably proceeding from the same point near the horizon, succeeds in a few hours. (R1)

X2. General observation. There is a phenomenon of common occurrence in this part of the country, connected with the atmospheric influences that move about the earth during the seasons of mild temperature, which I had never been able to explain according to any laws with which I was acquainted. I allude to the lights frequently seen in a clouded sky at night in advance of storms. These lights, some suppose, are caused by fire on the prairies; while by others they are called storm lights, and are said to indicate snow. They differ somewhat from the Aurora Borealis, but still are analogous." (R2)

X3. General observation. "An old friend of mine told me 40 years ago that a relation of his, who commanded for two or three years a cruiser in the Channel, had observed that whenever an aurora appeared the wind would blow hard from the S. W. in 48 hours. I have myself verified this as the unfailing result of my own observation ever since." (R3)

X4. General observations. "Amongst peoples living in the auroral zone there has always been some association between the aurora and the weather, much of it confused and contradictory, but none the less with what seems to have been sound observation. A Finnish name for the aurora is Vindlys = Windlight. Scottish farmers of the last century held that the first great storm of the year was heralded by an aurora. Observations between 1843 and 1870 at Uckfield, Sussex (lat. 51°N), made during an aurorally 'rich period', showed that after the brilliant displays occasionally seen, very stormy weather followed almost invariably after 10 to 14 days. The Southern Cross Expedition, which wintered at Cape Adare, Ross Dependency (72° S), in 1899-1900 found that brilliant and active aurora, observed to the north, was followed, too often for coincidence, by a violent storm from the southeast." (R4)

X5. Circa 1873. Tennessee. Weather lights seen to southwest. Violent whirlwind passed over town. (R5)

X6. February 26, 1884. Fayetteville, Arkansas. Weather lights on horizon were the precursor of a violent snowstorm. (R6)

X7. General observation. Major auroral displays followed by gale in 48 hours. (R7)

X8. General observation. Auroras often followed by wind and rain. (R8)

References

R1. Atwater, Caleb; "Facts and Remarks Relating to the Climate,...." American Journal of Science, 1:11:224, 1826. (X1)

R2. Hall, J.; "Storm Lights," Scientific American, 9:274, 1854. (X2)

R3. Slatter, John; "Aurorae and Storms," Symons's Monthly Meteorological Magazine, 33:182, 1898. (X3)

R4. Botley, C. M.; "The Aurora and the Weather," Weather, 20:117, 1965. (X4)

R5. Carnes, A. C.; "The Cause of the Aurora Borealis," Scientific American, 28: 373, 1873. (X5)

R6. Conrad, Cuthbert P.; Americal Meteorological Journal, 1:81, 1884. (X6)

R7. Denning, William F.; "Aurorae Boreales and Gales," English Mechanic, 15: 644, 1872. (X7)

R8. Clarke, W. B.; "Notice of a Singular Electrical Phenomenon on the Night of Feb. 18, 1837," Magazine of Natural History, 1:220, 1837. (X8)

# GLA17 Curious Folklore: Auroras and Silken Threads

Description. Folklore and some questionable testimony claim that silken threads sometimes fall during auroral displays.

Background. This charming thought seems to owe its origin to the analogy between the aurora's motions and patterns and the operation of a loom, perhaps reinforced by the occasional swishing sounds accompanying active auroras (GSH).

Data Evaluation. Obviously the testimony is highly questionable. This "phenomenon" is included only because of its similarity to other phenomena, as described below. Rating: 4.

Anomaly Evaluation. If true, a most remarkable event. Rating: 2.

Possible Explanations. None.

Similar and Related Phenomena. There are so many seemingly related phenomena that this little tale about auroras and silken threads, though ridiculous on the surface, must be included in this catalog. (1) Gelatinous meteors or pwdre ser events involve luminous phenomena and the fall of strange substances (GWF); (2) The folk tales relating the anomalous appearance of fibers after earthquakes, which of course may be accompanied by luminous phenomena(GQE); (3) Many remarkable falls of silken threads or gossamer, as produced by several species of spiders; and (4) The belief by some scientists in the past that auroras are "material" things (GLA12).

Example

X1. General observation. "The auroral light sometimes is composed of threads like the silken warp of a web; these sometimes become broken and fall to the earth, and possess exquisite softness and a silver luster, and I denominate these as the products of the silkery of the skies. I once obtained a small piece, which I preserved." (R1)

References

R1. Merriam, Mr.; "Brilliant Atmospheric Phenomena," Scientific American, 1: 178, 1859. (X1)

# GLA18 Correlation with Lunar Phase

Description. The dependence of auroral frequency upon the phase of the moon, with minimum auroral frequency occurring at full moon. Analysis with modern instruments confirm that this phenomenon is real and not subjective due to the moonlight masking auroral light.

Background. For over a century, observers have maintained that auroras are less frequent during full moons. (R1)

Data Evaluation. This correlation, originally based upon simple visual observation, has now been fairly well established by optical instrumentation. Rating: 2.

Anomaly Evaluation. Some reasonable physical mechanisms have been proposed to explain this phenomenon; only the details are in question. Rating: 3.

Possible Explanation. The passage of the moon through the solar plasma streaming toward the earth may delay or deflect some of the particles that would otherwise contribute to the formation of auroras.

Similar and Related Phenomena. The correlation of precipitation with lunar phase (GWP). The effect of the moon on geomagnetic parameters (GMS).

Examples

X1. General observations. "A significant periodicity in the occurrence frequency of aurora has been reported by Stoffregen. Not only does the length of the period (29.5 days) correspond to the synodical period of the moon, but there is consistently less aurora during periods of full moon." (R2)

References

R1. Oliver, J. A. B.; "Influence of the Moon upon Aurorae," English Mechanic, 33: 259, 1881.

R2. Henriksen, K., et al; "Lunar Influence on the Occurrence of Aurora," Journal of Geophysical Research, 82:2842, 1977. (X1)

# GLA19 Auroras Interacting with Lunar Halos

Description. Apparent physical interactions between auroral displays and lunar halos. One type of interaction occurs when the two luminous phenomena overlap and the points of intersection exhibit enhanced luminosity or, in the case of dark spaces between auroral streamers, reduced halo luminosity. A second type of relationship is seen in the appearance of halos of unusual radii during auroral displays or shortly thereafter.

Data Evaluation. A few rather vague general observations, mostly rather old, plus a single well-observed example from the Twentieth Century. Rating: 3.

Anomaly Evaluation. One would expect to observe some enhancement of light where the two displays overlap, but the dark spaces between auroral streamers should not mask any lunar halos. The reason for this is that auroras are basically very high altitude phenomena that are separated from the realm of lunar halos by hundreds of miles of altitude. Likewise, one would not expect auroras to be correlated with the appearance of halos of any radius, much less those of unusual radii. Rating: 2.

Possible Explanation. Only very low auroras (GLA4) could obscure portions of a lunar halo display, and even here obscuration is difficult to understand if low-level auroras are caused by precipitating particles, which should not absorb the halo's light. If, however, low-level auroras involve a different mechanism, say one dependent upon ice crystals or other nuclei, the aurora-halo interaction phenomena might be explained.

Similar and Related Phenomena. The relationship between auroras and clouds/haze (GLA12);

low-level auroras (GLA4), auroral fogs and mists (GLA21), black auroras (GLA22), and halos of unusual geometries (GEH).

Examples of Interactions between Auroras and Lunar Halos

X1. March 28, 1928. Cambridge, Massachusetts. "The lunar halo first attracted my attention about 11 P. M., Eastern Standard Time. A few minutes later the halo bore a fringe strikingly suggestive of a solar corona. By 11:30 a well-defined auroral fan was centered at a point on the horizon directly under the moon (at first quarter) with a streamer extending from the horizon directly past the moon and vertically upward. Oblique streamers arranged themselves approximately symmetrically about this line. One of these passed above Capella and at 11:50 P. M. extended across Polaris, and between Vega and Hercules to the eastern horizon. A similar brilliant streamer mounting from the auroral center extended southward, passing Procyron and Regulus in Leo. These two streamers crossed the lunar halo very nearly on the same circle of altitude as the moon and the crossing points were marked by exceptionally brilliant patches of auroral light. The outstanding feature of the phenomenon was the existence of horizontal streamers extending several degrees through the patches and diverted away from the moon as an apparent radiant point. These horizontal streamers with an apparent radiant at the center of the halo made angles of 20°-30° with the long auroral streamers but gave every appearance of a true auroral effect." (R1)

X2. 1894. Louth, England. During an auroral display. "At this time there was a brilliant halo around the moon, about 20° in diam., and a noticable feature was that the dark spaces between some of the streamers obliterated the halo where they crossed it. I paid most particular attention to this point, and am certain that it was not an effect of contrast alone, but an actual cutting out of the light of the halo." (R2)

X3. General observation. "Also in association with the aurora, halos of the unusual radii 10° and 15° were noticed by Richardson (1855) and on the 1820 Franklin expedition. Other halo observations include a greenish halo noted at Uppsala on 4 February 1874, a 'peculiar' corona recorded on 3 March in the same year at Peckloh in 'cirrus' of radius 20°. A particularly interesting observation was made at Bossekop on 2 November 1838 when an auroral ray reached the ordinary halo of 22° radius, 'this is wider and its light appears more condensed.'" (R4) "Baron Wrangel remarks, that, in Siberia, when shooting stars pass across the space occupied by polar lights, that fiery beams suddenly arise in the place traversed by the shooting star. Further, that, when

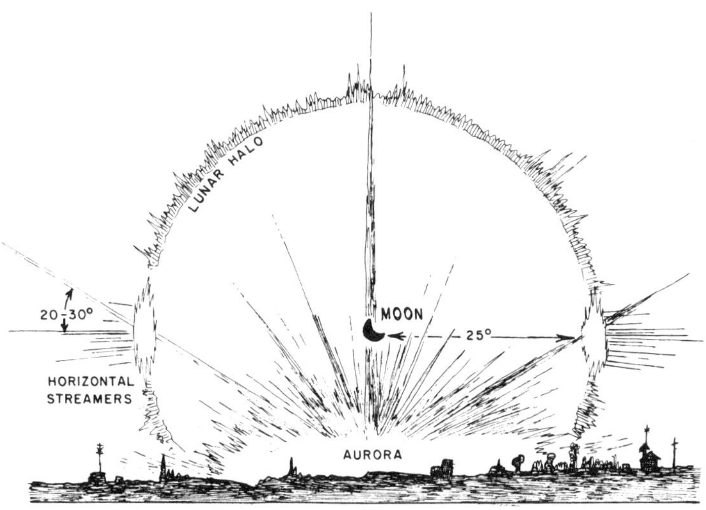


*Observed interactions between an aurora and lunar halo*

a polar beam rises high, it gradually forms a luminous circle around the moon, at a distance from her of from $20^{o}$ to $30^{o}$; remains in this form for a short time, and then disappears. (R3)

References

R1. Stetson, H. T.; "The Aurora of March 28, 1928," Science, 67:394, 1928. (X1)
R2. G., F. J.; "Auroral Display," English Mechanic, 59:198, 1894. (X2)
R3. "Polar Lights in Siberia," Edinburgh New Philosophical Journal, 3:381, 1827. (X3)
R4. Botley, Cicely M.; "Some Neglected Aspects of the Aurora," Weather, 18: 217, 1963. (X3)

# GLA20 Electrical Effects of Auroras at the Earth's Surface

Description. Surface electrical phenomena coincident with auroral activity, including physiological effects reported by observers, suppression of radio reception, fluorescence of minerals, strong perturbations of atmospheric electrical fields, and strong currents induced in long terrestrial conductors.

Background. The detection of appreciable electrical effects at the earth's surface during strong auroras supports the claims that visible auroras sometimes descend below the 80 kilometer limit usually assigned (GLA4), except in the case of induced currents in long conductors which are observed frequently during normal auroras.

Data Evaluation. Modern quantitative studies have firmly established the reality of induced currents in conductors and electric field changes, but the other potential effects mentioned above have not been well-substantiated. Overall rating: 2.

Anomaly Evaluation. When auroral action occurs there are normally powerful electric currents in the high atmosphere; thus, some induced surface electrical effects should be expected. The disruption of terrestrial communications, for example, is common when geomagnetic activity is high. The issue here is whether these surface electrical effects are strong enough to suppress radio reception, to be felt physiologically, to cause minerals to fluoresce and (especially) to help account for the numerous reports of low-level visible auroras (GLA4). Rating: 2.

Possible Explanations. As mentioned above, some surface electrical effects are to be expected, but it is not clear that they can be strong enough to create low-level auroras, the electrical odor occasionally experienced, and other electrical phenomena which seem best explained in terms of the slow, large-area discharge of electricity to space.

Similar and Related Phenomena. Low-level auroras (GLA4), the disruption of terrestrial communications during geomagnetic storms, the surface electrical effects of meteors (AYO), and the oft-reported suppression of radio reception and other electrical disorders during UFO sightings.

Examples of Auroral Electrical Effects

X1. Circa 1883. Human physiological effects. "There was an unusually brilliant aurora, and it was remarked by several that the streamers seemed to be very near to us; and presently, as we stood in the open air with heads uncovered, we began to feel the sensations produced by proximity to a body charged with electricity. The fact that such a sensation had actually been produced by the aurora was doubted by some scientific men to whom I mentioned it; and it was attributed to imagination, which, I fear, is guilty of much, and often accused of more." (R1)

X2. Winter 1929, Norwood, Ontario. Auroral effects on a radio receiver. "He observed that his radio set was 'dead', though the valves were alight and the aerial and ground wires were connected; no reason for the absolute quietness of the set was apparent. Looking out, he saw that a bright aurora was in progress; going outside to get a better view he found, to his astonishment, that a curtain of streamers extended from the sky to the ground and completely surrounded the house but at a distance of several feet. The curtain was many-coloured and unsteady, its scintillations being accompanied by 'visible sparking' and by 'snapping sounds.'" (R2) Radio suppression has also been noticed during milky seas (GLW9-X32).

X3. March 14, 1858. Armagh, Northern Ireland. Possible fluorescence during an aurora. "I availed myself of the opportunity to try whether this light was rich in those highly refrangible rays which produce fluorescence, and which are so abundant in the light of electric discharges; and I found it to be so. A drop of disulphate of quinine on a porcelain tablet seemed like a luminous patch on a faint ground; and crystals of platinocyanide of potassium were so bright, that the label on the tube which contained them (and which by lamplight could not be distinguished from the salt at a little distance) seemed almost black by contrast. These effects were so strong in relation to the actual intensity of the light, that they appear to afford an additional evidence of the electric origin of this phaenomenon, and as such I hope they may be of some interest to your readers." (R3)

X4. Auroral effects on terrestrial electric fields. "Evidence is presented here for coupling between the upper and lower atmosphere by electrical means related to aurorae. Electric field data obtained with rocket-launched parachute-borne probes are shown to be consistent with the hypothesis that aurorally-produced radiation can short out vertical electric field 'cells' in the mesosphere, with corresponding effects on the lower atmosphere electric field." (R4, R5)

X5. Auroral effects on power lines. "As human activity advanced northwards, its supporting systems, such as power transmission lines and oil/gas pipelines, extend across the auroral zone. As the auroral electrojet, a concentrated ionospheric current along the auroral zone, induces a potential of up to 1 V $km^{-1}$ on the Earth's surface, a significant amount of electric current can be induced in long conductors which are grounded at points separated by a long distance." (R6, R7)

References

R1. Quimby, E. T.; "Was It Imagination?" Science, 5:4, 1885. (X1)

R2. "A Low Aurora and Its Effect on a Radio Receiver," Nature, 127:108, 1931. (X2)

R3. Robinson, T. R.; "On Fluorescence Produced by the Aurora," Philosophical Magazine, 4:15:326, 1858. (X3)

R4. Hale, Leslie C., and Croskey, Charles L.; "An Auroral Effect on the Fair Weather Electric Field," Nature, 278:239, 1979. (X4)

R5. Freier, George D.; "Auroral Effects on the Earth's Electric Field," Journal of Geophysical Research, 66:2695, 1961. (X4)

R6. Akasofu, S.-I., and Aspnes, J. D.; "Auroral Effects on Power Transmission Line Systems," Nature, 295:136, 1982. (X5)

R7. Akasofu, S.-I., and Merritt, R. P.; "Electric Currents in Power Transmission Line Induced by Auroral Activity," Nature, 279:308, 1979. (X5)

# GLA21 Auroras and Surface Fogs and Mists

Description. Luminous fogs and mists observed at the earth's surface, usually during auroral activity. This phenomenon may be identical to low-level auroras (GLA4), with the observer being actually immersed in the active auroral region instead of viewing it from afar.

Data Evaluation. Enough reliable reports of luminous fogs and mists exist in the scientific literature to lend this phenomenon credibility. It is the interpretation of these observations that is controversial. Rating: 2.

Anomaly Evaluation. If luminous fogs and mists are actually created by the same physical process responsible for normal, high altitude auroras, they represent an important class of anomalies because auroral activity is supposedly limited to 80 kilometers and higher. See the discussion under low-level auroras (GLA4). Rating: 2.

Possible Explanations. As in the case of low-level auroras, some scientists suggest that luminous fogs and mists derive from light originating in normal auroras that is reflected and scattered by ordinary, nonluminous mists. A more radical interpretation is that the observer is actually in the midst of a large-area brush discharge (GLD).

Similar and Related Phenomena. Mountain-top glows (GLD1); brush discharge; large-scale St. Elmo's Fire (GLD3), those marine phosphorescent displays that seem to transpire well above the ocean's surface, sweeping past the observer in waves (GLW3). The older literature tells of luminous "dry fogs," blasting fogs, and strange mountain fogs (pogonips in which it is difficult to breathe, GWF). Low-level auroras (GLA4) may be closely related to luminous fogs, the latter being an example of an aurora at ground level.

Examples of Auroral Fogs and Mists

X1. General observations. "'I must yet add, says Gisler, 'that people who had travelled in Norway, informed me they have sometimes been overtaken, on the top of mountains, by a thin fog, very similar to the northern lights, and which set the air in motion; they called it Sildebleket (Haring's Lightning), and said that it was attended by a piercing cold, and impeded respiration.'" (R1)

X2. Winter of 1878-1879. The Arctic. I can maintain with full certitude, Baron Nordenskjold says, that the lighted segment of clouds which we saw during the winter of 1878-1879 had this origin; and most probably, several luminous mists which we saw during the nights of March 18 and 20, close by our ship, close by the ice, were due to the same cause; but I cannot affirm that quite certainly." Nordenskjold thought the luminous mists were perhaps ordinary mists lit up by the aurora. (R2)

X3. No date. Cape Breton, Canada. "It was my good fortune to observe an aurora, which to my eyes was embodied in and swept the earth with successive banks of Cape Breton fog....In this fog bank hung, as it were, a brilliant curtain of light, with a wide fringe or flounce of maximum brilliancy, along the bottom edge, the light fading upwards along the curtain, but traceable to the very zenith, and the curtain stretching from the eastern horizon out at sea to the western horizon on the low hill tops. The curtain was evidently vertical, thin, straight, long enough to reach from one limit of the vision to the other, and floating broadside before the south wind towards the north. No reasoning could convince us (he had a companion) that these were not elements of the phenomenon, and, moreover, that the lower edge of the bright fringe was more than one or two hundred yards away at its nearest point when we first saw it." The observer here was Professor J. P. Lesley. (R3)

X4. Winter 1908-1909. Hartford, Connecticut. "During the winter of 1908-1909, while attending Trinity College at Hartford, Conn., I observed a magnificent aurora. The light effects gave me the impression that the atmosphere was filled with fog, and that someone was illuminating it by playing a searchlight back and forth. The effect was very striking because the display was so close to the ground that I seemed to walk right through the illuminated fog." (R4)

X5. Winter 1917-1918. No location given. "Some of the accounts of low aurorae seen between the observer and terrestrial objects are very striking and circumstantial. One writer states that he and his party (members of a government radio station, in the winter of 1917-18) were enveloped in 'a light mist or fog-like substance in the aurora'; a hand extended could be seen as if in a coloured fog and a kaleidoscope of colours was visible between the hand and the body. It was impossible to feel this visible fog or mist, and there was no dampness. By stooping close to the ground it was possible to see under this light, which did not go below four feet from the ground." (R5)

X6. 1783. No location given. Dry fog luminous at night. (R6)

X7. 1831. No location given. Dry fog luminous at night. (R6) January 7, 1831. Italy. A luminous mist so bright that print could be read at midnight. (R7)

X8. February 19, 1852. Wyoming, New York. Auroral vapor near ground. (R8)

X9. July 7, 1928. Big Sauble Point, Lake Michigan. Auroral display associated with fog and high wind. Bright enough to read fine divisions on watch. (R9)

X10. No date. Canada. Engineer sees some

of his men walk into an aurora. (R10)

References

R1. "On the Noises That Sometimes Accompany the Aurora Borealis," Edinburgh New Philosophical Journal, 1:156, 1826. (X1)
R2. K., P.; "The Aurora," Nature, 25: 321, 1882. (X2)
R3. Abbe, Cleveland; "The Height of the Aurora," Nature, 60:130, 1899. (X3)
R4. Kelley, Floyd C.; "Audibility of Auroras and Low Auroras," Nature, 133: 218, 1934. (X4)
R5. Chapman, S.; "The Audibility and Lowermost Altitude of the Aurora Polaris," Nature, 127:341, 1931. (X5)
R6. von Humboldt, Baron; "On the Luminousness of the Earth," Edinburgh New Philosophical Journal, 39:339, 1845. (X6, X7)
R7. "Luminosity," Nature, 125:33, 1930. (X7)
R8. Olmsted, D.; "Great Aurora Borealis of February 19th, 1852," American Journal of Science, 2:13:426, 1852. (X8)
R9. Lemon, Harvey B.; "The Auroral Display of July 7," Nature, 122:167, 1928. (X9)
R10. Eve, A.S.; "Northern Lights," Smithsonian Institution Annual Report, 1936, pp. 149. (X10)

# GLA22 Black Auroras

Description. Intensely black, well-defined arches and streamers analogous to the normal luminous features of auroral displays.

Background. Occasional black patches appear in auroral displays, but "negative" displays are very rare and may, in fact, not be objective phenomena.

Data Evaluation. The observations of black auroras on record are few in number and all more than a half century old. Further, contrast effects and illusions may play important roles in this phenomenon. Rating: 3.

Anomaly Evaluation. If black auroras are not subjective and/or illusory, they may indicate the presence of geometrically arrayed streaks of opaque material, supporting the often-suggested link between auroras and daytime clouds (GLA12). Such a notion is contrary to modern theories of the aurora. Rating: 2.

Possible Explanation. Subjective contrast effects may be to blame, although experienced observers deny this. Radical though the idea may be, the presence of bands of absorbing material (possibly ice crystals) could help explain the similar phenomena cited below.

Similar and Related Phenomena. Banded skies (GLA23); the relationships between auroras and clouds (GLA12) as well as halo phenomena (GLA19); polar-aligned cirrus clouds (GWC); and some of the strange sunset band effects (GES).

## Examples of Black Auroras

X1. March 25, 1894. Croydon, England. "It consisted of five dark rays or streamers, which emanated from a point over the northern horizon at or near the true north, and radiated like the spokes of a wheel towards the zenith, or it might be the magnetic zenith. The rays were clearly defined against a hazy sky, which was faintly illuminated by the glare of the metropolis, and resembled five blades of a black outstretched fan. As I looked, the whole system wheeled slowly through an arc of 30° or 40° towards the east, and then faded out. The rays did not seem to quiver or change colour, but remained black or smoky, and reminded me of streaks made by a hair pencil dipped in Chinese ink. I regard the phenomenon as a black aurora, because of its behavior and the collateral phenomena. Black or smoky pillars and patches have been observed in auroras, but although I have examined the records of hundreds of displays, I have seen no account of a similar effect." (R1)

X2. August 26, 1894. Maidenhead, England.

"From near the Pleiades, then just rising, two pairs of parallel rays of luminosity proceeded, one stretching towards the celestial pole, the other following approximately the course of the Ecliptic. Each luminous band was about 5$^{o}$ in breadth, very straight, the light patchy and without any well-defined edges. The dark spaces between the parallel rays were of about the same width, and of a most curious and intense blackness. The light of the luminous portion was quite steady, without any sensible shooting or flickering, such as is usually seen in auroral displays. Where the northern rays crossed the Milky Way near Cassiopeia, the light was very similar both in brightness and general appearance to that of this brilliant portion of the Galaxy. What most struck my attention was the singular blackness of the intervening spaces. This may have been an optical illusion, due to contrast; but I think not. It looked like an intensely black cloud; but on sweeping with the telescope over the regions where it was most conspicuous I observed that the definition and seeing were unusually good. I have frequently observed similar dark rays. Can these be anything in the idea of a 'black aurora'?" (R2)

X3. March 24, 1903. Somerset, England. "Has any member of the B. A. A. ever seen a 'Black Aurora,' or an aurora in the negative? One was seen here (Frome, Somerset) to-day March 24, at 1:40 p.m. This meteoric effect, situated in the western part of the sky, which, for want of a better name, I have called a 'black aurora,' consisted of a black arch with streamers radiating upwards from it; in short it was precisely like the commonest form of auroral display, as seen in high latitudes, only in negative, the streamers and arch being black instead of white, and the background blue sky instead of a night sky. The streamers to the north were fine and straight, and about double the length of those towards the south, which were thicker and bent at the top like the handle of a walking stick. The sun was shining at the time, and some bright white clouds were being driven in front of the aurora at a rapid rate by a fresh breeze." (R3)

X4. No date. England. "Now with regard to the Derby earthquake, some ten or fifteen years ago, I was coming into luncheon a little late and saw a most curious arch in the N. E., it was precisely like an aurora in negative, instead of the lines being white they were black. It seemed so remarkable that the people were fetched out from lunch to look at it. The next day the newspapers reported an earthquake at Derby at that hours." (R4)

X5. General observation. "The black rays of the northern lights, first noticed by Professor Hanstein, i.e. sharply bounded stripes, surrounded by masses of light, which, however, are quite detached from them, are a not unfrequent, but extremely striking phenomenon." (R5)

References

R1. Ranyard, A. C.; "A Black Aurora," Knowledge, 17:86, 1894. (X1)
R2. Harnett, Fras. R.; "Aurora (?)" English Mechanic, 60:111, 1894. (X2)
R3. Nelson, Edward M.; "A Black Aurora,"

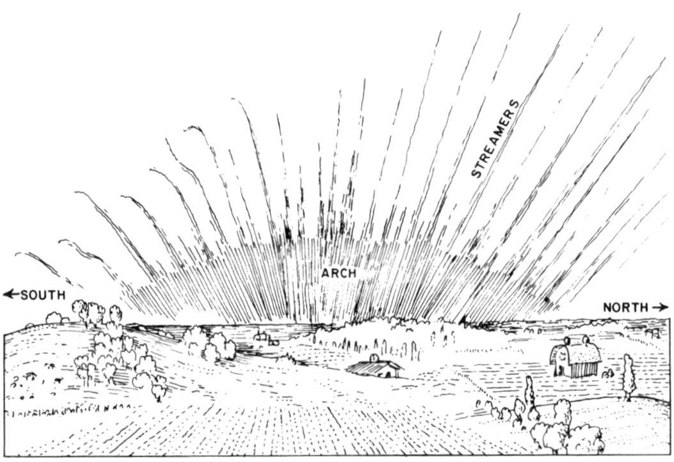


*Black arch and streamers appear over England in 1903*

English Mechanic, 77:168, 1903. (X3)
R4. Nelson, Edward M.; "Earthquakes and Electricity," English Mechanics, 119: 66, 1924. (X4)
R5. "Observations on the Aurora Borealis.." Edinburgh New Philosophical Journal, 35:384, 1843. (X5)

# GLA23 Banded Skies

Description. Arrays of luminous stripes or bands radiating from one horizon and converging on the opposite horizon. (They are, of course, actually parallel.) The filmy, hazy stripes may number from three to a dozen or more. Widths range from $2^{\circ}$ to about $5^{\circ}$, with dark intervening spaces of approximately the same width. The bands may be oriented in any direction. Slow movement perpendicular to the lengths of the lines is often observed. Stars shine through these gauze-like structures, which may pulsate, disappear, and reappear like many other aurora-like phenomena.

Background. Some obvious contradictions appear in the several specific examples presented below. This is a clue that we may be dealing with two or three different but geometrically similar phenomena.

Data Evaluation. Many highly qualified sky observers have seen and described banded skies. Rating: 1.

Anomaly Evaluation. If auroral in nature, the variable geometry and orientation are puzzling. If the banded sky is a cloud phenomenon, the origin of the pattern of ice nuclei presents a problem. If we are contending with an air-glow phenomenon, how does one account for such regularity over the hemisphere of the sky? Composite rating: 2.

Possible explanations. Many possibilities exist and they may not be mutually exclusive: (1) Systems of gravity waves create the glowing stripes; (2) Banded skies are simply remarkable auroral displays; (3) The bands are formed from meteoric or cometary debris arranged in windrows by gravity waves or circulation patterns (See R9 and X16.); (4) Aligned bands of cirrus clouds (GWC) are mistaken for an auroral display.

Similar and Related Phenomena. Related phenomena: auroral pillars (GLA1); auroral arches (GLA2); auroral meteors (GLA3), especially some of the air-glow events; possible identity of auroral displays and daytime clouds (GLA12); black auroras (GLA22). Similar phenomena: magnetically aligned cirrus clouds (GWC); radial spokes and similar atmospheric disturbances on Venus (AVO).

Examples of Banded Skies

X1. July 10, 1874. Halifax, Nova Scotia. "On looking towards the comet about a quarter to one o'clock on the morning of the 10th of July, long lines of dark and luminous clouds lay across the sky, diverging from the N.N.E. and S.S.W. points of the horizon. They lay on the west of the zenith chiefly, and their distance apart, or the width of the dark bands, was about $2^{\circ}$ near the zenith. After careful observation for some minutes the bright bands were seen to change in brightness two or three times a minute. Soon the dark spaces were crossed by scores of narrow bright bands, thus covering the sky as with the most delicate lattice work." (R1)

X2. December 5, 1892. Honeymeadbrook Station, New York. "I will add that at 10:45 P.M., Dec. 5, 1892, I saw, to me, an unique phenomenon. The moon was shining brightly, when diverging bands from the horizon in the north-north west spread at the zenith $60^{\circ}$ wide and converged again at the horizon in the south-south east. They were like thin clouds, through which the stars were easily seen. The belt of Orion was exactly then in their midst. I can liken their shapes to nothing more than the vibrations of a cord, stretched from horizon, over zenith, to horizon again. But they were stationary, and had so far disappeared at 11.30 P.M. standard time, that only curious traces and patches remained. I fancy that had not the moon been shining, these beautiful bands would have shown luminosity. I judged that the radiating point in the north north-west was a trifle west of the magnetic meridian there; but our western declination here is some nine degrees. These were, of course, parallel bands, the divergence and convergence points being the effect of perspective." (R2)

X3. October 11, 1903. England. "On the evening of October 11, 1903, after a fairly active display, the aurora disappeared, but its place was taken by a system of narrow bands of cirrus clouds stretching right across the sky, which, being illuminated by the bright moon, had all the appearances of the aurora. That they did not form part of the aurora could only be decided at first owing to no line appearing in the spectroscope when pointed at them; but later there could be no doubt, as they partly obscured the moon." (R3)

X4. October 2, 1921. England. "At 8.12 P.M. a fan-shaped white cloud was seen radiating from S.S.E. It spread into streamers with dark sky spaces between. The clouds crossed to W. of N.; the stars shown through them, and as they sank in the north they had all the appearance of auroral lights. These long parallel clouds with clear sky spaces between have been seen on many occasions and always moving to or from north. They move very slowly and seem to be at a higher level than ordinary clouds. They do not show the diffraction colours of a lunar corona, and they do not always cause a lunar halo." (R4)

X5. January 30, 1922. Cumberland, England. "Instead of the usual streamers, with radiant in N. moving across the sky, which were quite absent, faint parallel bands of luminosity crossed the heavens in a due E. and W. direction, and these never moved. I counted ten such bands between the Dipper and Ursa Major and Leo, with dark spaces between of about equal width, and they seemed to extend also N. and S. of this region indefinitely. They were of pale white colour, but at 1.30, suddenly, an exquisite crimson glow filled the sky, the colour occurring in large areas that seemed to shift continually, while the pale bands shone immovable over all." (R5)

X6. September 9, 1932. Cumberland, England. "There was a very definite display of banded sky on this date. Irregular bands, fainter and narrower than the Galaxy, radiated from horizon points B $50^{o}$ and $230^{o}$, but were almost entirely absent from W. area of sky. About 6 bands were counted at 00 30, the most conspicuous and broadest following closely the ecliptic. From Vega to Ursa Major the sky was clear and dark. From Aries to Aquarius, including the great square of Pegasus, was also dark, but a luminous patch was seen at times within the square of Pegasus. The rest of the sky was luminous and banded, of intensity estimated to be about half that of the Galaxy in Cygnus." (R6, R7)

X7. February 1, 1950. North Atlantic. "Three bands of light of a uniform width of about $2^{o}$ were observed apparently radiating from a point on the SW horizon and passing vertically overhead. The direction

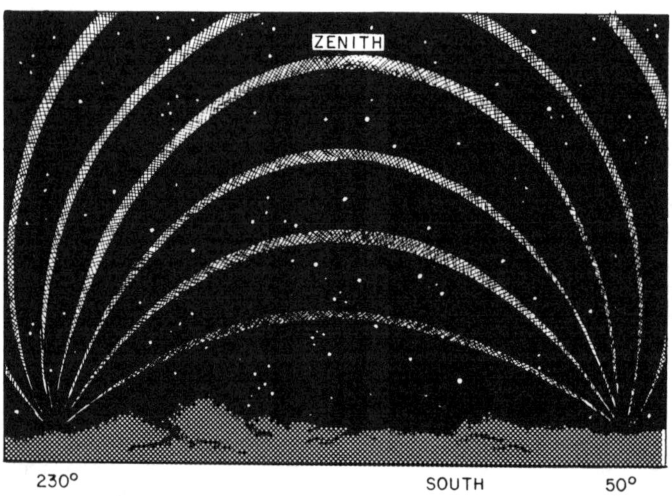


*1932 example of the banded sky*

in which they were lying was roughly 220°. The bands were quite distinct even though the moon was shining brightly. The two lower, or S'ly ones stretched from the point on the horizon to a position 20° to the south of the zenith, beyond which they were obscured by cumulus. The third one lay to the north of the others and was much shorter, starting from a position directly overhead and fading away in a 040° direction, leaving a band of light which gradually diminished in intensity, the whole representing an angular distance of 15°." (R8)

X8. General observation. "At times there is a strange luminosity in the sky at night. More or less extended areas glow faintly or stretch in bands across the sky. Such bands have been observed simultaneously at widely separated points, such as in Middle-Europe and in Scotland. Although occurring prevailingly in the northern sky, and often mistaken for some form of auroral effect, these bright areas are evidently not of that nature, because the light is steady, the streaks run prevailing from east to west, and the phenomenon is not accompanied with marked magnetic disturbances as in the case with the aurora." (R9, R10)

X9. General observation. "There is another phenomenon that has been visible on a number of nights of last year, and also in the present year, of which I have seen no record. This consists usually of long strips of diffused luminous haze. I believe that this is really ordinary haze, which for some reason becomes self-luminous. It is not confined to any particular region of the sky nor to any hour of the night. It always has a slow drifting motion among the stars. This motion is comparable with that of the ordinary hazy streaky clouds that are often seen in the daytime. They are usually straight and diffused and as much as 50° or more in length and 3° or 4° or more in width. In some cases they are as bright, or nearly as bright, as the average portions of the Milky Way---that is, they are decidedly noticable when one's attention is called to them. They apparently are about as transparent as ordinary haze. Sometimes, when near the horizon, where they may be quite broad, they have strongly suggested the 'dawn' or glow that precedes a bright moonrise. Their luminosity is uniformly steady." (R11-R13)

X10. General observation. Banded skies often called 'non-polar auroras.' Some displays seem unrelated to auroras. May be due to waves in the atmosphere. (R14)

X11. March 14, 1841. England. A splendid display of luminous cirri (?). The streaks would sometimes shoot across the sky. (R15)

X12. 1894. Louth, England. White streamers radiated from north through zenith to southern horizon. Disappeared and then reappeared. (R16)

X13. October 11, 1897. Balmain, U.K. Beams of light radiated from a point south of the moon. (R17)

X14. 1920. Stowmarket, England. Curious bands of clouds radiated from magnetic north, converging on SSE horizon. (R18)

X15. August 4, 1921. England. Luminous streaks with dark gaps in south. (R19)

X16. August 8, 1921. Germany. Luminous bands seen in sky. (R20) These bands appeared in Germany at the same time a mysterious bright object was seen near the sun from California. (R21) See AEO.

X17. October 7, 1956. Equatorial Pacific. Streaks of milky-white light ran east and west across the sky at an altitude of 50°. Horizontal width was 30°; bands 1-2° wide. Sky was completely overcast. The moon had not risen yet. (R23)

X18. July 7, 1914. Stowmarket, England. "About 9 P.M. in the north-west, where the light of the sun was still reflected from below the horizon, steady beams of white light were projected up into the sky like searchlights and remained stationary for nearly two hours. From the south-east the light of the rising Moon (Moon-rise 8.57 P.M. and Moon 'Full' at 2 A.M.) projected another series of white beams, whose focus remained the same after the Moon's altitude had increased. Between each distinctly defined beam the stars could be clearly discerned." Some of the rays from the northwest and southeast met at zenith. As the moon rose it became obvious that it was not the center of the southeast rays. (R24)

References

R1. Gledhill, Joseph; "Aurora?" Astronomical Register, 12:196, 1874. (X1)

R2. Hyatt, James; "Auroral Displays," Science, 20:374, 1892. (X2)

R3. Simpson, George C.; "Atmospheric Electricity in High Latitudes," Royal Society, Philosophical Transactions, A205:61, 1905. (X3)

R4. Cook, A. Grace; "Luminous Night Skies and Clouds," Observatory, 44:50, 1921. (X4)

R5. Housman, W. B.; "Crimson Aurora,"

English Mechanic, 115:39, 1922. (X5)
R6. "Luminous Sky," British Astronomical Association, Journal, 44:27, 1933. (X6)
R7. Housman, W. B.; "Luminous Sky," British Astronomical Association Journal, 43:48, 1932. (X6)
R8. Summers, C.; "Night Sky Phenomenon," Marine Observer, 21:17, 1951. (X7)
R9. "Luminous Night Skies Caused by Cosmic Dust," Science News Letter, 26:110, 1934. (X8)
R10. "Luminous Night Skies," Science, 80: sup 6, 1934. (X8)
R11. Barnard, E. E.; "Self-Luminous Night Haze," American Philosophical Society, Proceedings, 50:246, 1911. (X9)
R12. Barnard, E. E.; "Self-Luminous Night Haze," American Philosophical Society, Proceedings, 58:223, 1919. (X9)
R13. "Self Luminous Night Haze," Scientific American, 105:99, 1911. (X9)
R14. Hoffmeister, C.; "Investigations on Bright Night Sky and Luminous Bands," British Astronomical Association, Journal, 62:288, 1952. (X10)
R15. Stevenson, William; "Abstract of Observations on the Aurora....," Philosophical Magazine, 4:6:20, 1853. (X11)
R16. G., F. J.; "Auroral Display," English Mechanic, 59:198, 1894. (X12)
R17. McDonall, Francis K.; "Peculiar Atmospheric Phenomenon," British Astronomical Association, Journal, 7:145, 1897. (X13)
R18. Cook, A. Grace; "Moonlit Clouds," English Mechanic, 113:26, 1921. (X14)
R19. Cook, A. Grace; "A Luminous Night Sky," English Mechanic, 114:49, 1921. (X15)
R20. Wolf, M.; "Leuchtende Bander am Nachthimmel," Astronomische Nachrichten, 214:69, 1921. (X16)
R21. "Bright Object Near Sun," Popular Astronomy, 29:592, 1921. (X16)
R23. Jones, C. A.; Marine Observer, 27: 209, 1957. (X17)
R24. Cook, A. Grace; "Light Rays in the Sky," Observatory, 37:324, 1914. (X18)

# GLA24 Millisecond Brightness Pulsations of the Night Sky

Description. Atmospheric brightness pulsations lasting about one millisecond, consisting of damped oscillations at a frequency of approximately 10 kilohertz. The energy of a single pulsation is roughly 4 x $10^{-8}$ ergs per square centimeter and not visible to the naked eye. Long baseline observations have demonstrated that they have a terrestrial origin. (R1)

Background. Discovered accidently about 1970 during a search for pulses of fluorescence caused by cosmic-ray photons.

Data Evaluation. Apparently a well-established phenomenon, although only one reference has been discovered so far. Rating: 2.

Anomaly Evaluation. Since an explanation of these flickers is elusive in terms of terrestrial atmospheric physics, the phenomenon would seem to have a high degree of anomalousness. Rating: 2.

Possible Explanation. None.

Similar and Related Phenomena. Transient sky brightenings (GLA14) are similar but of much longer duration.

Reference

R1. "What Flickers So Briefly in the Night Sky?" New Scientist, 59:374, 1973.

# GLA25 False Dawn

Description. A transient brightening of the eastern sky one to two hours before the true dawn. Perhaps related is the popular observation that the night is darkest just before dawn.

Data Evaluation. Most support for this predawn sequence comes from folklore and popular belief. Little scientific observation has been discovered. Rating: 3.

Anomaly Evaluation. If, as some maintain, the false dawn is simply the zodiacal light, there is no real anomaly. However, one would not expect the zodiacal light to disappear suddenly and completely before the true dawn. Rating: 3.

Possible Explanations. The zodiacal light probably accounts for this phenomenon. Subjective considerations are important in such low levels of illumination. Little objective work has been done here. See AZO for zodiacal light anomalies.

Similar and Related Phenomena. The zodiacal light (AZO); transient sky brightenings (GLA14)

## Examples of the False Dawn

X1. General observations. "The first line of the second quatrain of the Rubaiyat of Omar Khayyam (Fitzgerald's translation) runs as follows: 'Before the phantom of False Morning died.' In the commentary by H. M. Batson on the Fitzgerald translation of the poem, the following remark is made regarding the line just quoted: 'Before the dawn comes the false dawn, a transient light on the horizon which Persians call blue dawn or morning grey; it appears an hour before daybreak, and is a well-known phenomenon in the East.'" Based on personal observations in South Africa, the author identified the false dawn with the zodiacal light. (R1)

X2. January 7, 1933. England. "I got out of the car and found the rain had ceased and strolled up and down for a few minutes, when I suddenly realised that it was getting light, and at first I thought it was the dawn, but knowing it was only about 5 a.m. I realised I was mistaken, but very unfortunately I cannot give exact times. But the time was almost certainly about 5.10 a.m.. It became so light that I was able to see the surrounding countryside. I could see the Dorsetshire hills sufficiently plainly to see that the clouds lay on their tops and that they were clear below, and I was able to turn off my headlights. The light was a general diffused light with no apparent focal point, but I should say it was lighter towards the south and east, although I could quite well see towards the north and west, and here again I failed to note times; but this condition lasted for probably about 20 minutes, after which period, and almost as suddenly as it appeared, it became quite dark again." True dawn commenced about 7.30 a.m. (R2)

X3. General observation. The following is attributed to W. F. Denning, the English meteor authority: "Before dawn a greater darkness seems to drop down like a mantle upon the immediate surroundings. Objects which were plainly observable during the previous hours of the night are blotted out, and a nervous feeling is sometimes induced by the dense opacity of the air." (R3)

## References

R1. Bird, John T.; "Omar Khayyam and the Zodiacal Light," British Astronomical Association, Journal, 40:336, 1930. (X1)

R2. Cumming, J. H.; "False Dawn," Meteorological Magazine, 71:189, 1936. (X2)

R3. "Darkness before Dawn," Scientific American, 111:331, 1914. (X3)

# GLB BALL LIGHTNING

## *Key to Phenomena*



## GLB0 Introduction

Ball lightning is one of nature's most mysterious manifestations. Basically, ball lightning is a mobile, luminous mass, usually spherical in shape, that accompanies such violent natural phenomena as thunderstorms, tornados, and earthquakes. It is one of those rare natural anomalies recognized by science as real but which eludes all attempts at explanation.

One reason ball lightning resists explanation is that it is so variable. It may be as small as a pea or larger than a house. It may be violet, red, or yellow, or even change colors during its brief life. Ball lightning is generally spherical, but rods, dumbbells, spiked balls, and other shapes have been sighted. Sometimes ball lightning appears to have a snake-like internal structure; long crinkled tails may be attached to the balls.

Ball lightning is a dynamic entity. It may glide silently and disinterestedly past an observer or it may inquisitively explore a room as if directed by intelligence. While a few of these enigmatic apparitions dematerialize silently, most explode violently. One has to hedge about the location of the sound for the actual explosion sometimes seems to occur elsewhere---outside the house in which the ball has appeared for example. This observation plus the frequent materialization of ball lightning in closed rooms and metal aircraft suggest that electrical induction may be involved in some cases. In other words, electrical forces from an enveloping thunderstorm may create a glowing ball of plasma in the air inside a closed container similar to the St. Elmo's Fire induced on pointed objects. When the electrical forces disappear (usually with a peal of thunder), the ball lightning vanishes, too. Observers, however, have no doubt that something palpable has visited them because the room is usually filled with the smell of electricity (ozone), and there may be smoke and much material damage.

No reasonable explanation exists for ball lightning. Glowing spheres of plasma created by natural electromagnetic forces have been proposed and found wanting. Antimatter meteorites, chemical reactions in the atmosphere, and intense cosmic radiation have also been proposed. None seems adequate. The sheer difficulty of accounting for ball lightning has led some scientists to assert that all observations of this phenomenon are illusory---aberrations of the eye and brain. But photographs of ball lightning do exist, and thousands of seemingly sound observations exist in the literature. Ball lightning seems very real.

# GLB1 "Ordinary" Ball Lightning

Description. Luminous spheres or near-spheres with diameters between roughly 1 inch and 5 feet, with 5-15 inches being most common. Ball lightning moves freely in the atmosphere, descending from clouds and following tortuous paths in the open air, sometimes against the wind, and wending its way around rooms in buildings almost as if it were inspecting the premises. The color is often yellowish, reddish, and bluish white, but purple and green fireballs have also been seen. The luminous spheroids may be sharply defined, fuzzy, or surrounded by a sort of halo. A hissing or buzzing sound sometimes accompanies them. Although ball lightning often appears silently it almost always disappears with an explosion. In its wake, it may leave a sulfurous or "electricity" smell. Lifetimes range from a few seconds to 10 minutes and more. Ball lightning is almost always associated with thunderstorm activity.

Background. Ball lightning or globe lightning was long doubted by scientists. Observations, even by reliable and competent individuals, were rejected as optical illusions or misidentifications of other natural phenomena. Yet, so many good observations have occurred in recent years and our knowledge of plasma phsyics has advanced so much farther that the reality of ball lightning is now generally accepted. The explanation of this phenomenon, however, still eludes us.

Data Evaluation. Hundreds of good reports, many by physical scientists, make ball lightning one of the best documented of all scientific anomalies. Rating: 1.

Anomaly Evaluation. Ball lightning is still an enigma to physicists. Scientists do not yet know how so much energy, as evidenced by its tremendous destructive power, can be confined so long in a detached, freely moving sphere. Rating: 2.

Possible Explanations. The explanation of ball lightning's shape and stability probably resides in the relatively new field of plasma physics. In other words, some sort of electromagnetic bottle confines the essential substance and energy. Early scientists thought that the substance of ball lightning was probably only combustible gas, but calculations show that much higher energy densities than those in, say, methane, are needed. It is fashionable now to talk in terms of highly energetic metastable states of molecules. So energetic are some examples of ball lightning that one investigator even proposed antimatter meteoroids! Some of the examples classified as ball lightning below may actually be only unusual cases of St. Elmo's Fire and corona discharge---especially those that seem to simply fade away without a detonation.

Similar and Related Phenomena. All of the other types of ball lightning described in this section are probably generically related. Also related are: St. Elmo's Fire (GLD); bead lightning (GLL), and corona discharge. Low-level meteors (GLM) and will o' the wisps (GLN) sometimes emulate ball lightning.

Examples of "Ordinary" Ball Lightning

X1. March 1731 or 1732. No place given. "On Sunday the 12th of March, 1731 or 1732 (the exact year seems to be uncertain) while walking in his garden between one and two o'clock in the afternoon he (Rev. A. Vievar) heard as it were a loud clap of thunder from the northeast. While looking into the air the noise was repeated very loud, but seemed more like the violent fall of a house, so that he expected every moment an outcry from the town; but he was soon undeceived when it began again, and he found it made towards him, with a different noise from the former, being like the grinding of flint stones, but very loud. The object seemed to be about three feet wide. He found it sink in the air and as it seemed to point directly at his head, he laid himself down on a grass slope to let it pass over. However, at the upper end of the walk it fell to the ground and came rolling down the grass walk and he can compare it to nothing better than to that of a violent grinding of flint stones, or a coach and six at full speed on a causeway of loose stones. He lay attentive, expecting to see something, and saw a piece of wood come running before it. When the phenomenon came to the water side it twisted up a large stake that stood in its way and tossed it toward him with much violence, and immediately fell into the water, with the violence and noise of a red-hot millstone."(R1) Unlike most examples of ball lightning, this one relates little about the luminous aspects. The phrasing and punctuation are typical of the period. (WRC)

X2. 1843. France. "After a loud thunderclap, the tailor being finishing his meal, saw the chimney-board fall down, as if overset by a slight gust of wind, and a globe of fire, the size of a child's head, come out quietly and move slowly about the room, at a small height above the floor. The tailor said it looked like a good-sized kitten, rolled up into a ball, and moving without showing its paws. It was bright and shining, but he felt no sensation of heat. The globe came near his feet, like a young cat that wants to rub itself against its master's legs; but by moving them aside gently, he avoided the contact. It appears to have played for several seconds about his feet, he bending his body over it and examining it attentively. After trying some excursions in different directions, it rose vertically to the height of his head, which he threw back to avoid its touching his face. The globe, elongating a little, then steering towards a hole in the chimney above the mantlepiece, which hole received a stove pipe in winter, but was now pasted over with paper. 'The thunder,' he said, 'could not see the hole;' but, nevertheless, the ball went straight to the aperture, removing the paper without hurting it, and made its way into the chimney. Shortly afterwards, and when he supposed it had time to reach the top, it made a dreadful explosion, which destroyed the upper part of the chimney, and threw the fragments on the roofs of smaller buildings, which they broke through." (R2, R17)

X3. July 11, 1874. England. "On July 11th, 1874, whilst occupying my sitting room, which is at the back of the house and above the level of the wall abutting on the fields, I heard a clap of distant thunder, and noticed the sky to be darkened by clouds. At a quarter to four, I noticed that the sky was covered with dense dark thunder clouds, and heard a loud peal of thunder; the wind rose rapidly, and then occurred a most violent storm of wind, accompanied by a deluge of rain, extremely loud thunder, flashes of sheet and forked lightning, and large hail stones. ....After a short lull, I saw a ball of fire of a pale yellow colour rise behind the houses, on the other side of the fields, apparently from the Regent's Canal. The ball at first rose slowly, apparently about as fast as a cricket ball thrown into the air, rapidly incresing its rate of motion until it reached an elevation of about 30°, when it started off so rapidly as to form a continuous line of light, proceeding at first due east and then to the west, rising all the time; and after describing several zigzags, disappeared into a large black cloud to the west, from which flashes of forked lightning had come. There were two large black clouds near each other, and the ball when it formed the zigzags moved first towards one and then towards the other. In about three minutes another ball ascended, and in about five minutes afterwards a third, both behaving as the first and disappearing in the same cloud...." (R3, R38)

X4. March 21, 1877. Vence, France. "Toward midnight there was observed, about eleven miles north-east of Vence, a large black thundery cloud, in a state of extreme agitation, and continually raising and lowering its position. At the upper part of this cloud three or four balls of fire issued every two minutes, as if from the invisible centre of the cloud, diverging in all directions, and after running a course of from six to eight degrees, broke silently with effulgent brightness. Their apparent diameter, as seen at a distance of eleven miles, was about a degree. They were mostly of a reddish colour, a few, however, being of a yellowish tinge, but all of them assumed a white colour in the act of bursting. Their course, which was horizontal and parallel to the plane of the cloud, was relatively slow, not exceeding two degrees per second, and they bore a strong resemblance to immense soap bubbles, both as regards apparent lightness and general appearance." (R4)

X5. May 1878. Lower Cumber, England. "A singular phenomenon was witnessed here on Wednesday night last, about eleven o'clock. A large fir tree, growing at the gate of the church avenue, appeared all dotted over with twinkling little stars, of a whitish-blue colour, whose rays shed a faint light for some distance around, and filled the surrounding atmosphere with an oppressive and disagreeable odour. These atoms of light in a short time united into a globe of fire at the top of the tree, of a brilliant green, and immediately rose in the air, wafted slowly by a slight breeze---leaving a trail of luminous matter behind it---towards the church tower, against which it struck, smashing the ball into atoms. A few very large drops of rain at this juncture fell, accompanied by a brilliant flash of lightning, and followed by a terrific peal of thunder, when the glittering sparks evaporated." (R5) Apparently St. Elmo's Fire coalesced into ball lightning.

X6. August 10, 1888. No place given. "Capt. Grace, of the schooner Rebecca Smith, on August 10, 1888, during squally weather with thunder and lightning and heavy rain, observed a bolt of lightning in the shape of a ball fall in a perpendicular (zigzag) line from the zenith and strike the surface of the water less than a length from the vessel's beam. There was apparently no disturbance of the surface when the lightning struck (although this might have been missed in the glare of the lightning), but for a space of at least five minutes after the lightning struck, a thin blue smoke, similar in appearance to powder smoke, was seen to rise from the surface to a height of at least ten feet. The smoking surface was generally circular in form, about twelve or fourteen feet in diameter, and had the spot in which the ball struck for a center. The occurrence took place in broad daylight." (R1)

X7. April 11, 1894. Dunstable, England. "Whilst watching the incessant forked lightning in the east at 2.30 p.m., I suddenly saw a broad spout of fire drop almost vertically from the clouds to the earth. The band of fire was not at all like lightning, as it was ten or twenty times as broad, and formed a continuously, slightly curved line, without the slightest trace of zig-zag. It was like a large ball of ribbon being quickly unrolled, one end being retained in the clouds. The fall was less rapid than lightning, and was accompanied by a dazzling light. It was immediately preceded and followed by the crash of thunder, but the thunder was at the time continuous. The fall appeared to be close by, and I soon after learned that a 'fireball' had descended on the Dunstable side of Luton, near Dallow Farm, about four miles from here. On visiting the spot, and questioning two or three eye-witnesses of the fall, I was told that the 'thunderbolt' was seen as a large 'ball of fire,' that the fall was accompanied by a loud rushing sound and a dazzling intense light." Great damage was done to a barn. (R6)

*Photograph of possible ball lightning suspended above a garden. (R101. Courtesy J. D. Barry)*

X8. July 20, 1897. Suffolk, England. "During the thunderstorm of July 20, with which the drought broke up, an elderly man, Thomas Smith, residing in this parish about half a mile from the railway station, was watching the lightning from his cottage door, between 5 and 6 p.m., when he noticed a white ball, 'about the size of an egg,' dancing about in the air 'like rooks when at play.' He watched it through the intervals between two or three lightning flashes, therefore during several seconds. After some interval (perhaps a few minutes), he still standing at the door, his wife just coming down the stairs to him, something seemed to pass between them which felt hot to their faces. Simultaneously Miss Downes, schoolmistress, sitting on the landing above the stairs, felt something hot pass her hair behind, and then in a small bedroom, with open door adjoining, a loud detonation took place; whitewash from the ceiling covered bed and floor, the wall paper was torn, the plaster fissured, and the house filled with a 'sulphurous' smell." (R7)

X9. August 21, 1900. Telemarken, Norway. A storm was approaching. "I had been watching the particularly vivid lightning for some time, when suddenly I saw a streak of 'yellow,' apparently about one inch broad, dart from the sky just above the top of the hill opposite and, gradually falling, make straight for where I was sitting. I was too spellbound to move, and the length of time of the whole occurrence was too short to call the attention of those who were sitting with me. I remember having the sensation that it must hit my forehead, when just as it got in front of the hotel window it changed to a ball of dazzling yellow fire (about the size of a cricket-ball) and then burst with a frightful crash, emitting volumes of violet-coloured flame, which spread in all directions. This must have happened in one or two seconds, but for several minutes afterwards I was completely dazed." (R8)

X10. July 1921. Grindelwald, Switzerland."A dark cloud was noticed approaching from the east, but neither rain, hail, nor snow fell before the occurrence of the fireballs, and only distant lightning had been observed. Suddenly from the air inlet of the large stove, in which a small wood fire was burning, came, practically simultaneously, a large number of very bright round balls of various colours, the largest perhaps nearly as large as my head. Almost at the same moment, a dreadful deafening explosion occurred, and the balls had all vanished. The room was full of a grey smoke---perhaps disturbed dust---and a strong peculiar odour was observed.....The fire had not been disturbed, and none of the fuel was ejected. The guests who were nearest the stove momentarily experienced a slight electric shock, but no one was injured." (R9) These spheres resemble aerial bubbles (GLD7). (WRC)

X11. July 27, 1952. No place given. "During an electrical storm on the afternoon of July 27, 1952, several witnesses observed accurately the sudden appearance of a ball lightning in a closed room on the ground floor of my house. A few seconds after a lightning stroke in the neighborhood, we observed outside the window a brightly gleaming sphere the size of a fist which moved downward in short serpentine lines. Then this luminous ball penetrated through the closed window pane and entered our room. At a depth of about 1 m it performed a sudden turn of $90^{\circ}$ parallel to the wall and continued floating another meter further into the room. Thereupon it burst, and the luminous sphere disappeared with a brief deafening explosion. This ball lightning was purplish with a reddish cast which persisted during the entire duration of the phenomenon. It lasted about 3 sec. No damage whatever was caused either on the inside or outside of the room. After the bursting of the luminous ball, we could perceive the typical odor which occurs in the case of electrical discharge." (R10)

X12. August 12, 1956. Tambosk, Russia. "A commercial airliner (LI-12) was struck by ball lightning on 12 August 1956 while flying in the lower Tambosk region of the USSR. Before being struck, the aircraft had been flying at 3.3 km altitude through a slowly moving cold front which contained dense thunderclouds. During a penetration of one thundercloud, where the air temperature was about $-3^{\circ}C$, the crew saw a rapidly approaching dark red almost orange fireball 25 to 30 cm in diameter to the front and left of the aircraft. At a distance of not more than 30 to 40 cm in front of the nose, the ball swerved and collided with a blade of the left propeller, exploded in a blinding white flash, and left a flaming tail along the left side of the fuselage. The sound of the explosion was loud enough to be heard over the noise of the engine. No substantial damage could be found. One of the left propeller blades had a small fused area 4 cm along the blade and less than 1 cm in depth. Around the damaged region was a small area of soot, which was easily wiped off." (R11)

X13. 1958 or 1959. Cheektowaga, New York. During a typical summer thunderstorm. "At one point a single lightning stroke was observed coming from a cloud directly over the open field. Some distance above the ground, the stroke divided into two simultaneous branches, each of which moved toward one of adjacent telephone poles. At a distance of some meters above the pole tops the entire lightning stroke disappeared, and in its place two (one over each telephone pole) large luminous spheres appeared. The color was yellowish, much like the flames of a brightly burning wood fire. The diameter at least two or three pole diameters, possibly as much as five or six such diameters (0.4 to as much as 1.2m). The luminous spheres slowly descended toward the pole tops. When the spheres came very near or in direct contact with the pole tops, they exploded with a loud sharp bang. During the extremely brief 'explosion' phase, the spheres appeared to contract (much against my expectation for an event described as 'explosion') and turned brilliant bluish white, as if to indicate much higher internal temperatures. Following the explosion, dark smoke was seen to rise from the pole tops, but it did not persist to indicate a fire." (R12)

X14. May 21, 1973. Newbury, England. "On 21 May 1973 at 2150 GMT I was standing at the front door of my bungalow looking at the northern sky, which was mainly clear of cloud. Suddenly there appeared a luminous object, blue in colour, about the size of a tennis ball, slowly descending towards the earth. At first I thought it was a meteor, but it lacked certain characteristics of meteors; there was no rushing movement, nor was there a trail of luminous sparks often associated with such phenomena. As I watched it, it suddenly vanished completely, leaving no trace. There was at the time, thunderstorm activity in parts of south-east England, but none in the vicinity of Newbury." (R13)

X15. August 6, 1868. Ireland. "Another instance of a remarkable kind is recorded by Mr. M. Fitzgerald as having occurred in the Glendown Mountains in Ireland. He noticed a globe of fire in the air floating leisurely along, it decended from a ridge into the valley and reached a stream about 800 yards from the spot where the observer stood. It then struck the land and reappeared in about a minute, again disappearing and reappearing several times, until it flew across the stream and finally lodged in the opposite bank, leaving a hole where it buried itself. On examination of the track of this meteor, an aperture about 20 feet square was found in the peat where it had first touched the land, with the peat turned up on the lea as if it had been cut with a huge knife; it next made a trench about 20 perches in length and 4 feet deep, afterwards ploughing up the surface to about a foot. Next it tore away the bank of the stream about 5 perches in length and 5 feet deep, and then hurling the immense mass into the bed of the stream.....From its first appearance till it buried itself could not have been less than twenty minutes, during which it travelled lesiurely as if floating with an undulatory motion through air and land over a mile. It appeared at first as a bright red ball of fire about 2 feet in diameter, but became rapidly smaller, particularly after each dip in the soil, so that it looked not more than 3 inches in diameter when it finally vanished." (R14, R36)

X16. No date given. Omaha, Nebraska. During a thunderstorm. "A sharp crackling noise caused me to look toward a window screen to my left. Then I saw a round, iridescent (mostly blue) object, baseball size, coming toward me. It curved over my head and went through the isinglass (mica) door of the kitchen range, striking the back of the oven and spattering into brilliant streamers. There was no sound and no effect on me except a tingle as it passed over my hair. Later examination showed a tiny hole with scorched edges in the screen and isinglass, and scorch-like marks on the back of the oven." (R15)

X17. No date given. West Yorkshire, England. During a thunderstorm. After one lightning discharge there appeared a ball of light, 'electric' blue in the centre with an ill-defined yellow white fringe, the whole being about 18 in. in diameter. It first appeared from beneath the tree branches on my side of the road, perhaps 20 yards away, and approached slowly with a staggering undulating motion, but keeping approximately to the same height under the trees. As it approached me, it crossed the road gradually and then, when almost opposite me, turned back across the road and slightly backwards, passed through the lodge gates where I lost sight of it. I estimate it moved at a moderate walking pace for 30 yards before I lost view of it and in this time, no changes in size or brightness were noted." (R16)

X18. No dates or places given. Six examples of ignition of inflammable fluids by ball lightning, two of which are presented here. "Distillate was being loaded into the open top of an 8,000 gallon transport truck. 'A ball of light travelled along the fill pipe and down the droptube. It entered the truck compartment and the truck blew up.'" "An observer on a ship several miles from an empty VLCC (very large crude carrier) tanker. 'A ball of light travelled along the deck of the ship and disappeared. Then the ship blew up.'" (R18)

X19. 1556. Gloucestershire, England. Eight people killed by a fiery, sulfurous globe that rolled in through a door. (R19)

X20. 1557. France. An oscillating flame moves about a bedroom. (R11)
X21. July 2, 1665. Norfolk, England. Great gray ball descended during a storm. Did great damage and killed one, lamed others. (R20)
X22. September 1713. Italy. A bluish-white flame rose from a floor, expanded, then disappeared suddenly. (R21, R22)
X23. November 4, 1749. No place given. A large ball of blue fire rolled along water surface, hitting ship and exploding. (R1, R22, R23) Same as GLB7-X5.
X24. July 23, 1763. Hertford, England. Fireball hits church, explodes. (R19)
X25. February 14, 1809. Portsmouth, England. Three lightning balls kill and injure sailors. (R24)
X26. 1841. Milan, Italy. People chase a fireball down the street. (R2, R25)
X27. 1850. North Atlantic. Fireball shatters ship's mast, but no explosion or noise. (R26)
X28. July 5, 1852. Paris, France. Same description as X2, but different date. (R27)
X29. February 13, 1853. Mount Desert, Maine. Many fireballs and purple lightning seen during snowstorm. (R28, R29)
X30. May 23, 1855. Chesapeake Bay. Numerous balls of fire roll quietly around a wharf. Some just faded away; others disappeared with a slight sound. (R1) This may have been St. Elmo's Fire. (WRC)
X31. 1861. North Atlantic Ocean. Three luminous bodies issue from the sea during a squall. In view 10 minutes. (R30)
X32. August 1866. Putney, England. (R31)
X33. 1866. Shreveport, Louisiana. Ball lightning splinters a tree. (R32)
X34. June 30, 1867. Hudson Bay, Canada. Fireball falls on ship from a cloud, shatters a mast, causes some temporary paralysis. (R33, R34)
X35. July 12, 1868. Guildford, England. Fireball 3 inches in diameter excavates a trench 10 feet long. (R35)
X36. August 14, 1872. Poughkeepsie, New York. Many balls of lightning enter buildings and explode. (R37)
X37. 1876. Palestine, Texas. Woman killed by ball lightning. (R39)
X38. October 12, 1877. Wandsworth Common, England. Several fireballs roll along before the wind. (R40)
X39. July 23, 1878. Southampton, England. Delicate, rose-pink, pear-shaped ball lightning, 6-8 inches long, fell upon a yacht and detonated. (R41)
X40. 1870s. Pyrenees, France. Fireball 6 inches in diameter travelled several miles. (R42)
X41. July 17, 1880. Kent, England. A fireball fell from the clouds and exploded. (R43)
X42. August 19, 1880. Paris France. Fireballs rose from the ground and vanished. (R44) This example is similar to the luminous spheres in GLD7. (WRC)
X43. September 1880. Hendon, England. Ball lightning falls into a pond killing over 100 fish. (R45)
X44. July 20, 1881. England. A fireball 0.5-meter in diameter destroys a house. (R46)
X45. August 23, 1881. Leamington, England. Ball lightning explodes creating much smoke. (R47)
X46. 1881. Scotland. Chain of unconnected balls of lightning observed over a ship. Each was about a foot in length. More fireballs seen running around the ship. (R48)
X47. Summer 1882. England. Mottled whitish ball of fire bounced off the ground and disappeared. (R49)
X48. July 28, 1883. Hartford, Connecticut. Fireball big as a man's head paralyzes man. (R22, R23, R50)
X49. July 1883. Buckinghamshire, England. Ball lightning hit a house. (R51)
X50. September 20, 1883. Banbury, England. Yellow globe the size of a cricket ball. (R52)
X51. October 9, 1883. No place given. Fireball injured three seamen. Exploded into flying sparks. (R53)
X52. July 26, 1884. Cologne, France. A scarlet fireball was visible for several minutes. (R54)
X53. January 28, 1885. No place given. A fireball on a vessel. May have been St. Elmo's Fire. (R1)
X54. August 19, 1886. New Harmony, Indiana. Ball lightning rolled down steps and across a carpet (without hurting it), and out a door. Considerable damage done inside and outside the house. (R55)
X55. August 25, 1886. During a hurricane (no location given), numerous balls of fire shot up from the northern horizon to an altitude of about 20°. (R1)
X56. March 14, 1887. No place given. Sea was phosphorescent, numerous fireballs flying in the air. (R1)
X57. March 22, 1887. No place given. Heavy gale, fireballs flying through the air, St. Elmo's Fire everywhere, sea phosphores-

cent. (R1)

X58. May 30, 1887. English Channel. Yellowish-white globe the size of the moon. (R56)

X59. November 12, 1887. Off Cape Race. A large ball of fire rose from the sea, moved against the wind. Lasted 5 minutes. (R1, R57, R58)

X60. November 20, 1888. No place given. Two blue balls of fire pass over a ship followed by sounds like artillery. (R1) May have been a pair of meteors. (WRC)

X61. April 6, 1889. No place given. Fireball ran down rigging and burst above the deck. (R1)

X62. July 20, 1889. No place given. Ball lightning falls on ship, glides around the rigging and deck, explodes. (R1)

X63. August 1889. France. Three fireballs fell into a courtyard. (R59)

X64. August 1889. No place given. Luminous ball passed through a house, bounded along the ground, disappeared without an explosion. (R60, R61)

X65. January 2, 1890. Spain. Ball lightning the size of an orange. (R62, R63)

X66. March 1890, Louisville, Kentucky. A tornado was accompanied by numerous fireballs rolling in the streets. (R64)

X67. Day of the Pentecost, 1890. East Prussia. As a man drove his carriage along a road lined by wire fences, two balls of fire appeared, one on each fence. They followed the carriage along until the end of the fence was reached, where they collapsed with a noise like rumpling paper. (R27) This may have been an unusual manifestion of St. Elmo's Fire. See GLD6-X2. (WRC)

X68. August 28, 1891. Martinique. A great hurricane during which many examples of ball lightning were seen. Some lasted several minutes. (R65)

X69. March 6, 1892. North Atlantic Ocean. Luminous ball bursts on steamer, spreading lightning in all directions. (R66)

X70. July 3, 1892. Liverpool, England. Ball lightning falls into a lake, creating a column of water 60 feet high. (R67)

X71. August 7, 1892. Germany. Large, tapered globe of fire explodes in church devastating it, and leaving a sulfurous odor. (R68)

X72. July 26, 1893. Epping, England. Ball lightning bursts inside a house. Prior to the appearance of the lightning, a dark space fringed with crimson flame appeared in the kitchen. (R69)

X73. December 12, 1893. Devon, England. Ball lightning descends, shedding sparks. (R70)

X74. June 21, 1895. Karachi, Pakistan. Ball lightning explodes, leaving a sulfurous smell. (R71)

X75. August 24, 1895. Donegal, Ireland. Boy injured by ball lightning. (R72)

X76. December 18, 1895. Devon, England. A fireball travels against the wind, kills a man, and tears up the ground for upwards of 100 yards. (R73)

X77. 1895. No place given. Fireballs 2-3 feet in diameter roll about the streets. (R74)

X78. August 1896. Dublin, Ohio. Ball lightning hits a carriage with a hissing roar. Smell of sulfur afterwards. (R75)

X79. August 1898. No place given. About 20 globes of fire roll, one after the other, down the side of a valley. (R76)

X80. 1898. Springfield, Massachusetts. A ball of lightning floats down like a balloon, hits ground, bounces once, then disappears at the same time ordinary lightning hits a house. (R75)

X81. August 4, 1899. Salt Lake City, Utah. Ruby red fireball, one foot in diameter, enters a house by one window, leaves by another. (R77)

X82. June 9, 1901. France. Ball of fire drops from a cloud like a stone. (R78)

X83. July 25, 1901. Dalston, England. Ball lightning, 6 inches in diameter, falls and explodes. (R79)

X84. 1903. England. Ball lightning falls onto a bicycle's handlebars. Rider smells sulfur. Ball falls to ground and explodes. (R80)

X85. 1904. Deux-Sevres, France. Child touches ball lightning with foot. It explodes, killing 11 cattle but not hurting the child. (R81)

X86. April 1905. Surrey, England. Ball lightning explodes over a church. (R82)

X87. Summer 1905. Harrow, England. (R83)

X88. August 1, 1907. Alpena, Michigan. A fireball drops to the floor, moves around the room in circles, smashes holes in walls. (R84)

X89. September 7, 1907. West Australia. Lurid red ball, 4 inches in diameter, with a halo of yellow flame 18 inches in diameter. (R85)

X90. July 1909. England. Globe of light, 4-5 inches in diameter, bursts inside a house with slight report. (R86)

X91. Summer 1909. Buzzards Bay, Massachusetts. Luminous ball descends. Many luminous bodies seen on roof of house. (R87, R88)

X92. Summer 1911. Newport, England. Bright green fluorescent ball floats in a door and then out again. Explosion followed.

(R89, R90)

X93. December 15, 1911. France. Ball lightning falls off a wire and explodes. (R91)

X94. July 13, 1912. Sussex, England. White ball, size of a full moon, travels slowly across the sky. In view for 55 seconds. (R92)

X95. January 25, 1915. Rome, Italy. Reddish globe, 40-45 centimeters in diameter, falls on church and explodes. (R93)

X96. August 1, 1915. Norwich, England. Bright, greenish-yellow spots of light seen moving against a cloud. (R94)

X97. April 1919. Kent, England. (R95)

X98. January 13, 1920. Sydney, Australia. Ball of light tinged with violet descends slowly in an oscillating fashion. (R96) This "falling leaf" motion has been reported in many UFO sightings. (WRC)

X99. Spring 1923. Chevy Chase, Maryland. Ball lightning the size of a toy balloon hits a tree and explodes. (R75)

X100. April 1924. Evanston, Illinois. (R97)

X101. June 8, 1924. Virginia. Fist-sized ball lightning explodes with sparks and smoke. (R75)

X102. September 1, 1924. Gold Coast. Large red globe, 4 inches in diameter, came through open shutter and bursts in a room. No storm at time. (R98)

X103. April 29, 1925. Saxony. Barrel-like object falls from a cloud and broke into small strokes of lightning. Much damage done. St. Elmo's Fire seen at the same time. (R10, R27)

X104. August 11, 1925. Yarmouth, England. Brilliant red lightning ball, 1 foot in diameter, moves 100 yards along a road surface with a hissing noise. Explodes. (R99)

X105. 1925 or 1926. Pennsylvania. After a severe thunderstorm, a camper sees a golden-yellow ball of fire the size of a basketball enter an open window and exit via the opposite window. (R27)

X106. September 23, 1926. Bangor, Ireland. Ball lightning moves around a room, leaving a trail of broken bric-a-brac and the smell of sulfur. Leaves home and explodes. (R100)

X107. 1926. Norwich, England. (R101)

X108. March 23, 1927. Cattlewater, England. Greenish-blue fireball three times the size of an English football explodes. Others seen in area. (R102)

X109. August 25, 1928. Surrey, England. Ball of fire, 3 feet in diameter, explodes. Dull day but no storm. (R103)

X110. 1928 or 1929. Near Nelson, British Columbia. Two examples of ball lightning seen simultaneously. One falls on a lake wharf, shattering a pile. The second emerges from a cottage fireplace. This second ball vanished when the first one exploded outside. (R104)

X111. July 1929. North Atlantic Ocean. (R105)

X112. Summer 1929. East Hampton, England. Yellow flame, 5 inches in diameter, spins like a top. No explosion, but room full of smoke. (R106)

X113. December 12, 1929. Somerset, England. Ball lightning seen from a window; room filled with blue flame; ball explodes with much smoke. (R107)

X114. July 9, 1930. Chatham, New Jersey. (R108)

X115. August 30, 1930. Lincoln, Nebraska. Many balls of lightning photographed during storm. (R109)

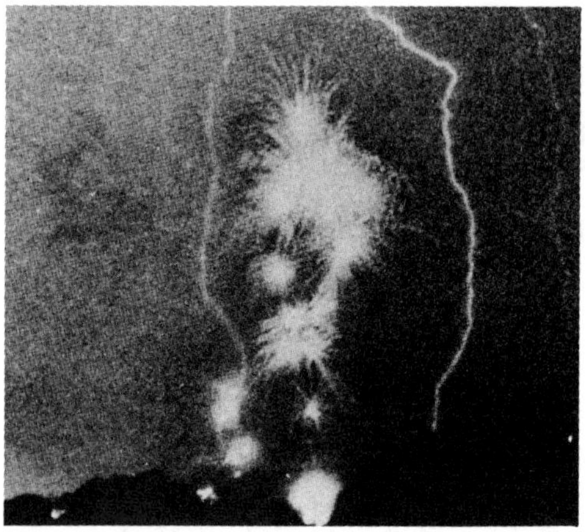

*A still camera photograph made during a thunderstorm showing several images identified as ball lightning. (R109. Courtesy J. D. Barry)*

X116. April 1, 1931. Atlantic City, New Jersey. Ball lightning enters a train and explodes. (R97)

X117. July 18, 1931. Dermott, Arkansas. Pungent odor and a swishing sound precede the explosion of head-sized ball lightning. (R75)

X118. July 9, 1932. Rock Rapids, Iowa. Fiery stream pours from a dust-laden tornado cloud; stream breaks up into separate balls of fire. (R109)

X119. August 31, 1932. Italy. (R110)

X120. March 7, 1933. Stoke Poges, England. (R111)

X121. January 11, 1934. Co. Antrim, Ire-

land. Three orange-red fireballs the size of cricket balls. (R112)
X122. March 5, 1934. Glasgow, Scotland. (R113)
X123. Summer 1934. Loch Tummel, Great Britain. (R114, R115)
X124. 1936. Lincoln, Nebraska. Several red fireballs roll and bounce along street. (R116)
X125. Summer 1938. Europe. Ball lightning comes in the open window of a flying boat at 8,500 feet altitude. Explodes. (R114)
X126. December 6, 1938. Co. Down, Ireland. (R117)
X127. April 26, 1939. Roche-fort-sur-Mer, France. Two balls of lightning, 15-20 centimeters in diameter, exploded doing considerable damage. (R11)
X128. October 1941. In aircraft over the English Channel. Bluish-green fireballs the size of apples hit and explode, leaving a smell of sulfur in the plane. (R118)
X129. 1943. Norwich, England. Orange fireball. 5-6 centimeters in diameter, drifts through a window and disappears with a loud pop. (R101)
X130. May 8, 1945. Maidenhead, England. Orange globe, the size of the sun, moves across the sky followed by a horizontal column of white vapor. (R119) Possibly a meteor although titled 'globe lightning.' (WRC)
X131. 1945. Oxford, England. (R120)
X132. July 9, 1947. Kirkcudbrightshire, Scotland. Fireball runs along electrical wire, blasts big oak to pieces. (R121)
X133. August 17, 1947. Cornwall, England. Misty-coloured fireballs come through a stand of trees. Where they struck the trees, zigzag lightning flashed out. (R121)
X134. 1952. Moody Air Force Base, Georgia. An aircraft collides with a big orange fireball. A sharp jolt was felt, but a fused radio compass was the only damage. (R11)
X135. August 12, 1953. Co. Durham, Ireland. Two blobs of light jump from one cloud to another. (R122)
X136. February 16, 1957. North Atlantic Ocean. Blue fireball falls from a cloud and shoots across vessel's bow at sealevel. (R123)
X137. August 2, 1960. Atlantic Ocean. (R124)
X138. September 1963. Austria. Whitish-yellow fireball, a little larger than a tennis ball, comes in a door and explodes. (R125)
X139. April 11, 1965. Toledo, Ohio. A tornado funnel emits orange fireballs from its point. (R126, R127)
X140. April 17, 1966. At sea, off Victoria, South Australia. (R128)
X141. March 9, 1970. Australian waters. Three occurrences of ball lightning in clouds. (R129)
X142. August 12, 1970. Sidmouth, England. Red fireball in sky explodes, sending jagged flashes of lightning toward the ground. (R130)
X143. August 1971. Minneapolis, Minnesota. (R131)
X144. 1972. England. A lightning strike generates a fireball 18 inches in diameter, with a red core and yellow halo. It moves several carlengths down the road and vanishes. (R101)
X145. April 3, 1974. Huntsville, Alabama. Tornado spews forth 10-15 red fireballs at a time. (R132)
X146. February 10, 1975. North Atlantic Ocean. Intense bluish-white fireball explodes after 5-7 seconds. (R133)
X147. April 12, 1978. Kent, England. (R134)
X148. December 1979. Surrey, England. Ball lightning appears with a bang, fades away, leaving a strong chemical smell. (R135)
X149. May 1, 1980. Arabian Sea. Orange fireball travels horizontally over the water. Much static electricity. (R136)
X150. October 8, 1980. Oxford, England. Big ball of white light explodes. (R137)
X151. July 9, 1981. Essex, England. (R138)
X152. No date given. Iowa. Yellowish-white fireball the size of a washtub bounces along a road with a rushing noise. It demolishes a shed and kills a horse. (R15)
X153. No date given. Ohio. Pig sniffs at ball lightning; it explodes. (R97)
X154. No date given. Georgia. A sharp click is heard, and ball lightning is seen floating along a hall. Odor of ozone detected. (R97)
X155. No date given. Fontella, Virginia. Globe, 6-8 inches in diameter passes through three locust trees. (R139)
X156. No date given. United States. Ball lightning enters through a window, sails around a room, scorching objects, exits through the same window. (R140)
X157. No date given. Leavenworth, Kansas. A fireball the size of a child's head hops across the floor like a soap bubble. (R141)
X158. No date given. British Columbia, Canada. Pilot and companion in a jet fighter at 48,000 feet see and photograph a luminous object, about the same size as the fighter, in a thundercloud. (R27)
X159. No date given. East Prussia. Reddish rotating ball, about 16 inches in diameter, passes by several people, runs along an electric bell wire and explodes. (R27)

X160. No date given. England. Ball lightning descends slowly and explodes. (R83)
X161. No date given. Denmark. Long erratic trail of purported ball lightning photographed. Trail ended when it went down a chimney and exploded. (R142)
X162. No date or place given. Ball lightning falls into a garden and vanishes quietly. (R143)
X163. No date given. Eastbourne, England. Balls of fire drop out of clouds. (R144)
X164. No date given. Norway. (R15)
X165. No Date given. Lake Geneva. Fireball, like the sun at noon, about 100 feet off the ground, drifts like a balloon for a half mile, settles on a railway pole, and vanishes noiselessly. (R120)
X166. No date given. India. A yellow ball, 6-7 inches in diameter bursts inside a house, leaving an orange-colored gas behind. (R145)
X167. No date given. New Zealand. Ball lightning floats through a shed; does not disturb cows. (R75)
X168. No date, no place given. Great ball of fire appears inside a house as it is struck by normal lightning. (R23)
X169. No date or place given. Glowing yellow bubble appears on a tile floor, explodes, leaving smell of burnt powder. (R10)
X170. October 1860. No place given. White or pale yellow globe of light, same angular diameter as the moon, fell to earth about a quarter mile distant. (R146)
X171. May 25, 1897. Szliacs, Europe. Ball lightning descends during thunderstorm, enters a building through an open window and leaves through a closed window, leaving a fist-sized hole. (R147)
X172. November 13, 1902. Queensland, Australia. Hot dry winds and duststorms. Fireballs hovered in the air, exploding and throwing out sparks. (R148)
X173. July 1865. St. Ives, England. Luminous bluish ball, hissing loudly, passes slowly overhead. (R149)
X174. Summer 1906. Folkestone, England. Ball of blue fire, 4 inches in diameter, rolls up a flight of stairs and disappears. (R150)
X175. April 30, 1925. London, England. Brilliant blue lightning, including blue, molten-looking, globular and elongated ball lightning. (R151)
X176. No date or place given. Bluish-white fireball with a fuzzy outline enters a room through openings in a wooden screen and exits through a pane of glass, melting a neat circular hole, 28 cm in diameter. The ball seemed to possess internal motion. (R152)
X177. June 1929. Morecambe, England. On a completely dry day, the doors of a slaughterhouse fly open, a ball of orange flame enters, then leaves via another door. (R101)
X178. August 2, 1927. Canton River, China. A 'thunderbolt' falls near a ship and explodes with a terrific report. The compasses were permanently affected. (R154)
X179. September 29, 1923. Mediterranean Sea. Ball lightning explodes 200 yards from ship with a brilliant flash and a strong concussion. A column of smoke or a geyser of water was formed. (R155)
X180. November 6, 1944. Kamaran Island, Red Sea. A fireball with the shape of a 'rugger' ball and twice as large explodes with a roar. (R156)

References.


R1. Mendenhall, T. C.; "On Globular Lightning," American Meteorological Journal, 6:437, 1890. (X1, X6, X23, X30, X53, X55, X56, X57, X60, X61, X62)
R2. "Curiosities of Thunderstorms," Eclectic Magazine, 38:458, 1856. (X2, X26)
R3. Tripe, John W.; "Ball Lightning Seen during a Thunderstorm," Royal Meteorological Society, Quarterly Journal, 13:301, 1887. (X3)
R4. "Ball Lightning," Nature, 15:539, 1877. (X4)
R5. "Singular Phenomenon at Lower Cumber," Royal Meteorological Society, Quarterly Journal, 4:167, 1878. (X5)
R6. Smith, Worthington G.; "Fireball," Nature, 49:577, 1894. (X7)
R7. Hill, E.; "Globular Lightning," Nature, 56:293, 1897. (X8)
R8. Crossley, Arthur W.; "Ball Lightning," Nature, 114:10, 1924. (X9)
R9. Reynolds, William C.; "Globular Lightning," Nature, 125:413, 1930. (X10)
R10. Powell, James R., and Finkelstein, David; "Structure of Ball Lightning," Advances in Geophysics, 13:141, 1969. (X11, X103, X169)
R11. Condon, Edward U.; "Scientific Study of Unidentified Flying Objects," Bantam Books, New York, 1969, pp. 729-735. (X12, X20, X127, X134)
R12. Gucinski, Hermann; "Observation of Ball Lightning," Weatherwise, 31:167, 1978. (X13)
R13. Ord, J. E.; "Ball Lightning," Weather, 29:77, 1974. (X14)
R14. Marcet, W.; "On Atmospheric Electricity," Royal Meteorological Society, Quarterly Journal, 14:197, 1888. (X15)

R15. Uman, Martin A.; "What Is Ball Lightning?" Science Digest, 70:21, October 1971. (X16, X152, X164)
R16. Clark, R.; "Ball Lightning," Weather, 20:134, 1965. (X17)
R17. "Curious Lightning Phenomena," Popular Science Monthly, 42:422, 1893. (X2)
R18. Ginsburgh, I., and Bulkley, W. L.; Nature, 263:187, 1976. (X18)
R19. "Fireballs---Some Fiery, Some Dark," Journal of Meteorology (UK), 1:312, 1976. (X19, X24)
R20. Schove, D. Justin; "Hail in History, A.D. 1630-1680," Weather, 6:17, 1951. (X21)
R21. Tomlinson, Charles; "On Certain Low-Lying Meteors," Knowledge, 16:96, 1893. (X22, X23, X48)
R22. Tomlinson, Charles; "On Some Effects of Lightning," Philosophical Magazine, 5:26:114, 1888. (X22, X23, X48)
R23. Elliot, Gertrude; "Ball Lightning," Symons's Monthly Meteorological Magazine, 39:133, 1904. (X168)
R24. "Thunderstorm at Portsmouth, February 14, 1809," Royal Meteorological Society, Quarterly Journal, 35:140, 1909. (X25)
R25. Marriott, W., and Ranyard, A. C.; "Photography of Lightning," Knowledge, 12:165, 1889. (X26)
R26. "Strange Event," Scientific American, 5:264, 1850. (X27)
R27. Ley, Willy; On Earth and in the Sky, Ace Books, New York, 1957, pp. 45-53. (X28, X67, X103, X105, X158, X159)
R28. "A Thunder and Lightning Story," Ellsworth Herald, March 4, 1853, p. 3. (X29)
R29. "Unusual Weather," Weatherwise, 29:84, 1976. (X29)
R30. "A Catalog of Observations of Luminous Meteors," Report of the British Association, 1861, 30. (X31)
R31. Hare, Charlotte; "Globular Lightning," Nature, 19:5, 1878. (X32)
R32. "Meteoric Explosion," Eclectic Magazine, 4:253, 1866. (X33)
R33. Seyhoth, Robert; "An Instance of Ball Lightning at Sea," Monthly Weather Review, 29:249, 1901. (X34)
R34. "Ball Lightning at Sea," Scientific American, 86:36, 1902. (X34)
R35. Capron, J. Rand; "Ball Lightning," Symons's Monthly Meteorological Magazine, 18:41, 1883. (X35)
R36. Fitzgerald, M.; "Notes on the Occurrence of Globular Lightning...," Royal Meteorological Society, Quarterly Journal, 4:160, 1878. (X15)
R37. "Remarkable Thunderstorms," Scientific American, 27:152, 1872. (X36)
R38. "Ball Lightning," Scientific American, 57:53, 1887. (X3)
R39. "Killed by a Meteor," Scientific American, 34:183, 1876. (X37)
R40. Michelski, M.; "Curious Phenomenon during the Gale," Royal Meteorological Society, Quarterly Journal, 4:166, 1878. (X38)
R41. Marsh, O. C.; "Globular Lightning," Nature, 53:152, 1895. (X39)
R42. Lyte, Maxwell; Royal Meteorological Society, Quarterly Journal, 13:306, 1887. (X40)
R43. Smith, W. F.; "Ball Lightning," Nature, 22:267, 1880. (X41)
R44. "Remarkable Effects of Lightning," English Mechanic, 32:107, 1880. (X42)
R45. Nature, 22:466, 1880. (X43)
R46. Tennant, John; "A Fireball," Nature, 24:285, 1881. (X44)
R47. Nature, 24:476, 1881. (X45)
R48. Hannay, J. B.; "Extraordinary Atmospheric Phenomenon," Nature, 25:125, 1881. (X46)
R49. Glatton; "Ball Lightning," English Mechanic, 90:306, 1909. (X47)
R50. "Lightning Attacks an Editor," Symons's Monthly Meteorological Magazine, 18:120, 1883. (X48)
R51. Millar, W. J.; "Fireballs," Nature, 30:312, 1884. (X49)
R52. Fortescue, C.; "Meteor," Nature, 28:541, 1883. (X50)
R53. Knowledge, 3:64, 1883. (X51)
R54. Petrie, W. M. Flinders; "Fireballs," Nature, 30:360, 1884. (X52)
R55. Owen, Richard; "Ball or Globe Lightning," American Meteorological Journal, 3:383, 1886. (X54)
R56. Buchanan, Wm., and Scudamore, Wm.; English Mechanic, 45:396, 1887. (X58)
R57. Nature, 37:187, 1887. (X59)
R58. "Globular Lightning," Science, 10:324, 1887. (X59)
R59. "Globular Lightning," Electrician, 26:5, 1890. (X63)
R60. Nature, 46:62, 1892. (X64)
R61. Bayley, C. C.; "A Fire-Ball," Science, 19:249, 1892. (X64)
R62. Nature, 41:303, 1890. (X65)
R63. "Globular Lightning at Pontevedra," Electrician, 24:414, 1890. (X65)
R64. "Weird Phantoms of the Air," Popular Mechanics, 48:979, 1927. (X66)
R65. Nature, 44:575, 1891. (X68)
R66. "A Lightning Story," Symons's Monthly Meteorological Magazine, 27:188, 1893. (X69)
R67. Lodge, Oliver J., and Hewitt, George H.; "A So-Called Thunderbolt," Nature,

46:513, 1892. (X70)
R68. Nature, 46:548, 1892. (X71)
R69. Brew, William; "A Peculiar Discharge of Lightning," Nature, 48:370, 1893. (X72)
R70. "Thunderstorm Phenomena," Royal Meteorological Society, Quarterly Journal, 20:73, 1894. (X73)
R71. Ryan, G. M.; "Globular Lightning," Nature, 52:392, 1895. (X74)
R72. Minchin, George M.; "A Fireball," Scientific American, 73:374, 1895. (X75)
R73. "Globular Lightning," Royal Meteorological Society, Quarterly Journal, 22: 75, 1896. (X76)
R74. Potts, Louis M.; "Ball Lightning," Science, 31:144, 1910. (X77)
R75. Humphreys, W. J.; "Ball Lightning," American Philosophical Society, Proceedings, 76:613, 1936. (X78, X80, X99, X101, X117, X167)
R76. R., C. F.; "Lightning Phenomena," Symons's Monthly Meteorological Magazine, 32:127, 1898. (X79)
R77. "Ball Lightning," Monthly Weather Review, 27:364, 1899. (X81)
R78. "Observation of Ball Lightning," Scientific American, 85:72, 1901. (X82)
R79. Dennett, Frank C.; "Ball Lightning," English Mechanic, 90:115, 1909. (X83)
R80. Hands, Alfred; "Fireballs," English Mechanic, 90:117, 1909. (X84)
R81. Botley, Cicely M.; "Ball Lightning," Nature, 126:919, 1930. (X85)
R82. Maclear, J. P.; "Was It Globe Lightning?" Symons's Meteorological Magazine, 40:66, 1905. (X86)
R83. R., C. F., et al; "Fireballs," English Mechanic, 90:95, 1909. (X87, X160)
R84. "Aug. 1, 1907. Ball Lightning at Alpena, Michigan," Nature, 126:153, 1930. (X88)
R85. Nature, 76:671, 1907. (X89)
R86. Plunkett, G. T.; "Ball Lightning," English Mechanic, 90:140, 1909. (X90)
R87. Hyatt, J. D.; "Electrical Fireballs: What Are They?" Scientific American, 101:215, 1909. (X91)
R88. Hyatt, J. D.; "Electrical Fireballs: What Are They?" English Mechanic, 89:618, 1909. (X91)
R89. Baggs, J.; "Ball Lightning," Electronics and Power, 27:98, January 1981. (X92)
R90. Crew, E. W.; "Ball Lightning Observation in the Summer of 1911," Journal of Meteorology (UK), 6:93, 1981. (X92)
R91. "Ball Lightning Due to a Bend in a Conductor," Scientific American, 104: 365, 1911. (X93)
R92. "Thunderstorms at Southwater, Sussex, July and August 1912," Royal Meteorological Society, Quarterly Journal, 38:313, 1912. (X94)
R93. Nature, 96:322, 1915. (X95)
R94. Xenon; "Ball Lightning," English Mechanic, 102:48, 1915. (X96)
R95. Cooke, Richard; "Thunderbolt at Detling," Symons's Meteorological Magazine, 54:54, 1919. (X97)
R96. Cleland, J. B.; "Ball Lightning," Nature, 110:40, 1922. (X98)
R97. Talman, Charles Fitzhugh; "Ball Lightning," American Mercury, 26:69, 1932. (X100, X116, X153, X154)
R98. Taylor, A. P.; "Occurrence of Ball Lightning at Tamale, Northern Territories, Gold Coast, September, 1924," Royal Meteorological Society, Quarterly Journal, 51:23, 1925. (X102)
R99. "Globe Lightning at Yarmouth, August 11, 1925," Royal Meteorological Society, Quarterly Journal, 51:399, 1925. (X104)
R100. Milligan, W. H.; "Globe Lightning," English Mechanics, 124:150, 1926. (X106)
R101. Charman, Neil; "Ball Lightning Photographed?" New Scientist, 69:444, 1976. (X107, X129, X144)
R102. Andrew, W. L.; "Ball Lightning at Cattewater," Meteorological Magazine, 62:186, 1927. (X108)
R103. Crawford, F. E.; "Ball Lightning," Meteorological Magazine, 63:234, 1928. (X109)
R104. Covington, Arthur E.; "Ball Lightning," Nature, 226:252, 1970. (X110)
R104. Covington, Arthur E.; "Ball Lightning," an Actual Meteorite," Popular Astronomy, 39:107, 1931. (X111)
R106. Wood, R. W.; "Ball Lightning," Nature, 126:723, 1930. (X112)
R107. Bromwich, Josephine E.; "Thunderbolt in Somerset," Royal Meteorological Society, Quarterly Journal, 56:66, 1930. (X113)
R108. Craigie, Hamilton; "Observations of Lightning," Science, 72:344, 1930. (X114)
R109. Jensen, J. C.; "Ball Lightning," Physics, 4:372, 1933. (X115, X118)
R110. Jones, W. Neilson; "Ball Lightning," Nature, 130:545, 1932. (X119)
R111. "Ball Lightning at Stoke Poges?" Meteorological Magazine, 68:68, 1933. (X120)
R112. Holmes, Marshall; "Three Discharges of Ball Lightning," Nature, 133:179, 1934. (X121)
R113. Wishart, E. M.; "A 'Fireball'," Meteorological Magazine, 69:65, 1934. (X122)
R114. Gold, E.; "Thunderbolts: The Electric Phenomena of Thunderstorms,"

Nature, 169:561, 1952. (X123, X125)
R115. "Ball Lightning Is Real!" Science Digest, 32:80, July 1952. (X123)
R116. "Ball Lightning Observed by Nebraska Family," Science News Letter, 30:52, 1936, (X124)
R117. "Curious Effects of Globular Lightning," Nature, 141:722, 1938. (X126)
R118. "Ball Lightning in a Cloud?" American Meteorological Society, Bulletin, 23:88, 1942. (X128)
R119. "Globe Lightning," Nature, 155:752, 1945. (X130)
R120. Charman, Neil; "The Enigma of Ball Lightning," New Scientist, 56:632, 1972. (X131, X165)
R121. "Ball Lightning," Meteorological Magazine, 76:210, 1947. (X132, X133)
R122. Whillis, J.; "Ball Lightning," Weather, 8:353, 1953. (X135)
R123. Johnston, L. H.; "Ball Lightning," Marine Observer, 28:11, 1958. (X136)
R124. Pratt, C. E.; "Ball Lightning," Marine Observer, 31:127, 1961. (X137)
R125. Wagner, Gunther A.; "Optical and Acoustical Detection of Ball Lightning," Nature, 232:187, 1971. (X138)
R126. Botley, Cicely M.; "The Widecombe Calamity," Weather, 21:318, 1966. (X139)
R127. Vonnegut, B., and Weyer, James R.; "Luminous Phenomena in Nocturnal Tornadoes," Science, 153:1213, 1966. (X139)
R128. Sheldon, G. S.; "Ball Lightning," Marine Observer, 37:59, 1967. (X140)
R129. Stock, P. J.; "Lightning," Marine Observer, 41:12, 1971. (X141)
R130. Gordon, A. H.; "Fireball Cuts Off 2500 TV Sets," Weather, 25:85, 1970. (X142)
R131. Anderson, F. J.; "A Report on Ball Lightning," Journal of Geophysical Research, 77:3928, 1972. (X143)
R132. Vaughan, Otha H., Jr., and Vonnegut, Bernard; "Luminous Electrical Phenomena Associated with Nocturnal Tornadoes in Huntsville, Ala., 3 April 1974," American Meteorological Society, Bulletin, 57:1220, 1976. (X145)
R133. Todd, R. E.; "Ball Lightning," Marine Observer, 46:9, 1976. (X146)
R134. "A Ball Lightning Report from Kent," Journal of Meteorology, (U. K.), 3:210, 1978. (X147)
R135. "Ball Lightning," Journal of Meteorology, (U. K.), 5:123, 1980. (X148)
R136. Chipperfield, B. V.; "St. Elmo's Fire and Ball Lightning," Marine Observer, 51:57, 1981. (X149)
R137. Webb, J. D. C.; "Ball of Lightning on 7 October 1980," Journal of Meteorology, (U. K.), 6:150, 1981. (X150)
R138. "Ball Lightning in Essex, 9 July 1981," Journal of Meteorology, (U. K.), 6:325, 1981. (X151)
R139. "Not Ball Lightning," Monthly Weather Review, 27:156, 1899. (X155)
R140. Horner, D. W.; "Ball Lightning," English Mechanic, 102:71, 1915. (X156)
R141. Meigs, M. C.; "Ball of Electric Fire," Science, 6:338, 1885. (X157)
R142. Petersen, Helge; "Ball Lightning," Weather, 9:73, 1954. (X161)
R143. Horner, D. W.; "Ball Lightning," Symons's Monthly Meteorological Magazine, 39:111, 1904. (X162)
R144. Russell, A.; "Ball Lightning," Nature, 126:809, 1930. (X163)
R145. McMillan, Walter G.; "On the Phenomena of the Lightning Discharge....," Nature, 40:295, 1889. (X166)
R146. Churchill, W.; Royal Meteorological Society, Quarterly Journal, 13:309, 1887. (X170)
R147.."Globular Lightning," Royal Meteorological Society, Quarterly Journal, 23:307, 1897. (X171)
R148. Jensen, H. I.; "Remarkable Meteorological Phenomena in Australia," Nature, 67:344, 1903. (X172)
R149. Skeeles, John; "Fireballs," English Mechanic, 90:66, 1909. (X173)
R150. "Globe Lightning," English Mechanic, 90:211, 1909. (X174)
R151. Wilby, Frances B.; "Globular Lightning," Royal Meteorological Society, Quarterly Journal, 51:285, 1925. (X175)
R152. Powell, James R., and Finkelstein, David; "Ball Lightning," American Scientist, 58:262, 1970. (X176)
R153. Talman, C. F.; "A Case of Ball Lightning," American Meteorological Society, Bulletin, 11:110, 1930. (X10)
R154. Campbell, Alex; "Thunderbolt," Marine Observer, 5:157, 1928. (X178)
R155. Owens, A. L.; "Ball Lightning," Marine Observer, 1:117, 1924. (X179)
R156. Thompson, D.; "Explosion of a 'Fire Ball' at Kamaran, Red Sea," Royal Meteorological Society, Quarterly Journal, 71:39, 1945. (X180)

# GLB2 Ball Lightning with Projections or Spikes

Description. Ball lightning adorned with spikes or protuberances.

Background. "Ordinary" ball lightning is often pear-shaped suggesting the presence of internal forces distorting the normal spheroid. The appearance of spikes, knobs, and projections may be another expression of internal forces.

Data Evaluation. Only a few reports of a testimonial nature have been found. Rating: 3.

Anomaly Evaluation. Projections and other complex shapes, if they are stable, would seem more difficult to explain than a simple spherical shape. Rating: 2.

Possible Explanation. All observations of ball lightning with projections and spikes have been fleeting, suggesting that deviations from a spheroid are transitory and may, in fact, be a prelude to detonation.

Similar and Related Phenomena. Ball lightning with rays (GLB3), dumbbell-shaped ball lightning (GLB5), fragmenting ball lightning (GLB9), ball lightning with long tails (GLB16)

Examples of Ball Lightning with Projections or Spikes

X1. August 22, 1904. South Devon, England. "During a sharp thunderstorm experienced here on the 22nd inst., between 11.30 a.m. and noon, after a sharp peal of thunder, globular or ball lightning was observed at 11.40 a.m. in the S. E. The ball was estimated to be about 1/2$^{\circ}$ in diameter and of a molten glowing yellow colour. It was situated about 5$^{\circ}$ above the surface of the sea, and remained visible for nearly five seconds. There was a narrow pear-shaped appendage attached to the top of the ball extending to about 1 1/2$^{\circ}$ above it." (R1, R5)

X2. May 14, 1919. Dublin, Ireland. "On the night of May 14 a thunderstorm took place over Dublin. A shower of rain fell after 9 p.m., but between about 9.25 and 9.40 there was practically no rain, only a few drops falling. At about 9.50 I went outside, and when I had gone about two steps from the door I suddenly saw a luminous ball apparently lying in the middle of the street. It remained stationary for a very brief interval---perhaps a second---and then vanished, a loud peal of thunder occurring at the same time. The ball appeared to be about 18 in. in diameter, and was of a blue colour, with two protuberances of a yellow colour projecting from the upper quadrants.....The thunder was heard just at the disappearance of the ball, but the sound seemed to come from overhead rather than from the place where the ball was." (R2)

X3. December 31, 1924. Cardiganshire, U.K. A brilliant fireball with luminous projections. (R3)

X4. May 17, 1981. Norfolk, England. Ball lightning appeared inside a house. It was about the size of a tennis ball with four sharp points. (R4)

References

R1. Russell, S. C.; "Globular Lightning," English Mechanic, 80:88, 1904. (X1)
R2. Gilmore, G.; "Globular Lightning," Nature, 103:284, 1919. (X2)
R3. Davies, Ben; "Ball Lightning Phenomena," Nature, 115:640, 1925. (X3)
R4. "Ball of Light inside Home," Journal of Meteorology, (U.K.), 6:189, 1981. (X4)
R5. Russell, Spencer C.; "Ball Lightning," Symons's Monthly Meteorological Magazine, 39:153, 1904. (X1)

# GLB3 Ball Lightning with Diverging Rays

Description. Luminous spheres with long, radial, fan-shaped rays shooting from their surfaces. These rays appear most often during detonation or disintegration of the balls, but occasionally they are long-lived features.

Background. The appearance of long radial streamers may be indicative of ball lightning's internal structure and/or its disintegration scenario.

Data Evaluation. Observations of rayed ball lightning are rare, but enough examples have accumulated to give us confidence that the rays are real features of some ball lightning occurrences. Rating: 2.

Anomaly Evaluation. Essentially nothing is known about the diverging rays. Are they intense beams of electromagnetic radiation or streams of incandescent material. Rating: 2.

Possible Explanation. The structure providing ball lightning with its external appearance, presumably some sort of electromagnetic bottle, may begin its dissolution by developing holes or weaknesses in the "bottle," allowing energy and material to escape. The rays may thus be precursors of detonation.

Similar and Related Phenomena. Protuberances and spikes on ball lightning (GLB2) may reflect the same sort of internal instability as the rays. Fragmenting ball lightning (GLB9) may also be closely related.

Examples of Ball Lightning with Diverging Rays

X1. Circa 1860. Lambeth Walk, England. Experience of a Mr. Munro during a very heavy thunderstorm. "While looking out from his position of shelter, he saw a ball of fire, about 2 inches in diameter, thrown into the air from some invisible source, with streamers of fire or light diverging from it, the ball eventually bursting with a loud report." (R1)

X2. Circa 1915. London, England. During a powerful thunderstorm. "Looking down toward Eccleston Bridge, he saw a ball of fire, resembling a brilliant full moon, its apparent diameter about 6 ins. at first, descending at about 45°. It approached rapidly and struck and dislodged 15 or 20 ft. of a heavy moulded cornice on a house at the eastern corner of the cross roads, which fell with a crash into the area below. He was uncertain whether there was also the sound of an explosion. At the same moment fanlike rays of what he believed to be lightning darted from the ball, and he experienced a painless but very heavy momentary pressure, followed apparently by a sudden depletion, which left him breathless...." (R2)

X3. August 10, 1937. Fitzwilliam, New Hampshire. "I was seated on a second story porch enclosed with glass watching the storm. A radio aerial extends from a distant tree to a point on the side of the house some distance from the porch. Coincident with a crash of thunder, the fire-ball appeared. I can not say that it followed the wire or came from the sky. It just came out of space and seemed to move directly toward the window and then fell as though to enter the cellar of the house. It was a round, bronze, glistening ball with gleaming rays shooting from the top and sides; by its beauty and brilliance reminding one of an ornament at the top of a Christmas tree." (R3)

X4. No date given. Mt. Bohul, Russia. Report by a geodesic party. "An extremely bright violet ball, surrounded with rays which were, the party says, about two yards long, struck the top of the peak. A second and a third followed, and the whole summit of the peak was soon covered with an electric light which lasted no less than four hours." (R4)

X5. August 23, 1878. London, England. Ball lightning bursts with a splendid mass of whitish-blue rays. (R5)

X6. April 15, 1916. Puy de Dome Observatory, France. Several fireballs explode, throwing out tongues of fire in all directions. (R6)

X7. No date given. Nioheim, Prussia. Ball lightning the size of a cannon ball explodes in a bundle of rays. (R7)

References

R1. Royal Meteorological Society, Quarterly Journal, 13:308, 1887. (X1)
R2. Glatton, W. H.; "Ball Lightning," English Mechanic, 102:13, 1915. (X2)
R3. Hunneman, Mary Ethel; "A Fire-Ball," Science, 86:244, 1937. (X3)
R4. Nature, 42:458, 1890. (X4)
R5. Stone, J. Harris; "A Meteorite?" Nature, 18:464, 1878. (X5)
R6. "Ball Lightning," American Journal of Science, 4:43:248, 1917. (X6)
R7. "Globular Lightning," Royal Meteorological Society, Quarterly Journal, 23:307, 1897. (X7)

# GLB4 Rod-Shaped Ball Lightning

Description. Luminous masses of cylindrical shape but otherwise possessing most of the characteristics of ordinary ball lightning (GLB1). Sizes from 5 x 15 inches to 3 x 8 feet have been reported. Bright flames sometimes issue from the surfaces of the cylinders. In general, the phenomenon seems somewhat more bizarre and energetic than ordinary ball lightning, as the examples below will confirm. Rod-shaped ball lightning has been seen to change into spherical ball lightning.

Background. Ordinary ball lightning is often pear-shaped, but cylinders with length-to-diameter ratios of two or greater seem to represent a distinctly different variety with, perhaps, a different mode of formation and another type of internal structure.

Data Evaluation. Rod-shaped ball lightning is much more rare than the spherical variety. The few extant cases do, however, display good internal consistency. Rating: 2.

Anomaly Evaluation. It is more difficult to imagine an electromagnetic containment scheme for cylinders with high length-to-diameter ratios than for simple spheroids. Rating: 2.

Possible Explanation. Rod-shaped ball lightning may be an intermediate state between ordinary ball lightning and dumbbell-shaped ball lightning, although this possibility certainly does not explain any variety of ball lightning. Interchangeable forms just heighten the enigma.

Similar and Related Phenomena. Ordinary ball lightning (GLB1), dumbbell-shaped ball lightning (GLB5), bead lightning (GLL).

Examples of Rod-Shaped Ball Lightning

X1. August 5, 1889. England. "On Monday, the 5th instant, at midday, this district was visited by a violent storm of rain, which lasted half an hour, and was accompanied by thunder and lightning. When the storm had passed over and the sky was getting bright, a rod-like object was seen to descend from the sky. It is described as being of a pale yellow colour, like a hot iron, and apparently about 15 inches long by 5 inches across. .....The object descended 'moderately slowly,' 'not too fast to be followed by the eye' and quite vertically. On reaching a point about 40 feet from the ground, and in close proximity with a chimney-stack belonging to a house in Twickenham Park, the object seemed to 'flash out horizontally as if it burst,' showing an intensely white light in the centre and a rosy red towards the outer parts. At the same instant a violent explosion was heard, and soon afterwards a strong smell was perceived, which is described by the observers as 'resembling that of burning sulphur,' for which the smell of ozone and nitric oxide might easily be taken." (R1)

X2. May 28, 1901. Liverpool, England. "On afternoon of May 28 last, whilst I was in the kitchen of my house in the country during a most violent thunderstorm, my housekeeper sitting on one side of a window, the top sash of which was open and I on the other, simultaneously with a very heavy peal of thunder a red-hot bolt the size of a small rocket flew in through the open window, followed by a stream of sparks. When in the middle of the kitchen the bolt exploded with a bright flash and sharp crack as of a pistol. Stranger still, this was followed in about 15 seconds by two more similar bolts, one of which came through the back door and one through the front door, both of which were open. There was no smell or after-effect noticeable. I may say the house is not in a conspicuous situation, the adjoining one being much loftier." (R2) The terms 'bolt' and 'rocket' seem to indicate a cylindrically shaped mass.

X3. Circa 1907. Burlington, Vermont. "I was standing on the corner of Church and College streets, just in front of the Howard Bank and facing east, engaged in conversation with Ex-Governor Woodbury and Mr. A. A. Buell, when, without the slightest indication or warning, we were startled by what sounded like a most unusual and terrific explosion, evidently very near by. Raising my eyes and looking eastward along College Street, I observed a torpedo-shaped body some 300 feet away, stationary in appearance and suspended in the air about 50 feet above the tops of the buildings. In size, it was about 6 feet long by 8 inches in diameter, the shell or cover having a dark appearance, with here and there tongues of fire issuing from spots on the surface resembling red-hot unburnished copper. Altho stationary when first noticed this object soon began to move, rather slowly, and disappeared over Dolan Brothers' store, southward. As it moved, the covering seemed rupturing in places and thru these intensely red flames issued. My first impression was that it was some explosive shot from the upper portion of the Hall furniture store. When first seen it was surrounded by a halo of dim light, some 20 feet in diameter. There was no odor that I am aware of perceptible after the disappearance of the phenomenon, nor was there any damage done so far as known to me. Altho the sky was entirely clear overhead, there was an angry-looking cumulo-nimbus cloud approaching from the northwest; otherwise there was absolutely nothing to lead us to expect anything so remarkable." (R3)

X4. 1959. Kansas City, Missouri. "In 1959, Mrs. Lillian Mack, of Kansas City, Missouri, was in her home during an afternoon thunderstorm when she heard a sound like 'crushing glass.' An object about two feet long and one inch in diameter came flying like a spent arrow into the room where it hovered while forming the shape of a ball. After half a minute, it dissolved, and 'while dissolving we all heard the sound of breaking glass... but no glass was broken... sparks flew from the falling dust-like stuff and the (object) looked like hot metal.'" (R4)

X5. July 1879. Lambertville, New Jersey. A cylinder of fire, 3 feet by 6-8 feet fell with a whizzing sound. (R5)

X6. August 25, 1880. No place given. A brilliant cylindrical body with conical ends passes between clouds. (R6)

References

R1. Hare, A. T.; "Globular Lightning," Nature, 40:415, 1889. (X1)

R2. G.; "Thunderbolt," English Mechanic, 74:15, 1901. (X2)

R3. Alexander, William H.; "A Possible Case of Ball Lightning," Monthly Weather Review, 35:310, 1907. (X3)

R4. Gaddis, Vincent H.; Mysterious Fires and Lights, David McKay Co., New York, 1967, p. 54. (X4)

R5. Van Dyck, F. C., and Hayden, Everett; "Globular Lightning," Science, 11:110, 1888. (X5)

R6. Tomlinson, Charles; "Remarks on the Weathering of Rocks and Certain Electrical Phenomena," Philosophical Magazine, 5:26:475, 1888. (X6)

# GLB5 Double and Triple Ball Lightning

Description. Two or three balls of lightning connected by a luminous rod-like structure. The balls may be of different sizes but they move together.

Data Evaluation. Very few cases of double or triple ball lightning have been discovered so far. One example, however, was well-described by trained observers (X2 below). Rating: 3.

Anomaly Evaluation. Double and triple ball lightning cannot be less mysterious than ordinary ball lightning. The puzzling connecting structure only aggravates the problem of explaining these long-lived energetic manifestations. Rating: 2.

Possible Explanations. Double and triple ball lightning may be abbreviated forms of bead lightning (GLL), which itself is not well-understood. Alternately, ball lightning with protuberances might ultimately fission into two spheres, so might rod-shaped ball lightning. These surmises are not, of course, explanations but only thoughts about possible relationships.

Similar and Related Phenomena. Ball lightning with projections (GLB2), rod-shaped ball lightning (GLB4), and bead lightning (GLL).

Examples of Dumbbell-Shaped Ball Lightning

X1. July 1898. Scranton, Pennsylvania. "In July, 1898, in the city of Scranton, Pa., between the hours of 12 noon and 1 o'clock P. M., as the sun shone brilliantly from a cloudless sky, there appeared in front of the writer, apparently not more than 100 feet away and 50 feet elevated from the earth passing from an easterly to a westerly direction, three balls of candescent light linked together, descending like a bolt of lightning and accompanied with a most terrifying and hissing sound, ending in a report much like the sudden immersion of a large molten mass of iron into a body of water." (R1)

X2. May 12, 1912. Dresden, Germany. "A

remarkable example of ball lightning was observed on May 12, 1912, at 6.45 p.m. from the 200-m-high section of Plaven near Dresden on the adjacent side of the narrow Weisseritz Valley. Although the observation site itself lay beyond the range of the thunderstorm (the nearest of the three thunderclouds was at a horizontal distance of 3 km), suddenly, without there preceding a lightning of some other form, there appeared at a height, which must have been less than 100 m above the ground, a luminous spherical mass at a distance of approximately 1 km from the observation site. The instant that one of us first noticed the luminous phenomenon there was a second, smaller ball about 1.5 m below the first which was connected to the upper ball by a fine strip similar to a string of beads. The color of the balls was reddish yellow (about 600 mμ). Both moved slowly one directly above the other in the direction northeast, whereby the upper ball maintained a constant height above the ground, while the lower one slowly descended as the connecting luminous strip faded..... Both balls had a velocity which could hardly have exceeded 1 m/sec.... we concluded that the diameter of the lower ball could not have been greater than 1 m..... No noise was perceived, and the total duration of the event from the moment the phenomenon was noticed to the instant the upper ball faded amounted to two minutes....." (R2, R3)

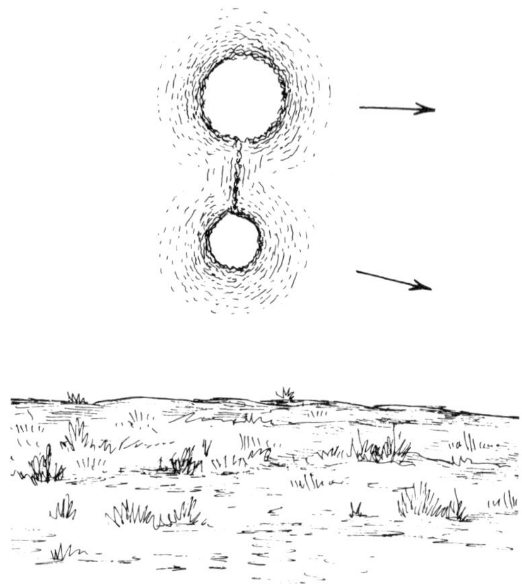

*Two balls of lightning connected by a luminous thread*

References

R1. Soper, C. H.; "A Daytime Meteor," Scientific American, 113:321, 1915. (X1)

R2. Meteorologische Zeitschrift, 29:384, 1912. (X2)

R3. "Double Ball Lightning," Literary Digest, 46:177, 1913. (X2)

# GLB6 Miniature Ball Lightning

Description. Spherical or nearly spherical luminous structures with diameters less than about 1 inch. Much smaller than ordinary ball lightning, this form has been seen only inside houses during thunderstorms or unsettled weather. The pea-sized lights sometimes disappears simultaneously with lightning and thunder outside, suggesting they may be induced by strong electric fields and possibly akin to St. Elmo's Fire (GLD). The energy content of miniature ball lightning is evidently very small.

Background. We know so little about the constitution of ordinary ball lightning that we cannot say that miniature varieties are impossible. However, the characteristics of miniature ball lightning seem somewhat different from ordinary ball lightning, indicating a basically different phenomenon, perhaps a mobile form of St. Elmo's Fire.

Data Evaluation. A very rare phenomenon, but the examples available show good consistency. Rating: 3.

Anomaly Evaluation. If miniature ball lightning is true ball lightning, the problem of explaining it is at least as difficult as it is for ordinary ball lightning. If miniature ball lightning is a mobile form of St. Elmo's Fire, as seems reasonable, the anomaly is less pronounced. Rating: 3.

Possible Explanation. A mobile form of St. Elmo's Fire or corona discharge.

Similar and Related Phenomena. St. Elmo's Fire (GLD).

Examples of Miniature Ball Lightning

X1. August 17, 1921. Eastbourne, England. "The two ladies were sitting at a table about 8 p.m., with the window open. It was raining heavily at the time, and there was no wind. Stormy clouds were about, but it was not un-unusually hot. Thunder and lightning at the same time were afterwards reported from London---a distance of, say, 50 miles---but there was no thunderstorm at Eastbourne. There had been no rain during the few preceding days. As one of the ladies took up a knife to cut bread the ball of light was seen to flash past the knife (without touching it) on to the table, travelling a distance of about 9 in. at an average height of about 3 in. from the table, but moving toward the latter. When the ball touched the tablecloth it 'went out with a spitting sound,' leaving no mark or trace of any sort.....As to the appearance of the ball itself, it was 'about the size of a pea, the light encircling it being about the size of a golf ball. The light was white and intensely bright, like electricity. Too dazzling to see through." (R1, R3)

X2. June 19, 1924. Baltimore, Maryland. "Mrs. Ames was standing on a rug during a thunderstorm with her hand at her waist, one finger more or less extended. I was about five feet away and noticed the air between her finger and the floor was quivering so that it looked just like the hot air over a field. I noticed something rise slowly from the floor up towards her finger and then there was for an instant a small oblong fireball about the size of a pecan attached to her finger. It was not very bright and appeared to shine through a haze. There came a flood of lightning outside and the fireball disappeared." (R2)

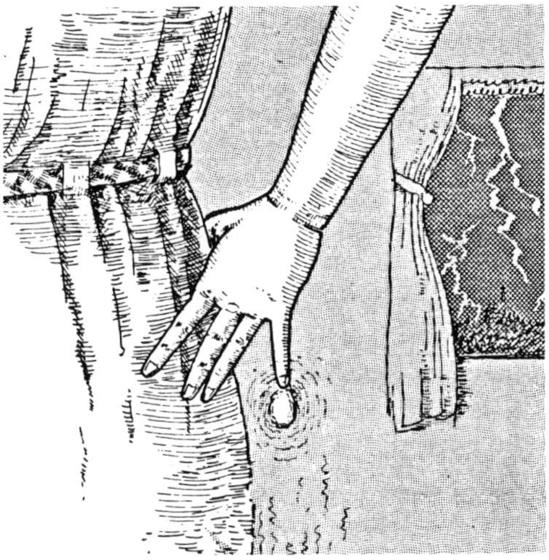

*Miniature ball lightning materializes in Baltimore house.*

References

R1. Chattock, A. P.; "Globular Lightning Discharge," Nature, 109:106, 1922. (X1)

R2. Humphreys, W. J.; "Ball Lightning," Americal Philosophical Society, Proceedings, 76:613, 1936. (X2)

R3. "Ball Lightning," Meteorological Magazine, 57:46, 1922. (X1)

# GLB7 Giant Ball Lightning

Description. A luminous, roughly spherical but sometimes shapeless mass ranging in size from approximately 3 feet to 60 feet. These glowing masses usually descend from thunder clouds, drift a bit, and fade away silently. There is often a strong resemblance to electric discharge phenomena (GLD).

Background. At the large end of the ball lightning size spectrum, as well as at the small end, the phenomena seem to take on the characteristics of such phenomena as electrified clouds and St. Elmo's Fire (GLD). Classical ball lightning is much more energetic and destructive than either giant or miniature ball lightning. Thus, the explanations may vary with the size of the luminous mass.

Data Evaluation. Good observations and photographs exist. Rating: 1.

Anomaly Evaluation. Although giant ball lightning does not seem to possess the energy density of ordinary ball lightning, no explanation exists for its ability to preserve its shape and motion for the several seconds it persists. Rating: 2.

Possible Explanation. Giant ball lightning may be a large, poorly organized plasmoid held together by a weak magnetic bottle.

Similar and Related Phenomena. St. Elmo's Fire and other electric discharge phenomena. (GLD)

Examples of Giant Ball Lightning

X1. November 1, 1864. Florence, Italy. "A white globe of fire many times larger than the full moon seemed hanging almost motionless in the air; a large portion of the surrounding heavens was lighted up by it, so much so that my head being turned away from it, I mechanically apprehended it was moonlight, though, had I thought at all, I should have known that no moon was at that time in sight; but a bright shimmering of light on a tall bay tree not far from me recalled my attention, and turning I saw this fireball, white for a moment, shades of orange and blue passing over its surface, the latter gaining on the former. After a full minute's time (after I had turned towards it) it disappeared suddenly---vanishing, not appeared to move from where it was; only just before its disappearance a smaller ball was seen immediately below it, of a fiery orange colour, the first one appearing at that moment of the same hue." (R1)

X2. February 8, 1906. Beccles, England. During a thunderstorm. "Then almost immediately, in the north-west, there appeared a huge circle of light, giving the impression of the heavens being opened, most vivid, and in size, as far as one could judge, 2 or 3 times the diameter of a setting sun. It appeared just above the cedar in height, but not near it in distance from me. At the same moment an appalling crash came like the bursting of a big shell immediately overhead. This was, I suppose, the moment at which the east end of our church, 100 yards from this house, was wrecked. The circle of light was visible for some appreciable time..... My wife, in a room above this sitting away from the window, describes the appearance almost exactly as I have done, except that she says the colour was very yellow, like gas, whereas I should say it was white or nearly so. My daughter, aged sixteen, looking out of window from the same upstairs room, speaks of it as a huge ball of light like a great sun, and says that the whole garden seemed to be a blaze of light. These two add that the globe of light seemed to travel over the house." (R2)

X3. July 9, 1975. Albany, New York. "Of particular interest was the observance of apparent ball lightning associated with, but not connected to, a long diagonal stroke which hit a factory a distance of just under

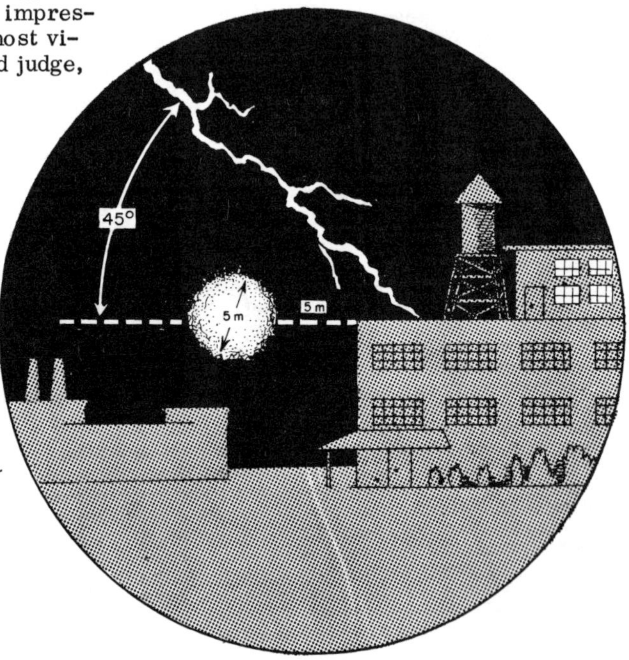


*Ball lightning about 5 meters in diameter observed near Albany in 1975.*

a kilometer to the north of the observatory. Over a three-second interval, four distinct lightning strokes, each following the same path, hit the building, and during this time an unmistakable fireball was seen near the ground an estimated 5 m to the west of the point of impact of the lightning stroke. The ball was very bright, fuzzy in appearance, equal in brightness to the lightning stroke, and orange-yellow in colour. Apparently, the ball was quite large, its size estimated to be approximately 5 m in diameter. It appeared at the moment the first stroke reached the factory and vanished when the final stroke disappeared. The ball maintained a constant brightness for the duration of the entire flash, a full 3 secs; it did not flicker on and off in unison with each individual stroke." (R3)

X4. June 8, 1977. Dyfed, West Wales, U.K. "The ball lightning phenomenon was very large and estimated to be about the size of a bus. It was described as a brilliant, yellow-green, transparent ball with a fuzzy outline which descended from the base of a towering cumulus over Garn Fawr Mountain and appeared to 'float' down the hillside. Intense light was emitted for about three seconds before flickering out. Severe static was heard on the radio. The object slowly rotated around a horizontal axis, and seemed to 'bounce' off projections on the ground. It was noticed that cattle and seabirds in the immediate vicinity became disturbed." (R4)

X5. November 4, 1749. North Atlantic Ocean. Ball of fire, the size of a large millstone, was seen rolling over the water. It exploded when it hit the ship and did considerable damage. (R5)

X6. June 26, 1921. St. John's Wood, England. Large incandescent mass floats in the air below the clouds. It remained stationary for some minutes. (R6)

X7. August 30, 1930. Nebraska. A lavender colored shapeless mass floats slowly downward as a squall line approaches, and globular structures, 28 and 42 feet in diameter, run along power lines. (R7-R9)

X8. July 22, 1951. Dorset, England. A large mass of blinding light hovers over ground and fades out. Not spherical, about 1 yard in all dimensions. (R10)

References

R1. Baldelli, Madame; "Large Fireball," Astronomical Register, 3:53, 1865. (X1)

R2. "Lightning at Barsham, Suffolk, February 8, 1906," Royal Meteorological Society, Quarterly Journal, 32:170, 1906. (X2)

R3. Colucci, S.; "Ball Lightning Observation," Weather, 31:68, 1976. (X3)

R4. Jones, Ian; "Giant Ball Lightning," Journal of Meteorology, (U.K.), 2:271, 1977. (X4)

R5. Chalmers, Mr.; "An Account of an Extraordinary Fireball Bursting at Sea," Philosophical Transactions, 46:366, 1749. (X5)

R6. Nature, 107:733, 1921. (X6)

R7. Jensen, J.C.; "Ball Lightning," Scientific Monthly, 37:190, 1933. (X7)

R8. "Ball Lightning Apparently Connected with Dust," Science News Letter, 24:323, 1933. (X7)

R9. "Ball Lightning Phenomenon," Science, 78:sup 9, October 27, 1933. (X7)

R10. Davis, M.; "A Near Flash," Weather, 6:384, 1951. (X8)

# GLB8 Transparent Ball Lightning

Description. Transparent luminous masses that otherwise resemble ordinary ball lightning. Ball lightning is usually so bright that one cannot see through it, but if the ball's energy density is very low it might well be transparent.

Background. Although this phenomenon is classified with other types of ball lightning, it may in fact be more closely related to nocturnal lights (GLN), as exemplified by ghost lights and will o' the wisps. The line of demarcation between ball lightning and nocturnal lights is not at all clearcut. Ball lightning generally seems to be more violent and closely associated with storms.

Data Evaluation. The data are scant and soft. Rating: 3.

Anomaly Evaluation. Rating: 2.

Possible Explanations. A very low energy density form of ball lightning or a variety of will o' the wisp, neither of which has a good explanation.

Similar and Related Phenomena. All of the nocturnal lights (GLN).

Examples of Transparent Ball Lightning

X1. September 1910. Mazoc, South Rhodesia. "...at about 9 p.m., he observed a ball of fire, the colour of an electric light, about one foot six inches by ten inches in size, and similar in shape to a rugby football. The ball was luminous and transparent, but it cast no light beyond its own body. When first seen it was stationary over a road on Mr. Biggs' farm, and it remained in one position for about five minutes, during which time, Mr. Biggs approached within three yards. The light then moved away in a zig-zag fashion at a rate of about six miles an hour, and finally entered a belt of trees, where it disappeared, leaving no trace whatever of its flight. No noise was heard, no trace of damage could be found, and no smell could be detected." (R1)

References

R1. Curtis, John A.; "Globular Lightning," Knowledge, 9:343, 1912. (X1)

# GLB9 Fragmenting Ball Lightning

Description. Ball lightning that bursts into many smaller balls or incandescent pieces that fly off in all directions. The formation of multiple, smaller, long-lived balls is rare. The "pieces" of ball lightning may carry dangerous amounts of energy or be harmless.

Data Evaluation. Fragmenting ball lightning has been observed frequently. Rating: 1.

Anomaly Evaluation. It is not clear how ball lightning's structure, whatever it may be, can split into smaller, luminous, apparently self-sustaining (temporarily stable), pieces. Do the fragments have the same structure as the parent ball lightning? How is the original ball's energy subdivided? Rating: 2.

Possible Explanations. The structure of ball lightning may be multicelled and fissionable when stimulated. Another possibility is that fragmenting ball lightning may generate electric fields intense enough to create corona discharge and St. Elmo's Fire, appearing as multiple spheres.

Similar and Related Phenomena. St. Elmo's Fire (GLD) has also been observed to split and even recombine. Ordinary lightning (GLL) frequently splits into many bolts and, more rarely, bead lightning. Luminous air bubbles. (GLD)

Examples of Fragmenting Ball Lightning

X1. November 16, 1857. Charleston, South Carolina. From the testimony of a Mr. Scriven. ",...he saw a red, fiery ball of the size and shape of an orange, slowly descending through a distance apparently of 20 or 30 feet, to the ground. Its fall was scarcely more rapid than that of a soap bubble, giving him time to call his sister, a little girl, to see it strike a high wooden fence, distant about fifty or sixty feet from the portico, and which separated the door-yard from a church enclosure adjoining. It seemed to adhere for an instant to the board against which it struck, and then separated into three parts and disappeared." (R1)

X2. July 17, 1890. Surbiton, England. During a severe thunderstorm. "The room I was standing in was situated about 30 ft. from the ground, when all at once at the time mentioned it seemed to me as if a huge ball of reddish yellow fire was suspended in mid air, about the level of the window. (The colour was that of an electric arc as seen in a dense fog). It did not last a moment, then it rapidly changed colour, and turned to an intense whiteness, move vivid than any other light I have ever seen, and burst with a terrific explosion, increasing in size many times over, and dozens of small splashes darted out of it on all sides; its brilliancy dazed me for several moments after." (R2)

X3. April 13, 1904. Essex, England. "A thunderstorm occurred during the early morning hours, at about 3 a.m. there was a blinding flash, lighting up the whole neighborhood for miles around, followed immediately by a crashing explosion. One person stated that he saw what appeared to be a cylinder, and another person, a ball of fire, descend and then explode, 'casting darts' in all directions. On careful examination in daylight it was found that in an oatfield which had recently been dredged there were three distinct sets of holes ranging from 9 in. down to about 1 in. in diameter. The holes, which were perfectly circular, diminished in size as they went downwards, and remained so on to the perfected rounded ends at the bottom. Upon digging sectionally into the soil, which is stiff yellow clay, it was found that the holes were 'as clean cut as though bored with an auger.'" (R3) Note that ordinary fulgerites have branching structures. (WRC)

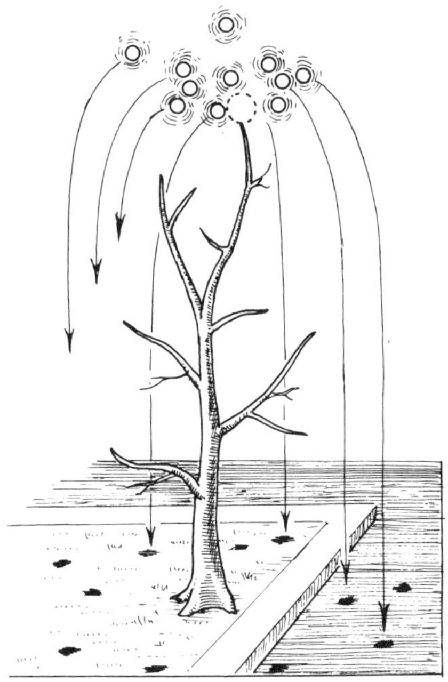

*Ball lightning fragments into smaller balls upon hitting a tree.*

X4. October 8, 1919. Salina, Kansas. "At about 6.30 p.m., October 8, 1919, a brilliant display of ball lightning occurred at Salina (sic), Kans., on one of the most frequented street intersections of the town. Eyewitnesses described it as 'a ball of fire as large as a washtub floating low in the air.' It struck the northwest corner of the Campbell Building, corner Santa Fe and Iron Avenue, about midway to the top of the building, which is 35 feet high, tore out some brick, demolished a second-story window, and then exploded with a bang that resembled the noise made by the discharge of a large pistol, filling the air with balls of fire as large as baseballs, which floated away in all directions. Some of these balls followed trolley and electric-light wires in a snaky sort of manner and some simply floated off through the air independently of any objects near by." (R4)

X5. 1960s. Neustadt, Germany. During a thunderstorm. "Suddenly, at a height of about 16 m above the ground and at a short range of about 24 m, I saw a spherical plasma ball coloured bright yellow-white. This object appeared to have a diameter of 50 to 100 cm. It moved vertically downwards with a speed of about 4 m $s^{-1}$, and its path ended in the top branches of a tree, at a height of approximately 9 m. On touching the tree the ball instantly disintegrated into eight to twelve smaller spheres. These were the same colour as the large one and each had a diameter of 12 to 15 cm. They fell to the ground, guided by the outer contour of the tree, and moved vertically during the last few metres in the absence of branches. On reaching the ground (an asphalt roadway and a neighboring footpath) the spheres instantly disappeared. There was no noise apart from that of the rain and no lightning associated with the primary plasma-like sphere. Three to five minutes afterwards, the same phenomenon occurred again in precisely the same way as before, indicating that the conditions needed to produce and guide the primary sphere were still maintained or reestablished during the time that had elapsed. Immediately after the rain had stopped, I went out into the street to look for further evidence. There were circular patches of melted asphalt on the wet asphalt cover of the roadway which showed the interference colours of thin layers. Their diameters were each 12 to 15 cm and they obviously marked the impact areas of the smaller spheres. This event was described similarly by all of the witnesses." (R5)

X6. July 28, 1973. Rabaul, Papua-New Guinea. Clouds to the south illuminated by lightning. "From the top of the tallest cloud of the group (it appeared to emanate from the top) a 'ball' of lightning very swiftly traversed the sky, leaving behind it what could only be described as a meteoric tail. The angle of elevation from me to the 'ball' I estimated as approximately 30°, at point of commencement (this was a rough guess and could have been less). Still leaving its meteor-like tail the 'ball' travelled towards me, climbing to an estimated angle of 50° or so, then literally exploded into four or five 'fingers' like a roman candle firework, each separate finger of the same constitution as a meteor's tail. These fingers climbed to an estimated angle of 60°, then appeared suddenly to retract into themselves down the paths that they had followed, forming into a ball again and returning as such into the cloud of origin. (R6, R7)

X7. July 16, 1750. Surrey, England. A large fireball divides into many parts. (R8)

X8. August 1881. No place given. A bright elongated fireball emerges from a cloud; before reentering the cloud a portion of it breaks off and falls to earth. (R9)

X9. August 1881. Connersville, Indiana. Lightning bolt fragments into many balls that roll around the ground. (R10)

X10. February 24, 1884. Amiens, France. A brilliant mass divides into many glowing but harmless globules. (R11)

X11. June 10, 1891. Woodstock, England. Fireball bursts into glowing fragments. (R12)

X12. August 9, 1893. Kingstown, England. A fireball darts about the surface of the water, leaving smaller balls behind. (R13)

X13. Circa 1895. Culday, Ireland. A great shining mass approaches a boy and bursts into several fragments, mangling his hand. (R14)

X14. Circa 1907. Minnesota. A fireball strikes a tree and scatters in all directions. (R15)

X15. Circa 1920. Marietta, Ohio. Ball lightning, 9 inches in diameter, explodes and hurls brilliant particles in all directions. (R16)

X16. September 10, 1924. Finley, North Dakota. A dull, greenish-yellow ball of light appears and splits into many smaller parts. (R17)

X17. June 10, 1948. London, England. A reddish-white fireball, twice the size of a football, explodes, throwing pieces in all directions. (R18)

X18. No date given. Lanarkshire, England. Ball of fire settles on the ridge of a thatched roof and gradually disintegrates into pieces that stream down the thatch. (R19)

X19. November 1825. Newton, Ohio. "At the same time, and at the same place the meteor was seen to fall, two ladies were walking in the road in the same direction in which the meteor was moving, found themselves suddenly enveloped in a mass of light. A ball of fire or light, several feet in diameter, seeming to come from above and behind fell upon them and the ground, breaking, as it struck, into a thousand smaller balls, rolling upon the earth and breaking again into still smaller balls, till it disappeared. Looking up, they saw the other portion of the meteor just as it disappeared in the distance before them. The phenomenon was attended with no noise or heat, and their clothing exhibited no traces of having been in contact with any foreign substance." (R20)

## References


R1. Shepard, Charles Upham; "On a Shooting Meteor,..." American Journal of Science, 2:28:270, 1859. (X1)

R2. "Notes on a Curious Lightning Flash," Electrician, 25:445, 1890. (X2)

R3. English Mechanic, 79:453, 1904. (X3)

R4. "Ball Lightning at Salina, Kans.," Monthly Weather Review, 47:728, 1919. (X4)

R5. Wittmann, A.; "In Support of a Physical Explanation of Ball Lightning," Nature, 232:625, 1971. (X5)

R6. Piddocke, B.G.; "Ball Lightning," Weather, 29:314, 1974. (X6)

R7. White, R.; "Ball Lightning and Electrometers," Weather, 30:134, 1975. (X6)

R8. Mendenhall, T.C.; "On Globular Lightning," American Meteorological Journal, 6:437, 1890. (X7)

R9. English Mechanic, 33:135, 1881. (X8)

R10. Van Dyck, F.C., and Hayden, Everett; "Globular Lightning," Science, 11:110, 1888. (X9)

R11. "Globe Lightning," Knowledge, 5:479, 1884. (X10)

R12. Science, 18:5, 1891. (X11)

R13. Ranyard, A.C.; "Lightning Photographs and Some Photographic Defects, Knowledge, 16:193, 1893. (X12)

R14. "A Recent Irish Meteorite," English Mechanic, 62:248, 1895. (X13)

R15. "Fireballs," English Mechanic, 90:66, 1909. (X14)

R16. Kaiser, John; "Ball Lightning," Science, 60:293, 1924. (X15)

R17. Stephens, O.B.; "A Curious Kind of

Lightning?" Scientific American, 132:128, 1925. (X16)
R18. Slater, A. E.; "Ball Lightning at Whipsnade," Weather, 3:252, 1948. (X17)
R19. Muirhead, Henry; "Globe Lightning," Symons's Monthly Meteorological Magazine, 18:57, 1883. (X18)
R20. Olmsted, Denison; "Observations on the Meteors of November 13th, 1833," American Journal of Science, 1:26:132, 1834. (X19)

# GLB10 The Materialization of Ball Lightning in Enclosures

Description. The formation of luminous spheres, usually smaller than average ball lightning, inside closed houses, aircraft, and electrically shielded volumes. Ball lightning may penetrate glass windows without breaking them. It may emerge from electrical apparatus, including telephones and aircraft instruments. Materialization may be coincident with lightning strikes outside the closed volume. Otherwise, this phenomenon has all the attributes of ordinary ball lightning.

Background. Ball lightning, supposedly an electrical phenomenon, should not materialize inside electrically shielded volumes, such as modern aircraft. This puzzling anomaly may be very significant in the explanation of ball lightning of all types---assuming several types actually exist!

Data Evaluation. Many excellent observations of ball lightning materialization exist; some by experienced scientists. Rating: 1.

Anomaly Evaluation. As mentioned above, the materialization of ball lightning in closed volumes really deepens the enigma of ball lightning. Rating: 2.

Possible Explanation. Electrostatic induction may help explain the appearance of ball lightning inside closed houses. Intense electrostatic fields generated by storms outside may penetrate the walls of the house and cause ionization of the air inside. This does not, however, explain how a long-lived spheroid is formed or how so much energy is concentrated in a small volume. Of course, the materialization of ball lightning in metal aircraft by induction would seem impossible because the electrostatic lines of force should not penetrate the metal skin.

Similar and Related Phenomena. St. Elmo's Fire and corona discharge (GLD). Miniature ball lightning (GLB6)

## Examples of Ball Lightning in Enclosures

X1. June 22, 1915. Drifton, Pennsylvania, During a severe storm during which many examples of ball lightning were seen. "I was on the ground floor of my home, which is near the center of the area affected. I saw a flash in the hallway resembling the discharge between the points of a high power induction coil, and heard a sharp crackling sound above the noise of the thunder or else immediately preceding it. On both second and third floors of the house 'lightning balls' were seen passing down the hallways, as well as one in the kitchen in the rear of the house. For a few moments afterward there was a distinct smell of ozone in the air. No damage whatever was done except that the lightning arrester of the telephone was blown out. Exactly similar effects were noticed at houses a quarter of a mile to the north and a quarter of a mile to the south of my home, and at many places in the town of Freeland, about a mile to the east.... The telephone operator at the Freeland exchange said that 'balls of fire' issued from most of the plugs on the switchboard. Persons who were out of doors at the time describe 'balls of fire' dropping from the sky and exploding on hitting the ground." (R1)

X2. Summer 1938. Wiltshire, England. In the Meteorological Office, Boscombe Down.

"Immediately after the lightning flash, a brilliant white sphere, with a diameter of about 20 cm and diffuse edges, was observed hovering near the aneomometer supports. The diameter, symmetry and absence of structure agree with Johnson's description. No heat radiation was noted..... The object followed an irregular course; it travelled from one anemometer leg to the other, and then from the anemometer to a radio receiver transmitter which was being operated at the time. The distance from the anemometer to the radio set would be about 2 m, and the height above the floor about 1.5 m. The object disappeared as it reached the radio set, perhaps 3-5 s after its appearance." (R2)

X3. August 7, 1951. In an aircraft over Southern England, as a research instrument was being pulled in. "..... an extremely loud bang was heard, and the cockpit was illuminated by an intense light which seemed to last about 1 sec. Observers in different parts of the aircraft saw different effects as follows:--- (i) From the co-pilot's seat, the flash appeared to be between himself and the pilot, near the instrument panel and the aircraft controls, and appeared as a white floating ball of fire. Its disappearance and the bang were simultaneous. (ii) A second observer immediately behind this position also saw the flash in the same place, and also reported that the observer holding the impactor was surrounded by a bluish glow." (R3)

X4. July 27, 1952. Hamburg, Germany. "A few seconds after a lightning stroke in the neighborhood, we observed outside the window a brightly gleaming sphere the size of a fist which moved downward in short serpentine lines. Then this luminous ball penetrated through the closed window pane (without damage to the glass) and entered our room. At a depth of about 1 meter it performed a sudden turn of 90° parallel to the wall and continued floating another meter further into the room. Thereupon it burst, and the luminous sphere disappeared with a brief deafening explosion. This ball lightning was purplish with a reddish cast which persisted during the entire duration of the phenomenon. It lasted approximately three seconds. No damage whatever was caused either on the inside or outside of the room. After the bursting of the luminous ball, we could perceive the typical odor which occurs in the case of electrical discharge." (R4) See also GLB1-X11.

X5. 1960. On an Air Force tanker en route to Nevada. "As I was concentrating on the instruments on the panel (no outside visual references were visible) a ball of yellow-white color approximately 18" in diameter emerged through the windshield center panels and passed at a rate about that of a fast run between my left seat and the co-pilot's right seat, down the cabin passageway past the Navigator and Engineer. I had been struck by lightning two times through the years in previous flights and recall waiting for the explosion of the ball of light! I was unable to turn around and watch the progress of the ball as it proceeded to the rear of the Aircraft, as I was expecting the explosion with a full load of JP-4 fuel aboard and concentrated on flying the aircraft. After approximately 3 seconds of amazingly quiet reaction by the 4 crew members in the flight compartment, the Boom operator sitting in the rear of the aircraft called on the interphone in an excited voice describing a ball of fire that came rolling through the aft cargo compartment abeam the wings, then danced out over the right wing and rolling off into the night and clouds! No noise accompanied the arrival or departure of the phenomenon." (R5)

X6. March 19, 1963. In an aircraft over the U.S. East Coast. "The aircraft encountered an electrical storm during which it was enveloped in a sudden bright and loud electrical discharge (0005 h EST, March 19, 1963). Some seconds after this a glowing sphere a little more than 20 cm in diameter emerged from the pilot's cabin and passed down the aisle of the aircraft approximately 50 cm from me, maintaining the same height and course for the whole distance over which it could be observed. The observation was remarkable for the following reasons: (i) The appearance of the phenomenon in an almost totally screened environment; (ii) the relative velocity of the ball to that of the containing aircraft was $1.5 \pm 0.5$ m $s^{-1}$, typical of most ground observations; (iii) the object seemed perfectly symmetrical in all three dimensions and had no polar or toroidal structure; (iv) it was slightly limb-darkened having an almost solid appearance and indicating that it was optically thick; (v) the object did not seem to radiate heat; (vi) the optical output could be assessed as approximately 5 to 10 W and its colour was blue-white; (vii) the diameter was $22 \pm 2$ cm, assessed by eye relative to the surroundings; (viii) the height above the floor was approximately 75 cm; (ix) the course was straight down the whole central aisle of the aircraft; (x) the object seemed to be in perfect equilibrium; (xi) the symmetry of the object was such that it was not possible to assess whether or not it was spinning." (R6)

X7. November 6, 1963. Nottingham, England. "..... at approximately 11.5 p.m., my father saw in his bedroom, in the centre of the room, a small, egg-shaped ball of brilliant light. Within the space of a few seconds, this small ball of light spread itself to form a sheet of darkish green light as wide as the room itself (approximately 12 feet). This curtain of light then moved towards my father and turned greyish colour. The whole sight then vanished as suddenly as it appeared, with a very loud bang, similar to the report from a rifle." (R7)

X8. August 25, 1965. Dunnellon, Florida. Sky overcast, slight drizzle, roofed patio with fiberglass screens. "Mrs. Greenlee had just swatted a fly when a ball of lightning the size of a basketball appeared immediately in front of her. The ball was later described as being of a color and brightness comparable to the flash seen in arc welding, with a fuzzy appearance around the edges. Mrs. Riggs did not see the ball itself, but saw the flyswatter 'edged in fire' dropping on the floor. The movement of the ball to the floor was accompanied by a report 'like a shotgun blast.' The entire incident was over in seconds." (R8)

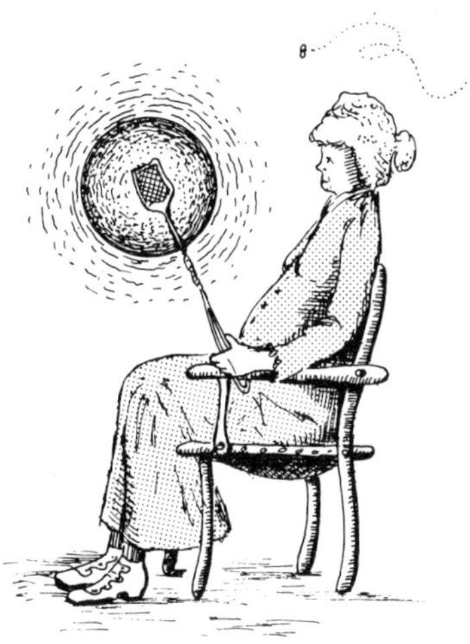

*Ball lightning materializes inside a screened porch in Florida. (Artist's interpretation)*

X9. August 8, 1975. Warley, England. During a vigorous thunderstorm; witness in her kitchen. "..... a sphere of light appeared over the cooker. The ball was ~10 cm across and surrounded by a flame-coloured halo; its colour was bright blue to purple. The ball moved straight towards the witness at an estimated height of 95 cm from the ground. Burning heat was felt, and there was a singeing smell. A sound something like a rattle was heard. The ball was in sight for only a second or so because its lifetime was cut short: 'The ball seemed to hit me below the belt, as it were, and I automatically brushed it from me and it just disappeared. Where I brushed it away there appeared a redness and swelling on my left hand. It seemed as if my gold wedding ring was burning into my finger.' Where the ball struck her, the clothing of the witness was damaged." (R9 R10)

X10. 1753. St. Petersburg, Russia. A pale blue, fist-sized ball of lightning emerges from the apparatus of the scientist, G. W. Richman, floats into his face, and kills him. (R4)

X11. July 16, 1905. Paducah, Kentucky. A large fireball appears in a closed house. (R11)

X12. June 19, 1924. Baltimore, Maryland. Miniature ball lightning materialized in a closed room. See GLB6-X2. (R12)

X13. December 1929. Liverpool, England. Orange fireball explodes in a closed house. leaving the odor of ozone. (R13)

X14. June 9, 1930. Auckland, New Zealand. Explosion and bright yellow flash in a closed kitchen. Attributed to ball lightning. (R14)

X15. 1956. Katsina, Nigeria. A fireball the size of a tennis ball appears inside a house with thick mud walls and a tiny remote window. (R15)

X16. No date, no place given. Ball lightning appears suddenly on the floor of the passenger cabin of an aircraft, runs down the aisle, and disappears with a bang. (R16)

X17. 1914. No place given. During a severe thunderstorm, a fireball emerged from electrical machinery. Diameter: 15-18 inches. It floated into an office, hit the ceiling, and splashed in all directions like water. Odor of ozone detected. (R17)

X18. April 20, 1915. Columbia, Missouri. Thunderstorm in progress. "Mr. Seaton had barely taken his position at the window (2.20 P.M.) when there was a sharp report, closely comparable with the report of a sporting rifle. There was an immediate answering click at the telephone. I instantly looked up, and saw a palish red, slightly corrugated ball, apparently 1 1/2 to 2 inches in diameter, or about the size of the outer rim of the mouthpiece of the telephone, moving across the space, about 6 feet, between the telephone and the window by which Mr. Seaton was standing. The fireball seemed to float as a liquid bubble does, though it seemed solid. It kept a fairly straight line for the window; it rolled over the windowsill, and disappeared, not into the outer air, but flickered out like a bubble. There was no explosion or sound of any kind following the click of the phone; there was no odor nor did it leave any marks on the windowsill. Mr. Seaton received no shock or injury of any kind. The telephone was not in the least damaged nor did it bear any marks to show that the ball actually came out of the mouthpiece....I should say that the ball took about two or three seconds to cross the space between the telephone and window. There was no more thunder after that one sharp report." (R18)

References

R1. Cone, E.J.D.;"Lightning without Rain," Scientific American, 113:141, 1915. (X1)

R2. Bromley, K.A.; "Ball Lightning," Nature, 226:253, 1970. (X2)

R3. Travers, V.J.; "Electrical Phenomenon Observed in Flight over Southern England," Meteorological Magazine, 81:121, 1952. (X3)

R4. Powell, James R., and Finkelstein, David; "Ball Lightning," American Scientist, 58:262, 1970. (X4, X10)

R5. Uman, M.A.; "Some Comments on Ball Lightning," Journal of Atmospheric and Terrestrial Physics, 30:1245, 1968. (X5)

R6. Jennison, R.C.; "Ball Lightning," Nature, 224:895, 1969. (X6)

R7. Falkner, M.F.; "An Occurrence of "Ball Lightning'," Meteorological Magazine, 93:95, 1964. (X7)

R8. Mohr, Frederick B.; "A Truly Remarkable Fly," Science, 151:634, 1966. (X8)

R9. Stenhof, Mark; "Ball Lightning," Nature, 260:596, 1976. (X9)

R10. "Woman Burned by Ball Lightning," New Scientist, 70:128, 1976. (X9)

R11. "Ball Lightning," Monthly Weather Review, 33:409, 1905. (X11)

R12. Humphreys, W.J.; Ball Lightning," American Philosophical Society, Proceedings, 76:613, 1936. (X12)

R13. Marchant, E.W.; "Globular Lightning," Nature, 125:128, 1930. (X13)

R14. Kidson, Edward; "Ball Lightning," Meteorological Magazine, 65:212, 1930. (X14)

R15. Charman, Neil; "Ball Lightning Photographed?" New Scientist, 69:444, 1976. (X15)

R16. Ingle, W.H.; "Lightning's Effects," New Scientist, 52:185, 1971. (X16)

R17. Uman, Martin A.; "What Is Ball Lightning?" Science Digest, 70:21, October 1971. (X17)

R18. Humphreys, W.J.; "Ball Lightning," American Meteorological Society, Bulletin, 1:97, 1920. (X18)

# GLB11 Black Ball Lightning

Description. Ball lightning that appears black or smoky, presumably because it is surrounded by an opaque layer. Black ball lightning usually explodes like ordinary ball lightning, producing the normal bright flash and frequently luminous fragments.

Background. Black lightning flashes, as detected visually and photographically, can generally be blamed on physiological effects and film overexposure, respectively (GLL). Black ball lightning, however, is more smoky than black and confirms its reality by exploding.

Data Evaluation. Observations of black ball lightning or smoke balls are rare. Rating: 3.

Anomaly Evaluation. The origin of the smoke or opaque gas of black ball lightning poses a serious problem if ball lightning is wholly an electrical phenomenon; that is, there is no combustion of matter. Rating: 2.

Possible Explanations. The smoke might be normal atmospheric gases made opaque (light-absorbing by electrical excitation. Possibly during it formation ball lightning accidently entrained some solid material that was subsequently vaporized.

Similar and Related Phenomena. Black lightning (GLL) and black auroras (GLA). Some fiery, smoky apparitions appearing in classification GLD may be related to or identical to black ball lightning.

Examples of Black Ball Lightning

X1. March 19, 1887. North Atlantic Ocean. "During a severe storm saw a meteor in the shape of two balls, one of them very black and the other illuminated. The illuminated ball was oblong, and appeared as if ready to drop on deck amidships. In a moment it became dark as night above, but below, on board and surrounding the vessel, everything appeared like a sea of fire. The ball fell into the water very close alongside the vessel with a roar, and caused the sea to make tremendous breakers which swept over the vessel. A suffocating atmosphere prevailed and perspiration ran down every person's face on board and caused every one to gasp for fresh air." (R1, R2) The sizes of the balls are not mentioned and this item might well be classified in GLD. (WRC)

X2. May 13, 1906. Morchard Bishop, England. Watching a thunderstorm. "I leaned forward to have a wider range through the open window, when I saw a large, dark egg-shaped ball swiftly falling straight down from the sky in the open space between two elm trees on the lawn and a fir tree. When the ball came to about the height of the latter, it suddenly collapsed and sparks of fire flew about in all directions like a magnificent firework..... When the ball split, a solid lump of rather light looking fire (like the colour of a blaze) issued forth, at first more from the bottom than the sides, and this emitted thousands of red-hot (bright deep red) sparks which flew in all directions, taking a circular shape." (R3)

X3. August 28, 1929. Peoria, Illinois. After a flash of lightning among some nearby trees. "About six feet from the trunk of one these hedge trees we observed a ball of smoke about two feet above the ground. The ball appeared to be about eighteen inches in diameter and perfectly spherical. The color of this smoke, if it was smoke, was a yellowish brown quite similar to the smoke given off by burning straw. The ball began immediately to diffuse into the surrounding air just as the smoke from an exploding shell." (R4)

X4. December 1596. Wells, England. A dark object the size of a football enters a church and explodes. (R5)

X5. December 9, 1897. Castelgandolfo. Italy. After close lightning, black globe 25-30 cm in diameter appeared in a kitchen. Seemingly formed of dense smoke, it scattered sparks and escaped through an open window. (R6, R7)

References

R1. "Rare Electrical Phenomenon at Sea," Monthly Weather Review, 15:84, 1887. (X1)

R2. "Rare Electrical Phenomenon at Sea," American Meteorological Journal, 4:98, 1887. (X1)

R3. G., S. E. L.; "Fireball," Symons's Meteorological Magazine, 41:191, 1906. (X2)

R4. Winchester, George; "A Peculiar Lightning Phenomenon," Science, 70:501, 1929. (X3)

R5. "Fireballs---Some Fiery, Some Dark," Journal of Meteorology, (U.K.), 1:312, 1976. (X4)

R6. Mathias, M.E.; "Les Globes Noirs et Blancs sans Lumiere Propre," Comptes Rendus, 182:32, 1926. (X5)

R7. "Black and White Globular Lightning," Nature, 117:211, 1926. (X5)

# GLB12 Ball Lightning's Electromagnetic Effects

Description. Ball lightning affecting ships' compasses, automobile ignition systems, etc.

Background. If ball lightning is an electrical phenomenon, electromagnetic effects are to be expected. If, however, ball lightning owes its existence to chemical combustion, antimatter meteorites, or many of the other suggested origins, electromagnetic effects should not occur.

Data Evaluation. Despite ball lightning's spark-emitting detonations, ozone odors, and close associations with thunderstorms and electrical equipment, good records of its expected electromagnetic effects are very scarce. Rating: 3.

Anomaly Evaluation. Ordinary lightning often causes electromagnetic effects, so perhaps it is the great rarity of electromagnetic effects from ball lightning that is anomalous. Rating: 2.

Possible Explanation. The energy of ball lightning may be temporarily stored inside an electromagnetic bottle which also confines electromagnetic fields.

Similar and Related Phenomena. The electromagnetic effects of ordinary lightning (GLL).

Examples of Ball Lightning's Electromagnetic Effects

X1. February 24, 1890. Lexington, Virginia. Balls of lightning do heavy damage and magnetize pens. (R1)

X2. No date or place given. Ball lightning strikes the sea 100 yards from the SS Sai On permanently affecting the vessel's compasses. (R2) See also GLB1-X178.

X3. October 7, 1811. Devon, England. Four balls of lightning seen in a church. The bell ringers complain that the bells seem "too heavy" to ring. A possible electromagnetic effect of the ball lightning? (R3) A tongue-in-cheek article? (WRC)

References

R1. Moreland, S. T.; "Lightning-Discharge," Science, 15:276, 1890. (X1)

R2. Barlow, E. W.; "Thunderstorms at Sea," Marine Observer, 12:49, 1935. (X2)

R3. Blair, A. J. F.; "Magnetic Fields, Ball Lightning and Campanology," Nature, 243:512, 1973. (X3)

# GLB13 Ball Lightning with Apparent Internal Structure

Description. Ball lightning with well-defined internal structure; usually a writhing mass of snake-like fibers of light.

Background. Ball lightning is usually so bright that internal structure, if it exists, cannot be discerned. There is no reason to suppose that all ball lightning possesses the structures reported here.

Data Evaluation. Reports of ball lightning with obvious internal structure are very scarce but also very consistent, especially considering the bizarre nature of this phenomenon. Rating: 3.

Anomaly Evaluation. No theory of ball lightning even attempts to deal with the snake-like structures reported. Rating: 2.

Possible Explanations. The luminous fibers might be transitory plasma currents or possibly wave-like structures on the surface of a plasma sphere.

Similar and Related Phenomena. Ordinary lightning has occasionally been described as running around ships and houses "like snakes."

Examples of Ball Lightning with Internal Structure

X1. 1887. England. While walking in a heavy thunderstorm under an umbrella. "Without any noise or notice, I saw what appeared to be a ball of live light fall just in front of me and slide along the road, and quietly disappear after going a few yards. The ball was about the size of a cricket ball, and appeared to be made up of a vibrating mass of strips of light, in rapid rolling motion. This, I suppose, was a sample of ball lightning, and at the time I felt very thankful it had fallen just clear of my umbrella rather than on it. What struck me most forcibly was the extraordinary appearance of the apparent fibres of light wriggling like a mass of eels." (R1)

X2. November 10, 1940. Coventry, England. "I was working at the far end of my garden, the weather was normal, no rain, no signs of thunder. Suddenly, I seemed to be in the centre of intense blackness and looking down observed at my feet a ball about 2 ft across. It was of a pale blue-green colour and seemed made of a mass of writhing strings of light, about 1/4 in. in diameter. It remained there for about 3 seconds and then rose, away from me, just missing a poplar tree about 8 ft away. It cleared the houses by about 20 ft and landed at the rear of the Weavers Arms on the Bell Green Road, a distance of about 1/4 mile. There was a loud explosion and much damage was done to the public house." (R2)

*Fibers of light wriggle on surface of ball lightning.*

References

R1. J., G.; "Fireballs," English Mechanic, 90:95, 1909. (X1)

R2. Matts, E.; "A Fire-Ball?" Weather, 19:228, 1964. (X2)

# GLB14 Unusual Physiological Effects of Ball Lightning

Description. Headache, nausea, neck pain, memory loss, disorientation, and related physiological symptoms suffered after encounters with ball lightning.

Background. Many of these symptoms occur after a severe electrical shock and are expected if ball lightning is truly electrical in nature. The thrust here, however, is the search for symptoms characteristic of nuclear radiation, which would imply a nuclear origin for ball lightning, as some have suggested. (R4, R5)

Data Evaluation. With the exception of one rather old example (X1), the physiological symptoms reported after exposure to ball lightning can well be attributed to simple electrical shock or eye-strain. Rating: 3.

Anomaly Evaluation. Symptoms of radiation sickness, if demonstrated, would suggest either a nuclear energy source for ball lightning (antimatter meteorites?) or the generation of nuclear reactions by high temperatures or electrically accelerated particles in ball lightning. All of these possibilities would be difficult to explain in conventional physical terms. Rating: 2.

Possible Explanations. Symptoms similar to those of radiation sickness may be induced by electrical shock, perhaps amplified by the traumatic experience of being close to ball lightning.

Similar and Related Phenomena. Nausea and disorientation experienced during earthquakes; the suffocating sensations and nausea felt near the vortex of a tornado (GWT). It is also important to point out that UFO observers frequently complain of headaches, nausea, memory loss, etc.; and some even exhibit burns and more overt effects.

Examples of Unusual Physiological Effects from Ball Lightning

X1. November 24, 1886. Maracaibo, Venezuela. "During the night of the 24th of October last, which was rainy and tempestuous, a family of nine persons, sleeping in a hut a few leagues from Maracaibo, were awakened by a loud humming noise and a vivid, dazzling light, which brilliantly illuminated the interior of the house. The occupants, completely terror stricken, and believing, as they relate, that the end of the world had come, threw themselves on their knees and commenced to pray, but their devotions were almost immediately interrupted by violent vomitings, and extensive swellings commenced to appear in the upper parts of their bodies, this being particularly noticable about the face and lips. It is to be noted that the brilliant light was not accompanied by a sensation of heat, although there was a smoky appearance and a peculiar smell. The next morning the swellings had subsided, leaving upon the face and body large blotches. No special pain was felt until the ninth day, when the skin peeled off, and these blotches were transformed into virulent raw sores. The hair of the head fell off upon the side which happened to be underneath when the phenomenon occurred, the same side of the body being, in all nine cases, the more seriously injured. The remarkable part of the occurrence is that the house was uninjured, all doors and windows being closed at the time. No trace of lightning could afterward be observed in any part of the building, and all the sufferers unite in saying that there was no detonation, but only the loud humming already mentioned. Another curious attendant circumstance is that the trees around the house showed no signs of injury until the ninth day, when they suddenly withered, almost simultaneous with the development of the sores upon the bodies of the occupants of the house." (R1) The symptoms mentioned in this item closely resemble those due to overexposure to nuclear radiation. The date of the phenomenon precedes the discovery of radioactivity, x-rays, etc. (WRC)

X2. July 2, 1893. Warwickshire, England. Large fiery globe of dazzling brilliancy. One observer is momentarily blinded and afterwards experiences pains in the head and neck. (R2) The bright light might have caused these symptoms. (WRC)

X3. No date or place given. An observer of ball lightning receives a shock and loses his memory for several hours. (R3) The electric shock alone may have been sufficient to cause memory loss. (WRC)

References

R1. Cowgill, Warner; "Curious Phenomenon in Venezuela," Scientific American, 55:389, 1886. (X1)

R2. Cumming, L.; "Thunderbolt in Warwickshire," Nature, 48:341, 1893. (X2)

R3. "Ball-Lightning," English Mechanic, 101:514, 1915. (X3)

R4. Altschuler, M.D., et al; "Is Ball Lightning a Nuclear Phenomenon?" Nature, 228:545, 1970.

R5. Hill, C.R., and Sowby, F.D.; "Radiation from Ball Lightning," Nature, 228:1007, 1970.

# GLB15 Artificial Ball Lightning

Description. The creation of detached luminous spheroids by high-current electrical apparatus.

Background. Small greenish plasmoids sometimes appear on submarines where high-current DC equipment is employed. These plasmoids may not possess all of the properties of natural ball lightning, particularly the high energy density and destructive power, but they do offer some insight into the nature and origin of the natural phenomenon. The subject of artificial ball lightning has been investigated at much greater depth by J.D. Barry (R4) and others.

Data Evaluation. The artificial production of luminous plasmoids is well-established. Rating: 1.

Anomaly Evaluation. While the production of artificial ball lightning is not too difficult, there is no consensus on just how the plasmoids are formed and maintained. Rating: 2.

Possible Explanations. See those for ordinary ball lightning (GLB1).

Similar and Related Phenomena. Ordinary ball lightning (GLB1) and artificial bead lightning (GLL).

Examples of Artificial Ball Lightning

X1. 1852. Ireland. "One of Sir William Hamilton's sons, who was fond of experimenting, had a telegraph wire running the length of a shrubbery, as far as the lodge; whether from accident or in the natural course of things I don't know, but an accumulation of the electric fluid, I suppose, passed along the wire from the Observatory and down to the lodge, where it left the wire, passed along the ground a distance of some yards, then rising over a hedge, on which some linen was bleaching, picked up a cravat and mounted high in the air with it, carrying it a distance of about half a mile. I saw it myself like a ball of fire with something white hanging to it; but I could not judge very well of its height." (R1)

X2. 1947. Philadelphia, Pennsylvania. Following a description of submarine electrical apparatus and the formation of arcs. "Usually the arc is 'blown-out.' However, when the amount of charge left in the batteries creates a sufficiently intense current through the arc, a green fireball will 'float' off the contacts into the engine room. The lifetimes of these plasmoids are of the order of a second. The green color is attributed to the copper from which the extensions of the silver electrodes are made. No appreciable damage has been reported from these 'engine-room plasmoids.'.....During the 'Guppy Reconversion Program' in 1947 at the Philadelphia Naval Shipyard, tests were performed on the reverse current gear of the U.S.S. Cutlass (Hull SS478), and a fireball was generated in the engine room using 260 volts and 156,000 amperes direct current. This gives a peak power of about 40 megawatts. Switching times on the order of 0.1 to 0.01 second are standard with mechanical switching, so that an energy on the order of 0.4 to 4 megajoules is the estimated energy range in which this plasmoid forms. The estimated diameter appears to vary between 4 to 6 inches or from 10 to 15 cm." (R2, R3)

References

R1. Williams, R. Packenham; "An Extraordinary Phenomenon," English Mechanic, 16:363, 1872. (X1)

R2. Silberg, Paul A.; "Ball Lightning and Plasmoids," Journal of Geophysical Research, 67:4941, 1962. (X2)

R3. Charman, Neil; "The Enigma of Ball Lightning," New Scientist, 56:632, 1972. (X2)

R4. Barry, James Dale; Ball Lightning and Bead Lightning, Plenum Press, New York, 1980.

# GLB16 Ball Lightning with Long Tails

Description. Luminous spheres similar to ordinary ball lightning but possessing long tails. The tails may be many times the diameter of the ball in length. Flat, tape-like, crinkled tails have been reported several times.

Background. Ordinary ball lightning is sometimes pear-shaped, and it is inviting to think that tailed ball lightning may just be pear-shapes with the stem end drawn out into a long strip. This may indeed occur but the most significant examples of tailed ball lightning are so bizarre---long, flat, wrinkled tails---that they cannot be stretched-out pear-shapes.

Data Evaluation. Bizarre though some of the examples may be, they possess an underlying consistency that gives one confidence that this phenomenon is real. Rating: 2.

Anomaly Evaluation. None of the proposed explanations for ordinary ball lightning seems to be able to accomodate tailed ball lightning. Rating: 2.

Possible Expanation. None.

Similar and Related Phenomena. The writhing, strip-like structure seen in some cases of ball lightning (GLB13) may be related.

Examples of Ball Lightning with Tails

X1. Circa 1920. Johannesburg, South Africa. "My first experience goes back to 1920 when, as a boy in Johannesburg, I observed a ball of lightning rolling slowly up an incline about 50 cm above and parallel to the sloping ground. This ball of light was about 50 cm in diameter with irregular saw-tooth streaks of light around it, and it had a comet-like tail. The ball continued until it met a dry-stone terrace wall, when it exploded in a flash of flame leaving an acrid smell. The wall was not damaged in any way. It had been possible to see that the ball of light was rolling in an anti-clockwise direction. (R1)

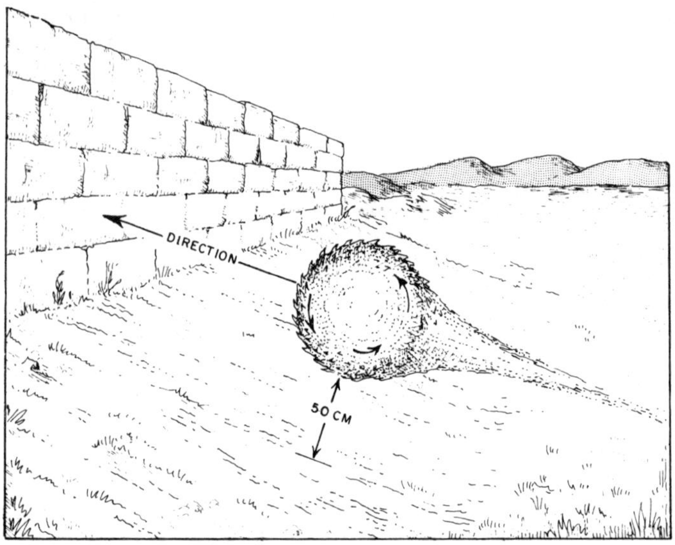


*Ball lightning with comet-like tail rolls along ground in Johannesburg.*

X2. September 22, 1965. Honolulu, Hawaii. In a garden after a light thunderstorm. "Suddenly, around the corner of the mock orange hedge, came the head of the lightning, crumpled with lots of black in the folds. As it came closer it was wider until at the end of its twenty-five foot length it was about two feet wide where it was chopped off clean. It was solid---not diaphanous or transparent or ethereal. It was the most brilliant, eye-dazzling electric blue without sparkle or scintillation. It moved fast just above the ground and about half-way between me and the birdbath. The bowl of the birdbath, but not the pedestal was visible. While it gave me the impression of being stiff and flat it might have been tubular. It showed no signs of elasticity, no flexibility; definitely it was not sinuous nor did it undulate. It had no sensation of heat. Had it been white hot, I would have been singed since it came so close to me. It was without a doubt dying lightning, but what made it die and in my garden?" It was like a strip of crumpled paper; dimensions were about 25 feet long and 1 1/2 to 2 inches thick. It made a sizzling sound and disappeared with a clap of thunder. (R2)

X3. August 15, 1975. Argyllshire, Scotland. "I must tell you of the magnificent thing I saw during the last storm of thunder and lightning which was in the autumn. It was what I called a bulb of lightning not a bolt. It was dome-shaped, opaque, like 8 inches in length, slim and came down slowly to within 9 inches of the ground in front of the house. This bulb lingered, and attached to it was a long tail which came down slowly after it. I would say the tail was 1 yard in length, seemed like 1/2 inch in width, like crinkly white silk ribbon, like the old-fashioned ribbon. The tail kept above the bulb which lingered, and then it all snuffed out like a candle, and left no trace." (R3)

X4. June 22, 1880. Leicester, England. "Fireballs fell repeatedly from a storm cloud. The balls were blunt on the bottom and drawn out into a tail above. (R4)

X5. Circa 1890. Upminster, England. Three lightning balls with tails wriggled across a meadow. (R5)

X6. April 30, 1959. No place given. Three fireballs with tails fall from a cloud. (R6)

X7. March 24, 1977. Capetown, South Africa. "At 2.50 a.m. (24 March) some hours after the passing of the storm, I was awa-

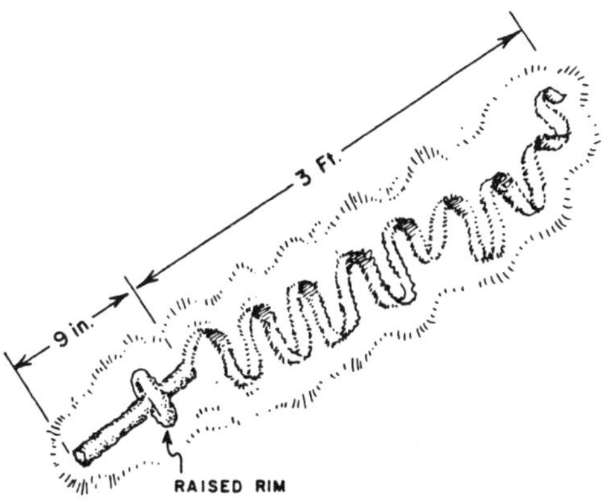


*In 1975, in Scotland, ball lightning was observed with a long, ribbon-like tail.*

kened by a sharp 's-s-s-spht' and to my amazement saw a small 'ball of fire', about 2 1/2 in-3 in in diameter, with a comet-like tail some 12 in-15 in, travelling north to south (or possibly north-east to south-west) about 2 ft away from the centre of a heavy double wardrobe, with quite heavy brass fittings and about the same distance from the floor and some couple of yards from my bed, the whole thing lasting only seconds. . ." (R7)

References

R1. Davis, R. J.; "Personal Experiences with Lightning in South Africa," Journal of Meteorology, (U.K.), 2:290, 1977. (X1)

R2. Price, Saul, and Carlstead, Edward M.; "Observation of a Curious Electrical Phenomenon in Hawaii," Monthly Weather Review, 94:272, 1966. (X2)

R3. Meaden, G. T.; "Remarkable Occurrence of Unusually-Shaped Ball Lightning, 15 August 1975," Journal of Meteorology, (U.K.), 1:310, 1976. (X3)

R4. Mott, F. T.; Curious Electrical Phenomenon," Nature, 22:193, 1880. (X4)

R5. Rowe, W. H.; "Ball-Lightning," English Mechanic, 101:470, 1915. (X5)

R6. Rollinson, A. C.; "Lightning," Marine Observer, 30:58, 1960. (X6)

R7. Seymour, Isabella P.; "Ball Lightning," New Scientist, 77:817, 1978. (X7)

# GLB17 Correlation of Ball Lightning Incidence with Solar Activity

Description. The increased incidence of ball lightning with increased solar activity.

Data Evaluation. Only one study has been uncovered so far, and it is based on scanty data from a small area (Holland). Rating: 3.

Anomaly Evaluation. Since thunderstorms are more frequent during periods of high solar activity, and ball lightning is almost always associated with thunderstorms, it would not be particularly surprising to find ball lightning incidence related to solar activity. The question of why, though, is a different one. Rating: 3.

Possible Explanation. Solar activity increases lead to more cosmic-ray bombardment of the terrestrial atmosphere. The increased ionization may lead to more electrical activity. More cosmic rays may cause more nuclear fission of atmospheric xenon, which may help create and sustain ball lightning. (R1)

Similar and Related Phenomena. The correlation of solar activity with thunderstorm frequency (GWS) and the incidence of auroras (GLA).

Examples of the Correlation of Ball Lightning with Solar Activity

X1. General Observation. "Time series analysis of ball lightning in Holland shows that from 1880 to 1956 the annual number of ball lightning increased with the intensity of solar activity as measured by Wolf average annual numbers." (R1)

References

R1. Arabadji, W. I.; "On the Problem of Ball Lightning," Journal of Geophysical Research, 81:6455, 1976. (X1)

# GLB18 Ball Lightning External to Aircraft (Foo Fighters?)

Description. Fireballs, singly or in formation, that pace aircraft despite high relative air velocities.

Background. Toward the end of World War II, pilots of all combatants, in all theaters, reported many cases of fireballs pacing their aircraft. The Americans called them Foo Fighters. Foo Fighters have seemingly become part of UFO lore and therefore ignored by scientists, but many of the sightings must have had some physical basis.

Data Evaluation. In addition to the large store of Foo Fighter observations in newspapers and the UFO literature, there are several excellent records of fireballs pacing modern aircraft. Rating: 2.

Anomaly Evaluation. In addition to the basic anomaly of ball lightning's nature, the ability of fireballs to maintain their positions relative to aircraft despite high relative air speeds poses a real puzzle to the scientist. Rating: 2.

Possible Explanation. St. Elmo's Fire may account for some observations but not for those instances where the fireball is well clear of the aircraft surface.

Similar and Related Phenomena. St. Elmo's Fire (GLD); UFOs.

Examples of Ball Lightning External to Aircraft

X1. General observation. "There are three kinds of these lights we call 'Foo Fighters,'" said Lieutenant Donald Meiers, of Chicago. 'One is a red ball which appears off our wing tips and flies along with us. No. 2 is a vertical row of three balls of fire, flying in front of us. No. 3 is a group of about fifteen lights which appear in the distance, like a Christmas tree up in the air, and flicker on and off." (R1)

X2. May 15, 1970. An aircraft 150 miles east of St. Louis, Missouri, in a thunderstorm. "At (what seemed) the moment of maximum turbulence and electric discharge, while the aircraft was still descending through the storm, a sequence of events took place that I list below. The whole took not more than 5 s, and I must admit that the order, save for numbers 4 and 6, is not necessarily chronological. 1. The turbulence ceased altogether. 2. The surrounding electrical discharges (glows) ceased altogether. 3. The wing stopped buckling altogether. 4. A white glowing sphere (ball lightning?) appeared on the port wing tip. I do not know if it was actually touching the wing. Its diameter was less than 1 m and more than 10 cm. Its boundary was 'fuzzy' and not distinct. 5. There was a soft 'pop'. 6. The ball lightning (?) vanished." (R2)

X3. No date or place given. "A further important observation was of a 20 cm ball which appeared at a height of about 50 cm over the trailing edge of the mainplane of an aircraft in flight. It moved parallel to the line of the mainplane at a speed of about 1 m $s^{-1}$ before being cast off the end and was not blown off in spite of the considerable air speed." (R3)

X4. April 23, 1964. An aircraft over Bedford, England. A loud bang and a whitish-blue flash of light. A ball of blue light the size of a football appeared on the starboard wing tip. It vanished in two seconds. (R4)

X5. No date or place given. "In the second case, a bright ball appeared on the top surface of the wing outside the aircraft, made rapid movements to and fro for an appreciable length of time and then disappeared. It seemed quite unaffected by the air passing through it at 250 mile/h, which does not accord very well with the theory that the balls may be composed of vaporised metal." (R5)

X6. No date or place given. Describing the types of St. Elmo's Fire. "The 'small ball' formation varies in size from two inches (5 cm) to a foot and a half (46 cm) in diameter and generally 'rolls around' the aircraft apparently unaffected by the movement of the aircraft. On one occasion a small ball (about six inches (15 cm) in diameter) of yellowish-white lightning formed on my left tiptank in an F-94B then rolled casually across the wing, up over the canopy, across the right wing to the tiptank and thence commenced a return, which I didn't note, but I was advised by my observer that it disappeared as spontaneously as it had arisen. I have seen this form several times but rarely for as long a period which I would estimate to be about two minutes in duration. Sometimes the balls are blue, blue-green, or white though it appears to favor the blue-green and yellow-white." (R7)

X7. General observation. Mysterious balls of fire follow Allied planes closely. No explosive or incendiary effects noted. (R8)

X8. General observations. The same sort of fireballs followed aircraft in both the European and Pacific theaters. B-29 pilots said they stayed about 500 yards off, were about 3 feet in diameter, had a phosphorescent glow, and could not usually be shaken off by maneuvering or flying into clouds. This report states that radar did not detect the foo-fighters. (R9)

References

R1. Creighton, Gordon W.; "Foo Fighters," Flying Saucer Review, 8:11, March/April 1962. Quotation credited to New York Herald Tribune, January 2, 1945. (X1)

R2. Felsher, Murray; "Ball Lightning," Nature, 227:982, 1970. (X2)

R3. Jennison, R. C.; "Can Ball Lightning Exist in a Vacuum?" Nature, 245:95, 1973. (X3)

R4. Vidler, G. T.; "Lightning Strike 23 April 1964," Meteorological Magazine, 93:254, 1964. (X4)

R5. Ingle, W. H.; "Lightning's Effects," New Scientist, 52:185, 1971. (X5)

R6. Aspden, H.; "The Enigma of Ball Lightning," Journal of Meteorology, (U.K.), 6:258, 1981.

R7. Condon, Edward U.; "Coronal Effects," Scientific Study of Unidentified Flying Objects, New York, 1969, p. 735. (X6)

R8. "'Foo-Fire' Reports Leave Scientists Guessing," Science News Letter, 47:24, 1945. (X7)

R9. Chamberlin, Jo; "The Foo Fighter Mystery," American Legion Magazine, 39:9, December 1945. (X8)

# GLD DIFFUSE ELECTRICAL DISCHARGE PHENOMENA

*Key to Phenomena*



## GLD0 Introduction

Everyone is familiar with fast, concentrated discharges of electricity, such as lightning. More rare are the slower, more diffuse flows of natural electricity. In these the passage of electrical currents is gentle and almost soundless in contrast to the violent thunderbolt. Even so, these slow discharges frequently give rise to luminous, often strangely beautiful phenomena.

Slow electrical discharges usually proceed from projections and sharp points, such as ship masts, radio antennas, and even human fingers held aloft. Pointed structures tend to concentrate electrical fields and, if the fields are intense enough, will encourage a slow flow of electricity from the point into the surrounding air. When these discharges become luminous, they are called St. Elmo's Fire. In damp and stormy weather, high voltage power lines may exhibit eerie luminous discharges around insulators and other structures. These bluish glows are termed 'corona discharges.' Reports from ships at sea and mountain-top scientific observatories tell of St. Elmo's Fire in the form of cold lambent flames and auras streaming from scientific instruments, guy wires, and even peoples' heads. St. Elmo's Fire also appears infrequently in snowstorms and sandstorms, as the falling and blowing particles transfer electricity between the earth and the air like those Wimshurst electrostatic machines so common in high school physics laboratories.

Corona discharge and St. Elmo's Fire are not particularly mysterious to physicists; they therefore constitute a good starting point for a journey into more controversial territory.

The next stop is a truly spectacular one: the so-called Andes Glow or, to be more general, the mountain-top glow. Since many mountains pierce the atmosphere with sharp projecting surfaces, the appearance of slow electrical discharges from their crests is not especially surprising. It is the scale of the process that is awe-inspiring. The sheets of flame and aurora-like beams of light projecting into the stratosphere may be visible for hundreds of miles. Where does this mountain electricity go? Into outer space? And why are mountain-top glows greatly enhanced during major earthquakes? No one has really studied this phenomenon carefully. It is in essence St. Elmo's Fire on a massive scale.

On a smaller, less-violent scale, intense electrical storms may create ground-level patches or waves of luminous electrical activity that may engulf humans in their paths, electrically shocking them or wrapping them in a garment of St. Elmo's Fire. In some ways, these surface displays resemble the marine phosphorescent displays, particularly the rotating phosphorescent wheels, described later in this volume of the catalog.

Considerably more mysterious are the glowing, enigmatic, floating spheres that observers often compare to toy balloons. These cavorting softly-lit bubbles are certainly not ball lightning and may not even be electrical at all. Whatever they are, modern science has paid little attention to them, perhaps because their explanation is so difficult and their existence so improbable.

These subjects bring us to those strange glows, flashes, and fireballs seen so often near earthquake epicenters. Earthquake lights also take the shapes of auroral beams, mountain-top glows, ball lightning, and moving waves of rock luminosity. Since modern scientific observations demonstrate that earthquake shock waves may penetrate the atmosphere into the ionosphere, it is possible these atmospheric disturbances may help create low-conductivity paths for earth-to-space electrical discharges. The ball lightning, the sheets of flame issuing from the ground, and other localized luminous phenomena may be generated by large-scale piezoelectric effects (i.e., the creation of electricity by stresses in rocks). An alternate explanation of earthquake lights involves the spontaneous ignition of natural gas liberated by the quaking earth.

Violent volcanoes and tornadoes also display unusual lights. Of course, normal lightning and ball lightning are to be expected in violent storms, but whence the peculiar shafts of light reported in tornado funnels and the strange glowing patches in and above storm clouds? The precise role of electricity in tornado action, if any, is highly controversial. Superficially at least, the funnel light columns may, like neon lights, arise from large-scale glow discharges in these naturally formed tubes.

That the earth-as-a-whole is a gigantic electrical machine cannot be doubted. The constant turmoil of the atmosphere, its never-ceasing bombardment by the solar wind, the electically charged wind-blown dust and snow, and the intense forces squeezing terrestrial rocks, all conspire to creat a wide spectrum of curious and poorly understood luminous effects.

# GLD1 Mountain-Top Glows

Description. Rays, undulating streamers of light, flashes, and steady glows appearing along mountain crests and ridges. The color is usually yellowish white, with green and orange being more rare. This phenomenon is observed in the Andes, the Alps, the Rockies, the Arctic, and probably many other places.

Background. Despite the widespread occurrence of mountain-top glows, science has taken little note of them. Instrumented studies are essentially unknown.

Data Evaluation. Many high quality observations, especially from the Andes and Alps. Rating: 1.

Anomaly Evaluation. Mountain-top glows are probably large-scale discharges on terrestrial electricity into the atmosphere---that is, greatly magnified St. Elmo's Fire. The anomalous aspects are: (1) The very large scale of the phenomenon; (2) The apparent heightening of the displays during earthquakes; (3) The possible periodicity of the flashes; (4) The close resemblance to auroras, which seems to underscore the reality of low-level auroras (GLA4). Rating: 3.

Possible Explanation. Large-scale discharge of terrestrial electricity.

Similar and Related Phenomena. Low-level auroras (GLA4); aurora-related fogs and mists (GLA21); earthquake lights (GLD8).

Examples of Mountain Top Glows

X1. September 25-30, 1868. Near Spitzbergen. "On returning from the island, where the instruments for the magnetic observations had been deposited, I perceived upon the ridge of the mountain, to the south, a brilliant polar light rising from 10° to 15° above the mountain in undulating rays, distinctly defined, at their base appearing as a diffuse yellowish light, but higher up as vertical orange beams, while at the top they formed a series of sharp points. The rays had an undulating motion, and the crest of the mountain was covered with a light fog, which the wind was moving from east-northeast to west-southwest. In a few moments the cloud of mist passed the mountain and the rays disappeared, but the crest of the mountain continued to be illuminated by a pale wandering light, which floated along the mountain.....On the 27th of September, after having observed in the morning a radiation of yellowish-white light proceeding from one edge of a cloud which stood out prominently from a wall of clouds, I perceived in the evening, at 11 o'clock 30 minutes, a pale, wandering light moving distinctly along the ridge of the mountain. The light appeared for a few moments in the form of rays of a clear and brilliant yellow, following every detail of the sinuosities of the mountain. The pale glimmer of light seemed to follow the ridge of the mountain, and I was convinced, from the movement of the mists, that the luminous phenomenon was formed upon the ridge itself. On the 30th of September, at 9h. 30 m., I witnessed on the island of Amsterdam a very intense luminous phenomenon, during which every peak and ridge, the most elevated, was illuminated with a pale light, particularly when covered with a vail (sic) of mist." (R1)

X2. August 14, 1903. Swiss Alps. Observing the peaks and ridges. "About 8 p. m. I noticed a faint quivering light overhead, supplemented by occasional flashes of greater brilliance and different colour. These manifestations rapidly increased in distinctness, and continued to play only along the opposite mountain-ridge, not extending into the regions beyond, so far as these could be seen from here, though I have since learnt that an independent series of flashes was seen around the Schilthorn on this side of the valley. Not a single peal of thunder was at any time audible.....As one watched the display it was easy to distinguish more definitely the two kinds of discharge. One of them took the form of a faintly luminous reddish or pink light, which shot with a tremulous streamer-like motion in horizontal beams that proceeded apparently from left to right, as if their starting point lay somewhere about the back of the Jungfrau. These streamers so closely resembled the aurora borealis that, had they appeared alone, one would have been inclined to wonder whether the 'northern lights' had not here made an incursion into more southern latitudes." (R2)

X3. August 21, 1911. Swiss Alps. "On the night of August 21, about ten o'clock, semicircular flashes of light shot up apparently behind the Monch, quivered for a few seconds, and then disappeared. I counted twenty-eight in a minute. The light was sometimes intense at a central point, which was steady, and from this quivering glow proceeded and lighted up from 15° to 20° of the horizon...The appearance seemed to me very like an aurora borealis which I saw in Scotland in the 'fifties, but the centre of the light here was to the south-west of where I stood." (R3)

X4. September 18-19, 1941. Mount Adams, New Hampshire. During the great auroral display to the south. "During the early stages of this display, the picture below (not reproduced) was taken by Howell, showing the green arc a few degrees above the horizon, and therefore in actual position some few hundred miles or so to the north, and bright enough to light up the peaks. Outlining the mountain can be seen a faint rim of light, greenish to the eye." (R4)

X5. Spring 1944. Algeria. "In the spring of 1944, one clear dark night, I observed the flash type of discharge being made repeatedly at the summit of what I believe is called Mont Leon to the north-west of Bone airport in Algeria. The night in question was clear and starlit and the top of Mont Leon was covered with a little cap of cloud; evidently it coincided with the inversion top, and it was in this cloud that the flash occurred. Enquiry seemed to show that this was commonly seen in the winter there." (R5)

X6. Early 1950s. Peru. While camped in the high Andes. "From our camp we had at least 50 km of the crest in direct view across the gorge. After supper, we noticed occasional lightning flashes and glows along the whole visible crest. The skies were fairly clear over us, and we heard no thunder. The nearest part of the crest was at least 10 km distant. As it grew darker, they became spectacularly frequent: there was hardly any instant without at least one flash somewhere, and often several blazed simultaneously. This Jovian show was still going undiminished when I went to sleep about midnight." (R6)

X7. June 7, 1954. Madeira. "On approaching Madeira from sw the island was completely covered with low cloud, St and Sc. On arriving within 16 miles of the island the cloud rapidly lifted and numerous brilliant white flashes were observed at frequent intervals on various mountain peaks. At the time of these occurrences the cloud was clear of the island, although there was some Sc to NW. After the flashes had continued for some 20 min a low rumbling was heard like distant thunder. (R7)

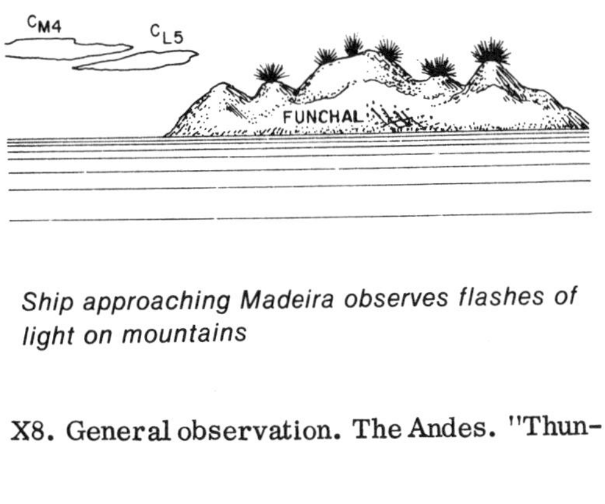


*Ship approaching Madeira observes flashes of light on mountains*

X8. General observation. The Andes. "Thunderstorms are rare in Chile, and this fact may possibly be explained on the assumption that the Andes act as a gigantic lightning rod, between which and the clouds silent discharges take place on a vast scale. The visible discharges occur during the warm season, from late spring to autumn, and appear to come especially from certain fixed points. According to Dr. Knoche they are confined almost exclusively to the Andes proper, or Cordillera Real, as distinguished from the coast cordillera. Viewed from a favorable point near their origin there is seen to be, at times, a constant glow around the summits of the mountains, with occasional outbursts, which often simulate the beams of a great searchlight, and may be directed westward so as to extend out over the ocean. The color of the light is pale yellow, or rarely reddish. One striking feature of these discharges is that they are especially magnificent during earthquakes. At the time of the great earthquake of August, 1906, throughout central Chile the whole sky seemed to be on fire;

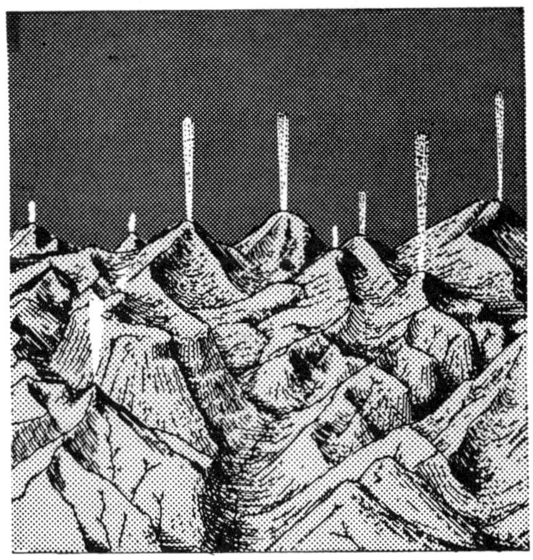

*Aurora-like pillars of light over the Andes*

never before or since has the display been so brilliant." (R8, R21)

X9. General observation. The Andes. "'Andes glow' or 'Andes lights' are terms used to describe illumination seen at night in the vicinity of certain mountain peaks. While the majority of reports have come from the Andes mountains of Bolivia, Chile and Peru, this phenomenon has also been reported in the European Alps, Mexico and Lapland and presumably could occur in many mountainous regions under favourable conditions. While sometimes thought to be lightning, for lack of a more obvious explanation, the interesting property of these light displays is that they can occur under cloudless skies. Sometimes they are but one single flash, while at other times they may persist intermittently for hours. On occasion, a periodicity has been noted in the time between flashes. At their most spectacular they have been described as '...not only clothing the peaks, but producing great beams, which can be seen miles out at sea.' They seem to favour particular mountain peaks where often they can be seen during the dry season." (R9)

X10. General observation. The Andes. Radial rays shooting up as far as the zenith. (R10)

X11. August 10, 1854. Mont Blanc, Switzerland. Mountain ridges seemed on fire. Climbers felt effects of static electricity. (R11)

X12. August 6, 1885. Jura Mountains, Switzerland. Mont Galene illumined by an inphosphorescent light. (R12, R13)

X13. December 17, 1937. Mt. Wachusett, Massachusetts. On top of the mountain everything tinged with violet. Tingling sensation felt. (R14)

X14. General observation. The Andes. Flashing along the crests. (R15)

X15. General observation. Andes glow may result as snow is blown over the peaks, creating strong electric fields. (R16)

X16. No date given. Bohul Mountain, Russia. See description GLB3-X4. (R17)

X17. General observation. Mountain-top glows seen in Spitzbergen and Lapland; Lemstrom's observations. (R18)

X18. No date given. Skye, Great Britain. Sir Edward Sabine noticed that the cloud caps on the highest hills were permanently luminous at night, occasionally producing aurora-like flashes. (R19)

X19. General observation. Andes lights can be seen far out at sea. (R4)

X20. General observation. Mount Ida, near the Dardanelles. Mountain-top glow is common. (R4)

X21. General observation. Rocky Mountains. Continuous, silent electric discharges. (R20)

X22. 1929. Pikes Peak, Colorado. Ghostly flares seen about the summit at night. (R22)

References


R1. Lemstrom, Selim; "Observations upon the Electricity of the Atmosphere and the Aurora Borealis, ...." Smithsonian Institution Annual Report, 1874, pp. 227-238, 1875. (X1)

R2. Geikie, Arch.; "Summer Lightning," Nature, 68:367, 1903. (X2)

R3. Brunton, Lauder; "A Pseudo-Aurora," Nature, 87:278, 1911. (X3)

R4. Botley, Cicely M., and Howell, W. E.; "Mystery on Mount Adams," Mount Washington Observatory News Bulletin, 8:9, March 1967. (X4, X19, X20)

R5. Young, J. R. C.; "The Andes Glow," Weather, 26:39, 1971. (X5)

R6. Cooke, H. R., Jr.; Personal communication, May 11, 1975. (X6)

R7. Robson, G.; Marine Observer, 25:95, 1955. (X7)

R8. "Curious Lightning in the Andes," Scientific American, 106:464, 1912. (X8)

R9. Markson, Ralph, and Nelson, Richard; "Mountain-Peak Potential-Gradient Measurements and the Andes Glow," Weather, 25:350, 1970. (X9)

R10. "Silent Discharges," Knowledge, 10:26, 1913. (X10)

R11. "Mont Blanc on Fire," Scientific American, 10:82, 1854. (X11)

R12. Nature, 32:426, 1885. (X12)

R13. "Electric Clouds and Phosphorescent Mountains," English Mechanic, 42:7, 1885. (X12)

R14. B., C. F.; "Glow Discharge on Mt. Wachusett," American Meteorological Society, Bulletin, 21:206, 1940. (X13)

R15. "Strange Light on the Andes," Literary Digest, 45:840, 1912. (X14)

R16. Latham, J.; "Mountain Peak Potential Gradients," Weather, 26:80, 1971. (X15)

R17. Nature, 42:458, 1890. (X16)

R18. "The Northern Lights," Scientific American, 56:135, 1887. (X17)

R19. "Artificial Aurorae," Symons's Monthly Meteorological Magazine, 18:33, 1883. (X18)

R20. "Electric Phenomena in the Euphrates Valley," Monthly Weather Review, 28:290, 1900. (X21)

R21. "Strange Kinds of Lightning," Literary Digest, 111:22, October 10, 1931. (X8)

R22. Talman, C. F.; "Static Electricity Flares on Pikes Peak," American Meteorological Society, Bulletin, 10:78, 1929. (X22)

# GLD2 Intermountain Electric Discharges

Description. Luminous discharges or lightning between mountain peaks accompanied by the sound of thunder.

Data Evaluation. The existence of this phenomenon is suggested primarily by testimony, some of questionable value. In addition, it is very difficult to judge the distance and location of a distant lightning flash in an active storm. Rating: 3.

Anomaly Evaluation. While ball lightning may be generated by high-current electrical apparatus (GLB15) and has been seen emerging from electrical equipment, it still seems unlikely that appropriate conditions exist for the creation of ball lightning or normal lightning between adjacent mountain peaks, although we are really steeped in ignorance in this matter. Rating: 2.

Possible Explanations. Possibly local geological conditions, say, ore bodies, might contribute to this phenomenon.

Similar and Related Phenomena. Mountain-top glows (GLD1); ball lightning (GLB).

X1. General observation. Taurus Mountains, Euphrates Valley. Ellsworth Huntington, the noted geographer, found the following phenomenon described in consistent terms by a dozen people in widely separated localities. "The facts, upon which all agree, are as follows: A ball of fire is sometimes seen to start from one mountain and to go like a flash to another. At the same time there is a sound like thunder. This occurs by day or by night, although by day no light is seen. It always occurs when the sky is clear and never when it is cloudy. It sometimes happens two or three times in a year, and then again is not seen for several years. For the last two years it has not been seen. It is most common (or possibly never happens except) in the fall at the end of the long, dry season of three months." One of the mountains involved, Karaoghlon, is 7,350 feet high and flat-topped. (R1)

X2. General observation. Comment on Huntington's item notes that interpeak lightning has not been scientifically observed though poets sometimes write of it leaping from crag to crag. (R2)

X3. No date given. Maryland. Lightning seemed to leap from peak to peak in the Appalachians. (R3)

References

R1. Huntington, Ellsworth; "Electric Phenomena in the Euphrates Valley," Monthly Weather Review, 28:286, 1900. (X1)
R2. "Electric Phenomena in the Euphrates Valley," Monthly Weather Review, 28: 290, 1900. (X2)
R3. "Notable Lightning," Monthly Weather Review, 28:290, 1900. (X3)

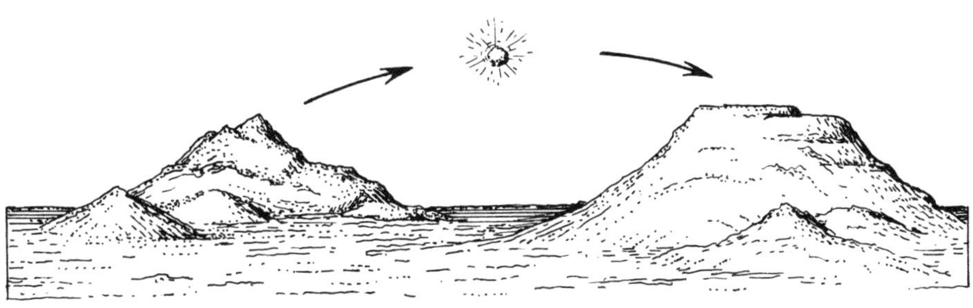

*Artist's conception of tales of intermountain fireballs in the Euphrates Valley*

# GLD3 Electric Fluids: Large-Scale St. Elmo's Fire

Description. The envelopment of people, buildings and large tracts by transient, flame-like electrical discharges. This phenomenon is usually preceded by a strong clap of thunder. Despite being completely surrounded by flames, people are not hurt by large-scale St. Elmo's Fire, although there are some minor electrical sensations and difficulties in breathing.

Background. Large-scale St. Elmo's Fire seems to fill a gap between ordinary St. Elmo's Fire (flames on isolated sharp points) and brush discharge (a large-area discharge of electricity more audible than visible).

Data Evaluation. Although this phenomenon is not particularly mysterious, it is supported by only a few rather old reports. Rating: 3.

Anomaly Evaluation. St. Elmo's Fire is, of course, well-known. It is the magnitude of the manifestations described below that merit consideration at least as a curiosity if not a significant anomaly. Rating: 3.

Possible Explanation. Extremely strong potential gradients in electrical storms are the undoubted cause, although the areas involved are very large.

Similar and Related Phenomena. Mountain-top glows (GLD1); low-level auroras (GLA4); electrified clouds (GLD4), auroral fogs and mists (GLA21).

## Examples of Large-Scale St. Elmo's Fire

X1. February 19, 1837. Orkney, Great Britain. Describing a boat moored to the shore. "....she was during this time attached to the shore by an iron chain, about 30 fathoms long, which did not touch the water, when, to my astonishment, I beheld a sheet of blood-red flame, extending along the shore for about 30 fathoms broad and 100 fathoms long, commencing at the chain and stretching along the shore and sea in the direction of the shore which was E. S. E., the wind being N. N. W. at the time. The flame remained about ten seconds, and occurred four times in about two minutes." (R1, R14)

X2. Circa 1842. Perthshire, Scotland. "I beg to observe that in the Highlands of Perthshire, some forty years ago, two men found themselves enveloped in flames, somewhat in the same style as Mr. Moir was on February 18 last. One Mr. John Stewart, who, for many years, drove the Mail gig between Dunkeld and Aberfeldy, told me that on a certain dark night, he and another man, climbing a rocky, heathery height in Rannock, were all at once set on flames by some mysterious fire, which appeared to proceed from the heather, which they were traversing, and the more they tried to rub the flames off the more tenaciously they seemed to adhere, and the more the fire increased in brightness and magnitude. Moreover, the long heather agitated by their feet, emitted streams of burning vapor, and for the space of a few minutes they were in the greatest consternation. They believed that they barely escaped a living cremation." (R2)

X3. August 5, 1879. Jura Mountains, Switzerland. As dark storm clouds approached. "At this moment the pine forest round St. Cergues was suddenly illuminated and shone with a light bearing a striking resemblance to the phosphorescence of the sea as seen in the tropics. The light disappeared with every clap of thunder, but only to reappear with increased intensity until the subsidence of the tempest." (R3, R4, R13)

X4. June 1880. Clarens, Switzerland. A thunderstorm in the distance; a heavy peal of thunder shakes the area. "At the same instant a magnificent cherry-tree near the cemetery, measuring a metre in circumference, was struck by lightning, Some people who were working in the vineyard hard by saw the electric 'fluid' play about a little girl who had been gathering cherries and was already 30 paces from the tree. She was literally folded in a sheet of fire. The vine-dressers fled in terror from the spot. In the cemetery six persons, separated into three groups, none of them within 250 paces of the cherry-tree, were enveloped in a luminous cloud. They felt as if they were being struck in the face with hailstones or fine gravel, and when they touched each other sparks of electricity passed from their finger-ends. At the same time a column of fire was seen to descend in the direction of Chatelard, and it is averred that the electric fluid could be distinctly heard as it ran from point to point of the iron railing of a vault in the cemetery. The strangest part of the story is that neither the little girl, the people in the cemetery, nor the vine-dressers appear to have been hurt...." (R5, R6)

X5. August 12, 1884. Scotland. During a thunderstorm on a moor. "As Sir Alexander described it to me, he was enveloped in a sheet of light which seemed to him like illuminated steam. He does not remember noticing any pricking sensation, and on my asking him what the feeling was, he said it was indescribable, but a more uncomfortable and weird feeling he never experienced. The gamekeeper, who was standing about ten yards from him, saw him enveloped in the luminous cloud....." (R7)

X6. February 10, 1888. Off the English coast. Thunderstorm in the distance. "At about 6 a tremendous report was heard, sounding like thunder, and the captain describes the appearance of the vessel as if it were shrouded in a mass of bright red flames, which lit up the surrounding waves. The phenomena was all the more surprising as the thunder and lightning appeared to be at some considerable distance from the steamer, and it could not be compared to an ordinary thunder clap and lightning flash, being far too violent and no regular flash of lightning being seen." (R8)

X7. November 8, 1930. Rapura, New Zealand. During a thunderstorm, an entire house was wrapped in a blue flickering flame for "an appreciable period." In a field, three horses were likewise enveloped in blue flames. (R9)

X8. May 30, 1869. Greiffenberg, Germany. People outside in a thunderstorm seemed wrapped in fire and deprived of air. (R10, R11)

References

R1. Traill, William; "St. Elmo's Fire Seen in Orkney," Franklin Institute, Journal, 24:362, 1837. (X1)

R2. Cameron, Donald; "A Strange Phenomenon," Nature, 25:437, 1882. (X2)

R3. Nature, 20:423, 1879. (X3)

R4. "The Fires of St. Elmo," Scientific American, 41:234, 1879. (X3)

R5. Nature, 22:204, 1880. (X4)

R6. Tomlinson, C.; "Atmospheric Electricity," Nature, 40:102, 1889. (X4)

R7. Tomlinson, C.; "Effects of Lightning," Philosophical Magazine, 5:27:208, 1889. (X5)

R8. "Remarkable Discharge of Atmospheric Electricity," Scientific American, 58: 210, 1888. (X6)

R9. "Unusual Lightning Phenomena," Meteorological Magazine, 66:41, 1931. (X7)

R10. "Extraordinary Phenomenon," Scientific American, 21:231, 1869. (X8)

R11. "Extraordinary Phenomenon," Eclectic Magazine, 10:507, 1869. (X8)

R12. Tomlinson, Charles; "Remarks on the Weathering of Rocks and Certain Electrical Phenomena," Philosophical Magazine, 5:26:475, 1888. (X4)

R13. English Mechanic, 29:622, 1879. (X3)

R14. Traill, William; "St. Elmo's Fire Seen in Orkney," Edinburgh New Philosophical Journal, 23:220, 1837. (X1)

# GLD4 Moving, Surface-Level, Electrified Light Patches

Description. Electrified luminous clouds or patches of light that drift over the surface of the ground. Sometimes the patches seem to move in waves over people and objects. Slight electrical shocks and sensations often accompany the enveloping patch of light. Usually, electrical storms are in the neighborhood.

Data Evaluation. This curious phenomenon has been well-described several times by competent observers. Rating: 2.

Anomaly Evaluation. The wavelike patterns and progressive motion of the luminous patches suggest an unrecognized regular structure for atmospheric electrical fields. Rating: 2.

Possible Explanation. Moving patterns of atmospheric electrical fields, a sort of St. Elmo's Fire with large-scale structure.

Similar and Related Phenomena. Large-scale St. Elmo's Fire (GLD3); auroral fogs and mists (GLA21); and, in appearance at least, the bar-type of marine phosphorescence display (GLW2). Some earthquake lights, especially GLD8-X5.

## Examples of Moving Electrified Patches of Light

X1. November 30, 1858. Boscastle, Great Britain. "Last night (Nov. 30), at 15 minutes to 9, it being very dark and raining heavily, I was ascending one of the steep hills in this neighborhood, when suddenly I was surrounded by a bright and powerful light which passed me a little quicker than the ordinary pace of man's walking, leaving it dark as before. This day I have been informed that the light was seen by the sailors in the harbour, coming in from the sea and passing up the valley like a low cloud." (R1, R2, R8)

X2. July 23, 1885. Midlothian, England. "About ten o'clock in the evening of July 23 a party of four of us were standing at the head of the avenue leading to this house, when we saw a feebly-luminous flash appear on the ground at a distance of some thirty yards down the avenue. It rushed towards us with a wave-like motion, at a rate which I estimate at thirty miles an hour, and seemed to envelope us for an instant. My left hand, which was hanging by my side, experienced precisely the same sensation as I have felt in receiving a shock from a weak galvanic battery. About three minutes afterwards we heard a peal of thunder, but, though we waited for some time, we neither saw nor heard anything further. The gardner who was one of the four, thus described what he saw:---

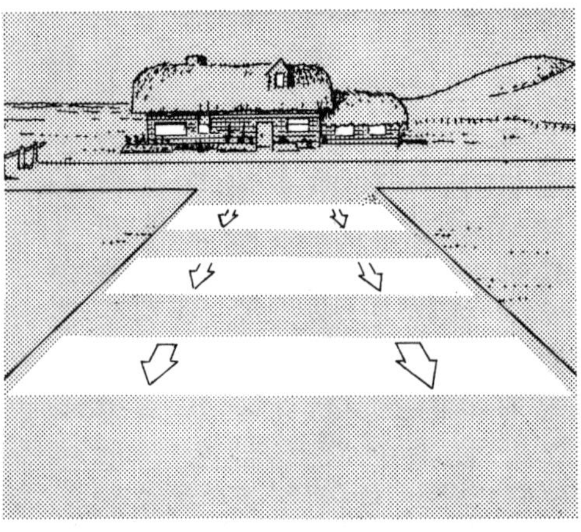

*Waves of light sweep across the ground in Midlothian, in 1885.*

I thought it was a cloud of dust blowing up the avenue, and before I could think how that could be when there was not a breath of wind, I saw you three gentlemen covered for a second in a bright light, and that was all. Another of the party says that he observed what seemed to be a luminous cloud running up the avenue with a wavy motion. When it reached the party it rose off the ground and passed over the bodies of two of them, casting a sort of flash on their shoulders. The distance traversed was about twenty yards, and the time occupied between two and three seconds." (R3)

X3. July 23, 1885. Dalkeith, England. Same time and general locality as X2. "When driving home from a professional visit in the country, and a mile south of this town, about ten o'clock I was suddenly startled by a peculiar sensation or slight shock, and immediately perceived, ten yards in front, on the road, a bright opalescent luminosity which travelled deliberately away in a northerly direction. This cloud or wave of light covered the whole breadth of the road, and was distinctly visible for some seconds. It seemed to rest entirely on the ground, and in character reminded one somewhat of the illumination resulting from the electric light. I should imagine it was travelling at the rate of twenty miles an hour.....I heard no thunder and saw no lightning or meteor to account for the strange and weird-looking light." (R4)

X4. February 22, 1912. Rochdale, England. Atmosphere close and heavy, with no breeze. "Suddenly, without the slightest warning, there appeared an area of faint electric-blue light, almost circular in shape and about 70 yards in diameter, which covered the plateau. The edge of this area was not more than 10 yards from where the observations were made. The whole electric field seemed to be three or four feet above the ground-level, and was in a state of intense agitation. Within the general blue ground there appeared flashes of a more decided blue, very similar in character to forked lightning, but not nearly so distinct. Sounds of two distinct types accompanied the agitation. The first consisted of whistling sounds, like that of numerous long-lashed whips swishing rapidly through the air, or perhaps that of the whistle of bullets. These sounds seemed to be associated with the general field of fainter blue. The other sounds consisted of the characteristic crackle of electricity, and these became so numerous as they approached the climax that they resembled a magnified rustle. These cracklings seemed to be associated with the forked discharges, and were probably due to the more distinct flashes coming into contact with the bushes which surround the plateau. The phenomenon lasted about fifteen to twenty seconds, and disappeared as spontaneously as it had arisen." (R5)

X5. September 1949. Yellowstone Park, Wyoming. While observing a violent thunderstorm. "It was then that I noticed a bluish light coming from over the low ridge to the west of Swan Lake. My first though was of a fire, perhaps caused by lightning. I watched the ridge for a moment and was amazed to see what can best be described as a hazy patch of blue light coming over the ridge and moving down the hill slope toward the flats around the lake. It was then that I observed a very low lead-gray cloud moving swiftly above the patch of light. The patch moved through the marshy north end of Swan Lake and caused several waterfowl to rise in hurried flight. The patch of light moved off the lake and onto the flats at a steady rate and proceeded directly toward my viewpoint. When the patch was but a few yards away, I noted a sudden calm in the air and a marked change in temperature, as well as what I believe was the odor of ozone. It was then that I realized that the display before me was some manner of static electricity, comparable perhaps to St. Elmo's fire and directly controlled by the low cloud moving above. The patch, which was actually a static field, enveloped my immediate area. To describe the weird feeling caused by viewing the progress of this phenomenon is difficult. It kept low to the ground, actually 'flowing around' everything that it came in contact with, coating it with a strange pulsating light. Each twig on the sagebrush was surrounded by a halo of light about two inches in diameter. It covered the automobile and my person but did not cover my skin. There was a marked tingling sensation in my scalp, and brushing my hair with the hand caused a snapping of tiny sparks." (R6)

X6. No date given. Natal. On a calm morning approaching a small airfield across farmlands. "But what had caught our attention in the misty dawn was an eerie reddish glow on the runway and about three hundred yards from the house. We were not more than two hundred yards away and walking down hill directly towards the phenomenon. My first thought was that it could be a very large bubble of irridescent marsh gas about 100 feet in diameter, bright pink in the centre and fading gradually towards the circumference..... By this time we were only a stone's throw away and the pinkish glow had started to rise eerily into the air. There was no sound, no rush of air, and with a sort of involuntary reaction we both broke into a run towards the phenomenon, Jock exclaiming, 'Just look at those sheep!' My eyes were following the rapid disappearance of the 'bubble' into the mist above, but as we reached the point where the glow had been I also looked at the sheep and noticed with amazement that they all appeared to be standing on tiptoe like ballet dancers with their heads held unusually high, just as if they were suspended in space with their hooves barely touching the grass. It was then that we both first experienced a peculiar feeling, almost of weightlessness." The Zulu name for this phenomenon (see article title) is translated as 'The red sun that rises straight up.' (R7)

X7. No date given. Constamon, France. Observation of sparks coursing over the prairies. (R8)

References

R1. Brown, Jabez; Report of the British Association, 1858, p. 156. (X1)

R2. Tomlinson, C.; "Atmospheric Electricity," Nature, 40:102, 1889. (X1)

R3. Watt, J. B. A.; "Electrical Phenomenon," Nature, 32:316, 1885. (X2)

R4. Lucas, Robert; "Electrical Phenomenon in Mid-Lothian," Nature, 32:343, 1885. (X3)

R5. M., J. McV.; "St. Elmo's Fire," Nature, 89:7, 1912. (X4)

R6. Sanborn, William B.; "An Electrical 'Bath' in Yellowstone," Natural History, 59:258, June 1950. (X5)

R7. Stocks, H. G.; "Hanga Elibovu Elisbonu Pezulu," Air Facts, 33:55, October 1970. (X6)

R8. Tomilinson, Charles; "Remarks on the Weathering of Rocks and Certain Electrical Phenomena," Philosophical Magazine, 5:26:475, 1888. (X1, X7)

# GLD5 Unusual Electric Discharge Phenomena during Duststorms and Snowstorms

Description. Ground-level lightning, subdued soundless flashes, dangerously electrified wire fences, widespread St. Elmo's Fire, stalled automobiles, sparking from projections, and other types of electrical discharges experienced during duststorms and snowstorms.

Background. Dust and snow particles carry electrical charges and can establish large potential differences at very low altitudes.

Data Evaluation. Many casual observations, backed in some cases (X4) by instrumented observations of atmospheric electrical gradients. Rating: 2.

Anomaly Evaluation. These manifestations of high electrostatic fields can be regarded only as curiosities. Rating: 3.

Possible Explanation. Frictional electricity created by blowing particles.

Similar and Related Phenomena. St. Elmo's Fire and corona discharge.

## Examples of Unusual Electrical Discharges during Duststorms and Snowstorms

X1. March 1813. In a boat on Lochawe, in Argyleshire; storm approaching. "In a few minutes, however, they were overtaken by a shower of snow; and immediately after, the lake, which was of glassy smoothness, with their boat, clothes, and all around, presented a luminous surface, forming one huge sheet of fire. Nor were the exposed parts of their bodies singular in this respect, for to the eye they all seemed to burn, although without any feeling even of warmth. When they applied their hands to any of the melting snow, the luminous substance adhered to them as well as the moisture, and this property was not lost by the snow for twelve or fifteen minutes." (R1)

X2. December 20, 1894. Ft. Laramie, Wyoming. During a fierce windstorm. "Reliable information, obtained from over a section of country extending 12 miles south and 20 miles east, states that the electrical current was freely felt in many localities embraced in that area. Mr. Silas Doty, a very reliable man living 8 miles south, reports that he discovered a fence post on fire 200 yards north of his house, and upon going to it to extinguish the fire found it burnt more than half through where one of the fence wires was fastened, and partly burned where another wire was in contact with it. Other fence posts were slightly marked by the electric sparks. Mr. John F. Barnes, another reliable man, living 11 miles south of here, states that in going from his stable to his house he caught hold of a fence wire to assist him in walking against the strong wind and received a severe electric shock from which he was some time recovering. He had a strong healthy cow in a lot inclosed with wire fence. The wind drifted the cow into a corner against the fence and held her there, where Mr. Barnes found her dead late in the evening." (R2)

X3. March 3, 1964. Tucson, Arizona. During a heavy snowfall of large, wet flakes. "Short flashes of 'lightning' were seen to be occurring at intervals estimated to be about 15-20 seconds and at random places around the town. The flashes continued to occur throughout the storm, until the snow ceased at about 11.30 pm. Visibility was about 1 mile at first, but later decreased during the heaviest snowfall to about 200 yards. The flashes of 'lightning' contained several unusual features: they appeared to be single, short flashes of light without the 'flicker' normally associated with lightning. They were less intense than normal lightning and were not observed to cast sharp shadows. Moreover, thunder was not heard at any time, nor could correlation between the flashes and 'static' over the radio be found. The 80-foot high observation tower affords a good view over the surrounding town, which is mainly occupied by single-story dwellings. From this elevation, the lightning was seen to originate from points at or very close to the ground. No brilliant, illuminated track such as is usually associated with lightning could be detected, although several discharges as close as 200 yards were observed. The illumination seemed to emanate from a single point which lit up the falling snow and surrounding cloud. The phenomenon was observed by several trained meteorological observers..." (R3)

X4. May 11, 1971. White Sands National Monument. "I made a fascinating observation of electrical activity on May 11, 1971, when a thunderstorm passed over the site. Strong winds were blowing large quantities of sand into the air, and atmospheric temperatures were lower and relative humidities were higher than usual for the month of May in that region.... Electric sparks were observed extending from the top of the sand dunes up into the air terminating at a height of a few metres. At least five sparks, all on different dune tops, were observed within the period 1510 to 1600 MDT. These sparks extended stright up, and no branches were observed." (R4)

*Long sparks rise from gypsum dunes in White Sands National Monument.*

X5. General observation. Kansas. "No thunder or lightning occurs during these electrified dust storms, nor is there any known relation between them and earth magnetism. Any metallic object insulated from the earth seems to become highly electrified as the fine dust blows past it. One rancher relates an instance of driving cattle in one of these odd electrical displays and seeing a glowing light at the tips of the horns of each steer. When autos were new to the country, drivers were puzzled to find their engines stalled in these disturbances and nothing apparently wrong with them." (R5)

X6. January 18, 1817. Vermont. During a rapid fall of wet snow, people were enveloped by light. Much St. Elmo's Fire on fence posts, hourses, etc. (R6)

X7. January 20, 1895. Oklahoma. Flashes of light seen during dust storm. (R7)

X8. November 19, 1903. Geneva, Ohio. Purple and milky-white flashes during a snowstorm. Faint rolling thunder. (R8)

X9. General observation. Lake Maracaibo, Venezuela. Constant silent lightning seen near the bar as the southern end of the lake. Called the Faro of Maracaibo since the Spanish Conquest. (R9) See also GLN1-X1.

X10. General observations. Khartum, Sudan and Kano, Nigeria. Observations of blue sparks. Attributed to wind-driven sand. (R10)

## References

R1. Smith, Colin; "Luminous Snow Storm in Lochawe," Edinburgh Philosophical Journal, 12:405, 1825. (X1)

R2. "Local Electricity in a Windstorm in Wyoming," Monthly Weather Review, 22:509, 1894. (X2)

R3. Matthews, J. Brian; "An Unusual Type of Lightning," Weather, 19:291, 1964. (X3)

R4. Kamra, A. K.; "Visual Observation of Electric Sparks on Gypsum Dunes," Nature, 240:143, 1972. (X4)

R5. "Electrified Dust Storms of the West," Popular Mechanics, 49:226, February 1928. (X5)

R6. Allen, J. A.; "On Luminous Appearances in the Atmosphere," American Journal of Science, 1:4:341, 1822. (X6)

R7. "A Silent and Electrical Dust Storm in Oklahoma,: Monthly Weather Review, 23:18, 1895. (X7)

R8. "Lightning Phenomenon," Monthly Weather Review, 31:534, 1903. (X8)

R9. "Infernito," Nature, 40:429, 1889. (X9)

R10. "Blue Sparks Explained," Fortean Society Magazine, 1:7, September 1937. Credited to the New York Times of August 1, 1937. (X10)

# GLD6 Unusual Manifestations of St. Elmo's Fire

Description. Unusual tongues of flame, luminous halos, and marching spheres of light, frequently accompanied by curious sound effects. Many odd occurrences of St. Elmo's Fire are experienced on mountain tops where climbers may become wreathed in luminous auras, while the surrounding rocks buzz noisily with electricity. If the atmospheric electrical fields are strong enough large areas of the surface seem to burst into flame. Usually, the luminous jets and spheres remain attached to conducting surfaces, but rarely they detach themselves and drift in the air like ball lightning.

Background. Sometimes it is difficult to differentiate between the less energetic forms of ball lightning and detached spheres of St. Elmo's Fire; if indeed there is a difference. Complicating the picture are the luminous air bubbles (GLD7) which seem to have some properties of both ball lightning and St. Elmo's Fire.

Data Evaluation. St. Elmo's Fire is a well-recognized phenomenon. Rating: 1.

Anomaly Evaluation. The manifestations of St. Elmo's Fire reported below pose at least two problems: (1) How do freely moving balls of St. Elmo's Fire maintain their existence? (2) Why do spheres of St. Elmo's Fire move along conductor surfaces rather than inside? Rating: 3.

Possible Explanation. Ordinary St. Elmo's Fire appears around sharply pointed objects when atmospheric electric field strengths are very high. It is simply the luminous flow of electricity through the air. These observations, however, do not explain detached masses of St. Elmo's Fire and its movement along conductor surfaces.

Similar and Related Phenomena. Ball lightning (GLB), luminous aerial bubbles (GLD7), earthquake lights (GLD8).

Examples of Pranks of Pranks of St. Elmo's Fire

X1. June 7, 1882. Pike's Peak, Colorado. St. Elmo's Fire everywhere. "In placing my hands close over the revolving cups of the anemometer---where the electrical excitment was abundant---not the slightest sensation of heat was discovered, but my hands instantly became aflame. On raising them and spreading my fingers, each of them became tipped with one or more beautiful cones of light, nearly three inches in length. The flames issued from my fingers with a rushing noise, similar to that produced by blowing briskly against the end of the finger when placed lightly against the lips, accompanied by a crackling sound. There was a feeling as of a current of vapor escaping, with a slight tingling sensation. The wristband of my woolen shirt, as soon as it became dampened, formed a fiery ring around my arm, while my moustache was lighted up so as to make a veritable lantern of my face. The phenomenon was preceded by lightning and thunder, and was accompanied by a dense driving snow, and disappeared suddenly at 8:55 o'clock, simultaneously with the cessation of the snow." (R1, R2)

*St. Elmo's Fire atop Pike's Peak.*

X2. No date or place given. "In the Zeitschrift fur Elektrotechnik a description is given of the following unique display of atmospheric electricity which was observed by a farmer while driving along a narrow private road having a wire fence on each side. When his vehicle had gone about 400 feet from the commencement of the wire fences his attention was attracted by a bright light behind him. On looking round he saw fire balls about the size of a man's hand travelling towards him along the wire on both sides. In a moment they were abreast of the carriage, and then travelled along with it

Balls of St. Elmo's Fire follow carriage moving between two wire fences.

pari passu, while brush discharges and audible crackling, as from a large electrical machine, were observed to proceed from the fire balls towards, apparently, the iron parts of the carriage. Vibrations of the wires could be distinctly heard, and a torrent of sparks sprang over from the fences to the carriage and horses. The frightened horses now ran away, but they could not shake off their fiery escort till the fences were interrupted by gaps on each side. No detonation or noise, other than the crackling already mentioned, was heard. The whole phenomenon lasted about 12 seconds." (R3)

X3. No date given. Worcester, Massachusetts. "While walking upon the Worcester and Norwich Railroad track about a mile south of Worcester Junction, I suddenly saw a ball of fire, or what looked like it, about the size of a large marble, running along on top of one of the rails just ahead of me. It was going at so slow a rate that I could have overtaken it in a few seconds, and my first impulse was to do so; but the sober second-thought warned me against making the attempt. I, however, watched it move until it came to the end of a rail separated from the adjacent one by something like half an inch, when it stopped, and in a second or two vanished, when there was a clap of thunder in a cloud overhead which I had not before noticed as being a thunder-cloud. The brightness of this small ball was not excessive, nothing to be compared with an electric arc. It was more like that of a red-hot bullet. It did not scintillate or make any noise, that I noticed. Now, while this was an accompaniment of a thunder-cloud, as are such manifestations generally, I think there is some reason for not calling the phenomenon itself an electrical one in the same sense as lightning is electrical. If electricity can gather itself up into a spherical form as if it was subject to some sort of cohesion, and if it can roll along on top of a good conductor instead of traversing the body of the conductor subject to Ohm's law, then there are some exceptions to this latter law." (R4)

X4. March 7, 1932. Hilversum, Holland. On a broadcasting antenna consisting of five horizontal wires. "Again, on March 7, at 17.00, when the antenna power was raised from 7 kw. to 20 kw., the corona on the antenna reappeared. It consisted of slowly moving luminous spheres of about 10 cm. diameter. They first appeared at one end of an outer wire, they slowly moved following the wires towards the down leads and disappeared at a distance of about 15 m. from their origin. Several of these luminous spheres were observed to occur simultaneously with a mutual distance of 0.5 m., so that the phenomenon made the impression of a string of beads, moving towards the centre with a speed of the order of 1 m./sec." (R5)

X5. January 1924. No place given. Seen on a barn roof on a clear, starry night. "There were six incandescent globes or disks or hemispheres, it is impossible to state exactly which, upon the ridge of the barn, seemingly about three feet in diameter and spaced exactly even with a distance of six inches between their edges---for a distance of four rods around the barn everything was clearly illuminated. Edward then called our rapt attention to the disk on the north end of the barn with the exclamation: 'It's falling!' Sure enough it did fall, leaving the ridge with a sort of jump and landing, without loss of brilliance, half way down the roof, then bouncing from there over the eave and disappearing before reaching the ground. Immediately upon extinction of the first light the second disk started to fall, and then the rest, one after another, in exactly the same way at regular intervals. Abruptly as the last one disappeared the light around the barn vanished and all again was dark." (R6)

X6. No date given. Giza, Egypt. "Towards eight o'clock in the evening, I noticed a light which appeared to turn slowly around the third pyramid almost up to the apex; it was like a small flame. The light made three circuits around the pyramid and then disappeared. For a good part of the night I attentively watched the pyramid; towards eleven o'clock I again noticed the same light, but this time it was of a bluish colour; it mounted slowly almost in a straight line and arrived at a certain height above the pyramid's summit and then disappeared." (R7)

X7. No date given. Virginia City, Nevada. "The city was interested, last evening, says my report, by the appearance on C-street of a strange phenomenon. At first it had the appearance of sparks of fire coming up through the pools of water beside the street. These sparks seemed to explode on reaching the surface, in many instances producing reports loud enough to be heard across the street, and being accompanied by a little cloud of smoke, and emitting a decidedly sulphurous smell. It was noticed that the phenomena occurred only on one side, under the telegraph wires. The sparks seemed to be caused by drops of water falling from the wires of the telegraph, which exploded when striking the pools of water. This solution was seeminly confirmed by the fact that when the wires became dry the phenomena ceased. It still remains to be explained, however, why, under the circumstances, such results should follow the falling of the water drops from the wires. (R8)

X8. August 18, 1876. France. "M. Trecul, in a note to the Comptes Rendus, states that on the 18th August 1876, while writing at an open window between 7 and 8 a.m., he observed simultaneously with some loud thunder, small luminous columns descend obliquely on his paper, about two metres long, and half a decimeter broad at the widest part; obtuse at the further end, but gradually thinning towards the table. They had mostly a reddish yellow tint, but near the paper the tints were more intense and varied. In disappearing they left the paper with a slight noise like that produced by pouring a little water on a hot plate." (R9-R12)

## References

R1. Gregory, R. A.; "Electrical Storms on Pike's Peak," Nature, 42:595, 1890. (X1)

R2. "Curious Electrical Phenomena on Pike's Peak," Scientific American, 47: 16, 1882. (X1)

R3. "Remarkable Electrical Phenomenon," Royal Meteorological Society, Quarterly Journal, 22:295, 1896. (X2)

R4. Dolbear, A. E.; "Globular Lightning," Science, 11:38, 1888. (X3)

R5. van der Pol, Balth.; "Bead-Corona on Radio Antenna," Nature, 130:662, 1932. (X4)

R6. Humphreys, W. J.; "Ball Lightning," American Philosophical Society, Proceedings, 76:613, 1936. (X5)

R7. Botley, Cicely M.; "St. Elmo's Fire in Egypt," Meteorological Magazine, 73:96, 1938. (X6)

R8. "Notes on Electrical Phenomena," English Mechanic, 25:435, 1877. (X7)

R9. Tomlinson, Charles; "On Certain Low-Lying Meteors," Knowledge, 16: 96, 1893. (X8)

R10. Omond, R. T.; "Atmospheric Electricity," Nature, 40:102, 1889. (X3)

R11. Tomlinson, Charles; "Remarks on the Weathering of Rocks and Certain Electrical Phenomena," Philosophical Magazine, 5:26:475, 1888. (X8)

R12. Lake, J. J.; "Beaded Lightning," English Mechanic, 24:234, 1876. (X8)

# GLD7 Luminous Aerial Bubbles

Description. Large assemblages of colored spheres or bubbles drifting randomly, quietly, almost playfully. Observers often compare them to toy balloons and soap bubbles. Their changing colors and irridescence strengthen the soap-bubble analogy. These softly glowing spheres do not detonate destructively like ball lightning but disappear silently like real soap bubbles. Hundreds of bubbles may fill the air during this phenomenon.

Background. Two possibly distinct phenomena are lumped together here: (1) Low-level objects that superficially resemble soap bubbles; and (2) Swarms of higher-altitude objects that appear primarily as moving points of light but seem to drift with the wind in a buoyant manner. Example X1 below mentions a gelatinous residue from one group of low-level bubbles; this characteristic brings to mind the even-more-bizarre gelatinous meteors or pwdre ser (GFG). Also, some cold flames seen issuing from the ground (GLN1) may leave a residue. The luminous aerial bubbles may in fact become nocturnal lights once the sun sets.

Data Evaluation. This bizarre phenomenon is difficult-to-believe, but it is confirmed by many reports. Rating: 2.

Anomaly Evaluation. Although very puzzling, aerial bubbles are probably explainable without gross modifications of chemistry and physics; but we know so little about the phenomenon that this is only a surmise. Rating: 2.

Possible Explanations. (1) An electrical origin similar to that of ball lightning; (2) A chemical origin similar to that proposed for some will o' the wisps; (3) An optical origin, say, the reflection of light off windborne specks of material, as in the case of telescopic meteors; (4) They may be real bubbles from some natural source.

Similar and Related Phenomena. Ball lightning (GLB); will o' the wisps (GLN); St. Elmo's Fire (GLD); telescopic meteors (ATO); pwdre ser (GFG), kaleidoscopic suns (GEL).

## Examples of Luminous Aerial Bubbles

X1. May 16, 1808. Biskopsberga, Sweden. "On the 16th of last May, being a very warm day, and during a gale of wind from southwest, and a cloudless sky, at about 4 o'clock, P.M. the sun became dim, and lost his brightness to that degree, that he could be looked at without inconvenience to the naked eye, being of a dark-red, or almost brick colour, without brilliancy. At the same time there appeared at the western horizon, from where the wind blew, to arise gradually, and in quick succession, a great number of balls, or spherical bodies, to the naked eye of a size of the crown of a hat, and of a dark brown colour. The nearer these bodies, which occupied a considerable though irregular breadth of the visible heaven, approached towards the sun, the darker they appeared, and in the vicinity of the sun became entirely black. At this elevation their course seemed to lessen, and a great many

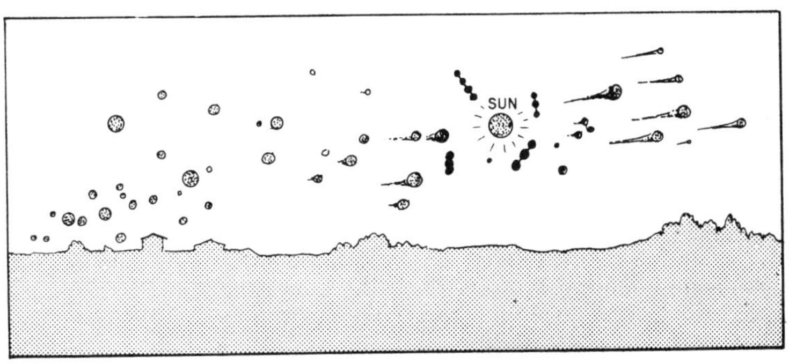


*Curious spherical bodies drift over Biskopsberga, Sweden, in 1808.*

of them remained, as it were, stationary; but they soon resumed their former, and an accelerated motion, and passed in the same direction with great velocity and almost horizontally. During this course some disappeared, others fell down, but the most part of them continued their progress almost in a straight line, till they were lost sight of at the eastern horizon. The phenomenon lasted uninterruptedly, upwards of two hours, during which time millions of similar bodies continually rose in the west, one after the other irregularly, and continued their career exactly in the same manner. No report, noise, nor any whistling or buzzing in the air was perceived. As these balls slackened their course on passing by the sun, several were linked together, three, six, or eight of them in a line, joined like chain shot by a thin and straight bar; but on continuing again a more rapid course, they separated, and each having a tail after it. apparently of three or four fathoms length, wider at its base where it adhered to the ball, and gradually decreasing, till it terminated in a fine point. During the course, these tails which had the same black colour as the balls, disappeared by degrees. It fortunately happened, that some of these balls fell at a short distance, or but a few feet from Mr. Secretary K.G. Wettermark, who had then for a long while been attentively looking at the phenomenon, in the aforesaid village. On the descent of these bodies, the black colour seemed gradually to disappear the nearer they approached the earth, and they vanished almost entirely till within a few fathoms distance from the ground, when they again were visible with several changing colours, and in this particular exactly resembling those air-bubbles, which children use to produce from soadsuds by means of a reed. When the spot, where such a ball had fallen, was immediately after examined, nothing was to be seen, but a scarcely perceptible film or pellicle as thin and fine as a cobweb, which was still changing colours, but soon entirely dried up and vanished." (R1)

X2, January 1871. Remenham, England. A cold, gray day, with low, rusty-colored clouds. "The wall paper and furniture of the room in which the lady was sitting were suddenly flushed with rose colour, which gradually deepened into crimson, passing through bright gold into orange, lilac and deep violet. It was then seen that from the centre of the level space of snow within view, a group of air bubbles, of the shape and apparent size of the coloured India rubber balls sold in the streets, rose to a considerable height and then began to move up and down within a limited area, and at an equal distance from each other, some descending others descending. The appearance lasted about two minutes, at the expiration of which the balls were carried away by a current of wind to the eastward and disappeared. Another group of balls arose subsequently from the same spot, and the phenomena were precisely reproduced. It was remarked that the balls assumed in succession the tints which had been observed on the walls of the room." (R2, R3)

X3. August 17, 1876. Ringstead Bay, England. "Between 4 and 5 p.m. two ladies who were out on the cliff, saw surrounding them on all sides, and extending from a few inches above the surface to two or three feet overhead, numerous globes of light, the size of billiard balls, which were moving independently and vertically up and down, sometimes within a few inches of the

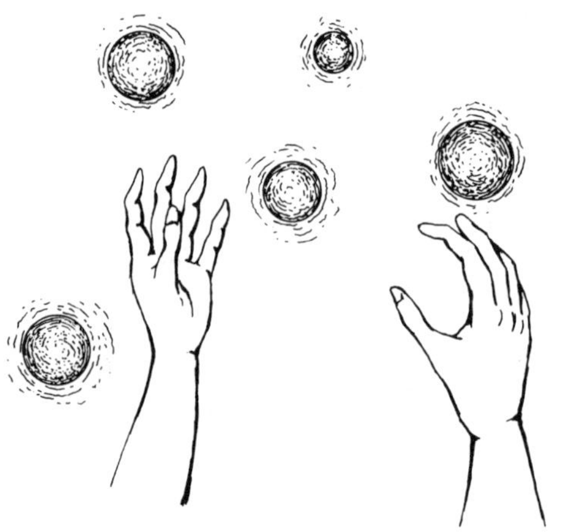

*Luminous spheres evade grasp at Ringstead Bay.*

observers, but always eluding the grasp; now gliding upwards two or three feet, and as slowly falling again, resembling in their movements soap bubbles floating in the air. The balls were all aglow, but not dazzling, with a soft, superb irridescence, rich and warm of hue, and each of variable tints, their charming colours brightening the extreme beauty of the scene. The subdued magnificence of this fascinating spectacle is described as baffling description. Their numbers were continually fluctuating; at times thousands of them enveloped the observers, and a few minutes afterwards the numbers would dwindle to perhaps as few as twenty, but soon they would be swarming again as numerous as ever. Not the slightest noise accompanied the display." (R2, R4-R6)

X4. March 22, 1880. Trakehnen, Germany. "A remarkable phenomenon was observed at Kattenau, near Trakehnen (Germany), and in the surrounding district, on March 22. About half an hour before sunrise an enormous number of luminous bodies rose from the horizon and passed in a horizontal direction from east to west. Some of them seemed of the size of a walnut, others resembled the sparks flying from a chimney. They moved through space like a string of beads, and shone with a remarkably brilliant light. The belt containing them appeared about 3 metres in length and 2/3 metre in breadth." (R7)

X5. Circa 1882. Glenora, New York. "I was walking home just at dusk during an evening of thunderstormy weather. The air was sultry and there were flashes of heat lightning on the horizon. When I came to what we called in New England the village green, I saw a remarkable exhibition of what I assumed to be static electricity. There were twelve or more round balls of luminescence bouncing around in the most undisciplined manner in all directions. They would apparently strike the ground, then bounce up into the air in the lightest and most buoyant fashion. They looked to be about the size of a football." (R8)

X6. April 1886. Cherry Valley, New York. Towards sunset on a warm, thawing day. "...as they rode down the long hill towards this village from the east, (they) saw what appeared to be innumerable spherical bodies floating in the air like soap bubbles. Both men saw and wondered at the appearance for some moments before either spoke. Capt. H. then said, 'I wonder whether I am dreaming?' The other rubbed his eyes and echoes the sentiment. 'Well,' said the captain, 'I wonder if you see what I see; what do you see?' They questioned each other, and both agreed as to their impressions. An orchard lay along the lower and northwesterly side of the road, and all in among the apple trees were thick, gently descending multitudes of these bubbles, pretty uniform in size, say, 8 or 9 inches in diameter, apparently; none less than six; no small ones being observed. The observers state that they carefully fixed their attention on particular bubbles, in order to compare notes, and saw them seem to rest on the bough of a tree, or the top board of the fence, and then gently roll off and disappear or go out of sight. The sun was sinking and dropped below the opposite hills as they reached the foot of the long descent and entered the village, and the appearance came to an end. But up to this time the air seemed to be filled with these transparent floating spheres. The position of the observers with regard to the light seems to have made some difference as to seeing well this or that large aggregation or swarm that one or the other pointed out. The bubbles were highly colored, irridescent, gave the same sort of reflections as soap bubbles, and apparently vanished individually in much the same way." (R9)

X7. June 1898. Kailana, Himalayas. Air hot and misty after the passage of a thunderstorm. "In the dense mist a succession of pale, round lights were seen floating in the direction of the storm; each was watched till it was out of sight. The balls appeared about 6 ins. in diameter, but the edges were not well-defined. The colour was similar to that of a firefly, but the behavior of the lights proved that they were not fireflies; moreover, no fireflies were found in the locality!" (R10)

X8. December 18, 1922. Salisbury, England. "There was in this locality on December 18th. between 6 and 8 p.m., an extraordinary display of the so-called ball lightning. The phenomenon took the form of balls of bluish, misty flame, which either hovered stationary or darted about the heavens in a curiously irregular manner. Vivid lightning of the ordinary kind accompanied the display, which must have lasted at least two hours." (R11)

X9. May 15, 1836. Havana, Cuba. During the solar eclipse, many luminous globules detached themselves from the sun and moved in various directions. (R12)

X10. May 21, 1877. Venice, Italy. Balls of fire rise from a dark cloud and burst. The balls looked and moved like soap bubbles. (R13)

X11. November 15, 1899. Wiltshire, England. The air seemed to be full of small luminous bodies. (R14)

X12. May 10, 1902. Devonport, England. At sunset the sky seemed to be filled with violet, yellow, and green "suns" that looked like toy balloons. (R15)

X13. September 21, 1910. New York, New York. A great number of round objects passed west to east over the lower part of the city. They were though to be little balloons. (R16)

References

R1. Acharius, E.; "Account of an Extraordinary Meteoric Phenomenon," North American Review, 3:320, 1916. (X1)
R2. "Ball Lightning," Scientific American, 57:53, 1887. (X2, X3)
R3. Bonney, A.; "Note on an Appearance of Luminous Bubbles in the Atmosphere," Royal Meteorological Society, Quarterly Journal, 13:306, 1887. (X2)
R4. "Electrical Phenomena near Weymouth," Nature, 126:262, 1930. (X3)
R5. Eaton, H.S.; "Note on a Display of Globular Lightning at Ringstead Bay..." Symons's Monthly Meteorological Magazine, 32:127, 1898. (X3)
R6. Eaton, H.S.; "Note on a Manifestation of Electricity at Ringstead Bay,...." Royal Meteorological Society, Quarterly Journal, 13:305, 1887. (X3)
R7. Nature, 22:64, 1880. (X4)
R8. Talman, Charles Fitzhugh; "Ball Lightning," American Mercury, 26:69, 1932. (X5)
R9. Swinnerton, Henry U.; "Aerial Bubbles," Science, 21:136, 1893. (X6)
R10. "Ball Lightning," Meteorological Magazine, 57:46, 1922. (X7)
R11. Smith-Gordon, Lionel; "Ball Lightning," Meteorological Magazine, 57: 336, 1922. (X8)
R12. Poey, M.A.; "Sur le Passage d'une Quantitie Considerable de Globules Lumineux...." Comptes Rendus, 56:88, 1863. (X9)
R13. English Mechanic, 25:132, 1877. (X10)
R14. "Spurious Daylight Observations of the Leonids," Observatory, 23:68, 1900. (X11)
R15. Markwick, E.E.; "Curious Phenomenon," English Mechanic, 75:417, 1902. (X12)
R16. New York Tribune, September 22, 1910. (X13)

# GLD8 Earthquake Lights

Description. Flashes of light in the sky, ordinary and ball lightning, aurora-like streamers and rays, flames issuing from the ground, sky glows, and St. Elmo's Fire observed in the general vicinity of an earthquake. Earthquake lights may appear before, during, and after the quake.

Background. Earthquake lights, like ball lightning, have been recorded since ancient times. Until recently, they have not been recognized as legitimate phenomena by most scientists. Some skepticism remains even today. One reason for this is the great variety of earthquake lights, many of which can be explained in terms of ordinary lightning in accompanying storms, the arcing of damaged power lines, and the ignition of ruptured gas lines. Furthermore, earthquake light observations may be contaminated by the psychological impact of strong earthquake shocks.

Data Evaluation. Scores of old and recent observations, including photographs, vouch for the reality of earthquake lights. Rating: 1.

Anomaly Evaluation. If all earthquake lights could be explained in terms of coincident thunderstorms, power- and gas-line phenomena, and even the spontaneous ignition of releases of natural gas, we would have no anomaly to explain. However, the fireballs, sky flashes, and aurora-like beams common to many earthquakes suggest that some other agent, probably electrical in nature, is at work here. Rating: 2.

Possible Explanations. The variety of earthquake lights may infer the existence of strong electrical fields, due perhaps to the piezoelectric effect as rocks are stressed by earthquake forces. The measured properties of rocks and the stresses exerted do not, however, seem to be adequate to explain the observations. Furthermore, earthquake lights have been seen at sea; a fact that weakens the piezoelectric theory while supporting the natural-gas-eruption theory. Another possibility is that the gases and dust injected into the atmosphere during a quake may alter the air's electrical properties, stimulating luminous phenomena and perhaps the hazes and thunderstorms that are reported so frequently. Earthquake lights may owe their existence to all of the mechanisms mentioned above.

Similar and Related Phenomena. Tornado lights (GLD10), volcano lights (GLD9), rock luminosities (GLD13), low-level auroras (GLA4), sky flashes (GLA14), transient lunar phenomena (ALO), nocturnal lights (GLN1).

Examples of Earthquake Lights

X1. Late 1811 through 1813. New Madrid, Missouri, and environs. "Dillard, in speaking of the shocks (not especially the first one), says: 'There issued no burning flames, but flashes such as would result from an explosion of gas, or from passing of electricity from cloud to cloud.' Lewis F. Linn, United States Senator, in a letter to the chairman of the Committee on Commerce, says the shock was accompanied 'ever and anon by flashes of electricity, rendering the darkness doubly terrible.' Another evidently somewhat excited observer near New Madrid thought he saw 'many sparks of fire emitted from the earth.' At St. Louis gleams and flashes of light were frequently visible around the horizon in different directions, generally ascending from the earth. In Livingston County, according to Mr. Riddick, the atmosphere previous to the shock of February 8 was remarkably luminous, objects being visible for considerable distances, although there was no moon. 'On this occasion the brightness was general, and did not proceed from any point or spot in the heavens. It was broad and expanded, reaching from the zenith on every side toward the horizon. It exhibited no flashes nor coruscations, but, as long as it lasted, was a diffused illumination of the atmosphere on all sides.' At Bardstown there are reported to have been 'frequent lights during the commotions.' At Knoxville, Tenn., at the end of the first shock, 'two flashes of light, at intervals of about a minute, very much like distant lightning,' were observed. Farther east, in North Carolina, there were reported 'three extraordinary fires in the air; one appeared in an easterly direction, one in the north, and one in the south. Their continuance was several hours; their size as large as a house on fire; the motion of the blaze was quite visible, but no sparks appeared.' At Savannah, Ga., the first shock is said to have been preceded by a flash of light." (R1-R3)

X2. June 16, 1819. Cutch, India. "At the moment of the shock, vast clouds of dust were seen to ascend from the summit of almost every hill and range of hills. Many gentlemen perceived smoke to ascend, and in some instances fire was plainly seen bursting forth for a moment. A respectable native chieftain assured me, that from a hill close to the one on which his fortress is situated fire was seen to issue in considerable quantities; and that a ball of large size was vomited as it were into the air, and fell to the ground still blazing on the plain below, where it divided into four or five pieces, and the fire suddenly disappeared. On examining the hill next day (the chieftain stated), it was found rent and shattered, as if something within had sunk; and the spot where the fire-ball was supposed to have fallen, bore marks of fire in the scorched vegetation." (R4)

X3. Circa 1872. Cerrogardo, California. "Immediately following the great shock, men whose judgment and veracity are beyond question, while sitting on the ground near the Eclipse Mines, saw sheets of flames on the rocky sides of the Inyo Mountains, but half a mile distant. These flames, observed in several places, waved to and fro, apparently clear of the ground, like vast torches. They continued for only a few minutes." (R5)

X4. April 18, 1906. San Francisco, California. Many reports of luminous phenomena from the region. "The display of blue flames before the onslaught of the red ones, and their final yellow sequences was very remarkable. 'The appearance of the blue lights,' says Professor Larkin, 'was over a wider area than first thought. At Petaluma ...blue flames eighteen inches in height played over a wide expanse of marsh land.' Before the earthquake only 'a flickering ominous haze was seen playing above the ground.' 'A dark funnel-shaped mass was seen in Fourth Street, San Francisco, suspended in the air, and it was illuminated by scintillating lights, like fire-flies.' 'Blue flames were seen hovering over the bases of foothills in western San Francisco.' In San Jose, on the street called the Alameda, looking eastward, at the time of the shock the whole street was seen 'ablaze with fire, it being of a beautiful rainbow colour but faint. This, no doubt,' observes Professor Larkin, 'was an electrical display, for had gas been on fire all along the street the houses would have been ignited. And letters from a point north of San Francisco describe blue lights as flickering like an aurora over a wide area of marsh land with a troubled surface of adjoining water." (R6)

X5. August 16, 1906. Chile. A great earthquake with hundreds of observations of various types of luminous phenomena. Diffuse illuminations, will o' the wisps, lightning, and meteors. (R8) The Andes mountain-top glow burst forth greatly enhanced. (R9) The sky all over central Chile flashed with a quivering light. (R11) Ships at sea observed waves of light moving on the hills. (R13) (R7-R13)

X6. November 26, 1930. Idu, Japan. Reports from some 1500 observers were collected. "The lights were very strong---in one place brighter than moonlight. They were usually described as bluish in colour, but sometimes as reddish yellow, or reddish blue. In shape, they resembled the rays of the rising sun, search-lights, and fireballs. The duration of each light was longer than that of lightning, and some careful observers report that the same light continued more than a minute. The

*Rays of light on skyline during a Japanese quake*

directions in which they were seen pointed usually, but not always, to the epicentral region of the earthquake." (R14-R21, R25)

X7. November 2, 1931. South Hyuga, Japan. 355 observations of luminous phenomena. "The directions in which the lights appeared are varied, but most observers on the coast saw them towards the district in which the submarine epicentre lies. The luminescence usually seemed to radiate from the horizon or to be like a search-light turned to the sky, in colour most often blue or bluish. The phenomena were clearly not due to houses on fire, or to lightning or landslides, though some may be assigned to electric sparking or to meteors." (R17, R19, R22)

X8. 1965-1967. Japan. The Matsushiro Earthquake swarm. Many luminous displays recorded around mountain peaks. The accompanying sketch shows a typical luminosity that appeared at the top of Mt. Noroshi on April 20, 1966, for 10 seconds. (R23)

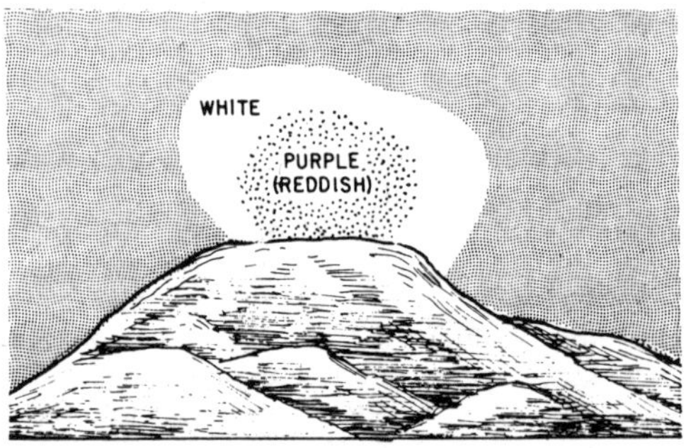


*Luminous display at top of Mt. Noroshi, Japan, during 1966 earthquake swarm*

X9. General observation. From a study of 156 earthquakes, the following numbers were found to be accompanied by: auroras (23); meteors (73); will o' the wisps (2); light flashes exclusive of thunderstorms (15); and flames issuing from fissures (10). (R24)

X10. 1257. Kamakura, Japan. Bluish flames emerged from fissures. (R25)

X11. February 5, 1663. Along the St. Lawrence River, Canada. Many varieties of luminous phenomena preceded the earthquake. (R26)

X12. 1672. Tokyo, Japan. Flying luminous objects seen. (R25)

X13. 1698. Tosa, Japan. Wheel-like luminous phenomena in sky. (R25)

X14. November 4, 1704. Zurich, Switzerland. Preceded by a brilliant meteor. (R27)

X15. December 31, 1730. Tokaido, Japan. Luminous bodies in sky and luminous air. (R25)

X16. October 21, 1731. England. Quake followed by vivid lightning display. (R27)

X17. April 2, 1750. England. Multitude of blood-red rays converged from all parts of the sky. (R28)

X18. November 18, 1795. England. Ball of fire passed over Derby when shock was felt. Sky flashes. Some people experienced electric shocks. (R28)

X19. December 29, 1820. Island of Zante, Greece. Fire on the surface of the sea observed 3-4 minutes before the shock. (R29)

X20. March 22, 1821. Fiume de Canera, France. Column of fire passed over the village at time of quake. (R30)

X21. December 13, 1823. Bellay, France. The sky seemed to be on fire. (R30)

X22. August 1830. Kyoto, Japan. On the night preceeding the earthquake the sky was very bright and the illumination emitted from the ground made it bright as day. (R25)

X23. November 17, 1831. Sweden. Extraordinary light on the northern horizon during shock. (R31)

X24. October 22, 1835. Pyrenees, France. A burning column of sulfurous air enveloped a cirque. (R32)

X25. February 9, 1836. Hungary. Flames issued from the ground. (R33)

X26. April 24, 1836. Italy. Great beams of fire in the sky. (R30, R33)

X27. March 22, 1841. Coblentz, Germany A blue flame rose from a hill and then subsided. (R33, R34)

X28. July 4, 1841. France. Ball lightning accompanies shock. (R34)

X29. April 2, 1851. Chile. Intense flash of light before an earthquake; no sound. (R35)

X30. December 16, 1857. Italy. Half an hour before an earthquake, a light as bright as the moon hovered over the area. Smell of sulfur. (R36)

X31. Circa 1867. Algeria. Atlas mountains enveloped in a luminous atmosphere. (R37)

X32. November 1879. West Cumberland, England. Vivid flash of light at instant of shock. (R38)

X33. April 3, 1881. Turkey. Broad flashes on light on the horizon. (R39)

X34. July 14, 1894. Fresno, California. Red cloud some 50 miles long settled over the Sierra Nevada. Electrical displays on edges of cloud reached maximum at time of quake. (R40)

X35. December 17, 1896. England. Many reports of vivid light flashes from region of earthquake. (R41)

X36. Circa 1901. French Alps. An avalanche of rocks was accompanied by brilliant sparks, forming a sheet of flame. (R42) Some earthquake lights may originate in such avalanches. (WRC)

X37. December 28, 1908. Calabria, Italy. Flash of light during quake. (R43)

X38. November 16, 1911. Germany and Switzerland. Widespread reports of diffuse illuminations, will o' the wisps, lightning, and meteors from quake area. (R8)

X39. September 1, 1923. Kwanto, Japan. Stationary fireball seen in sky. (R25)

X40. June 1932. Mexico City, Mexico. Dull red glow in sky and lightning during shocks. (R44)

X41. January 20, 1941. Cyprus. Bright flash in the direction of the epicenter. (R45)

X42. December 17, 1941. Taiwan. Widespread luminous phenomena. (R46)

X43. June 24, 1942. Wairarapa, New Zealand. Blue and green flashes seen in many places. (R47)

X44. 1961, Hollister California. Small sequential light flashes from random locations on a hillside. (R19, R21)

X45. January 5, 1968. Chiba, Japan. Fan-shaped light seen in sky. (R46)

X46. October 1, 1961. Santa Rosa, California. Flashes and streaks of colored light. (R19, R21, R48)

X47. February 9, 1971. San Fernando Valley, California. (R49, R50)

X48. September 5, 1975. Eastern Turkey. Sky brightening in the direction of the epicenter. (R51)

X49. July 28, 1976. Tangshan, China. A brilliant light seen for hundreds of miles around quake area. (R20, R52)

X50. 1977. Romania. Earthquake lights seen on western horizon, but the epicenter was in an easterly direction. Author believes this emphasized the regional character of the phenomenon. (R53)

X51. General observation. Sheets of flame reported during some earthquakes may be due to the release and spontaneous ignition of subterranean reservoirs of methane. (R54)

X52. General observation. Earthquake lights seem most frequent when the lunar tidal force is decreasing. (R55)

X53. No date given. Derby, England. Black aurora seen at time of quake. See GLA22. (R56)

X54. April 13, 1750. England. Red rays seen converging at zenith. (R33)

X55. April 15, 1752. Stavanger, Norway. An octagonal luminosity in the sky emitted fireballs from its angles. (R33)

X56. February 1692. Sicily. One evening during an episode of earthquakes, the entire village of Alari seemed to be in flames for about 6 minutes. (R57)

X57. December 30, 1820. Ionian Isles, Greece. Three or four minutes before the earthquake shock, a 4-6-foot fireball was visible over the sea for 5 or 6 minutes. (R58)

References

R1. Fuller, Myron L.; "The New Madrid Earthquake," U.S. Geological Survey

Bulletin 494, 1912. (X1)
R2. Broadhurst, Garland C.; "The New Madrid Earthquake," American Geologist, 30:76, 1902. (X1)
R3. Persinger, Michael A.; "Odd Luminosities (UFOs) and Other Fortean Events before Earthquakes: The New Madrid Test," Pursuit, 14:68, 1981. (X1)
R4. Macmurdo, Captain; "Account of the Earthquake Which Occurred in India in June 1819," Edinburgh Philosophical Journal, 4:106, 1820. (X2)
R5. Nature, 6:89, 1872. (X3)
R6. "The Earthquake and Professor Larkin," Science, 24:178, 1906. (X4)
R7. Nature, 90:550, 1913. (X5)
R8. de Ballore, Montessus; "The So-Called Luminous Phenomena of Earthquakes, and the Present State of the Problem," Seismological Society of America, Bulletin, 3:187, 1913. (X5, X38)
R9. "Curious Lightning in the Andes," Scientific American, 106:464, 1912. (X5)
R10. "Luminous Phenomena Associated with Earthquakes," English Mechanic, 96:7, 1912. (X5)
R11. "A Strange Light in the Andes," Literary Digest, 45:840, 1912. (X5)
R12. Milne, John; "Earthquakes and Luminous Phenomena," Nature, 87:16, 1911. (X5)
R13. Turner, H.H., et al; "On Seismological Investigations," Report of the British Association, 1907, p. 87. (X5)
R14. "Luminous Phenomena Accompanying Earthquakes," Nature, 128:155, 1931. (X6)
R15. "Luminous Phenomena Accompanying Earthquakes," Nature, 129:27, 1932. (X6)
R16. "Celestial Signs of Earthquakes," Literary Digest, 115:27, 1933. (X6)
R17. Davison, Charles; "Luminous Phenomena of Earthquakes," Discovery, 18:278, 1937. (X6, X7)
R18. Finkelstein, David, and Powell, James; "Earthquake Lightning," Nature, 228:759, 1970. (X6)
R19. Derr, John S.; "Earthquake Lights: A Review of Observations and Present Theories," Seismological Society of America. Bulletin, 63:2177, 1973. (X6, X7, X44, X46)
R20. "Earthquake Lights," Geotimes, 22:31, 1977. (X6, X49)
R21. Derr, John S.; "Earthquake Lights," Earthquake Information Bulletin, May/June 1978. (X6, X7, X44, X46)
R22. "Luminous Phenomena Accompanying Earthquakes," Nature, 130:969, 1932. (X7)
R23. Yasui, Yutaka; Report, Kakioka Magnetic Observatory, April 4, 1968. (X8)
R24. Parnell, Arthur; "Earthquakes and Electricity," Journal of Science, 20:697, 1883. (X9)
R25. Condon, Edward U.; Scientific Study of Unidentified Flying Objects, p. 740. (X6, X10, X12, X13, X15, X22, X39)
R26. Rothovius, Andrew E.; "Aerial Phenomena in Canada during the Great Earthquake of 1663," INFO Journal, 4:29, 1974. (X11)
R27. Mallet, Robert; "Report on the Facts of Earthquake Phenomena," Report of the British Association, 1852, pp. 128, 129. (X14, X16)
R28. Milne, David; "Notices of Earthquake Shocks Felt in Great Britain...," Edinburgh New Philosophical Journal, 31:92, 1841. (X17, X18)
R29. Mercati, Count; "Account of the Earthquake Which Desolated the Island of Zante,...." Edinburgh Philosophical Journal, 6:19, 1821. (X19)
R30. Perrey, Alexis; Comptes Rendus, 17:608, 1843. (X20, X21, X26)
R31. Mallet, Robert; "On the Facts of Earthquake Phenomena," Report of the British Association, 1853, p. 143. (X23)
R32. Philippe, M.; "Sur un Phenomene Singulier Qui, au Cirque de Troumouse, a Accompagne le Tremblement de Terre du Octobre Dernier," Comptes Rendus, 1:469, 1835. (X24)
R33. Mallet, Robert; "On the Facts of Earthquake Phenomena," Report of the British Association, 1854, pp. 57, 259. (X25, X26, X27, X54, X55)
R34. Milne, David; "Notices of Earthquake Shocks Felt in Great Britain....," Edinburgh New Philosophical Journal, 36:367-369, 1844. (X27, X28)
R35. Gillis, J.M.; "On the Earthquake of April 2, 1851, in Chile," American Journal of Science, 2:22:388, 1856. (X29)
R36. "Earthquake in Italy," American Journal of Science, 2:25:280, 1858. (X30)
R37. L'Annee Scientifique, 12:307, 1867. (X31)
R38. Nature, 21:19, 1879. (X32)
R39. Nature, 23:564, 1881. (X33)
R40. Holden, Edward S.; "A Catalogue of Earthquakes on the Pacific Coast," Smithsonian Miscellaneous Collections, vol. 37, 1898. (X34)
R41. "The Earthquake of December 17th. 1896," Symons's Monthly Meteorological Magazine, 31:177, 1897. (X35)
R42. Scientific American, 85:35, 1901. (X36)
R43. Nature, 79:255, 1908. (X37)
R44. "The Mexican Earthquake," Science, 75:sup 12, June 17, 1932. (X40)
R45. Aziz, Abdulazim; "Luminous Pheno-

menon Accompanying the Cyprus Earthquake, January 20, 1941," Nature, 149: 640, 1942. (X41)

R46. Yasui, Yutaka; "A Study of the Luminous Phenomena Accompanied with Earthquakes (Part II)," Memoirs of the Kakioka Magnetic Observatory, 14, no. 1, 1971. (X42, X45)

R47. "The Wairarapa (New Zealand) Earthquake of June 24, 1942," Nature, 151: 229, 1943. (X43)

R48. Yasui, Yutaka; "Seismo-Luminous Phenomena at Santa Rosa," Memoirs of the Kakioka Magnetic Observatory, 15, no. 1, 1972. (X46)

R49. "How Do Earthquakes Generate Lightning?" New Scientist, 58:328, 1973. (X47)

R50. Finkelstein, David, et al; "The Piezoelectric Theory of Earthquake Lightning," Journal of Geophysical Research, 78:992, 1973. (X47)

R51. "Phenomena Preceding Two Large Quakes," Science News, 112:408, 1977. (X48)

R52. "Tangshan Quake: Portrait of a Catastrophe," Science News, 111:388, 1977. (X49)

R53. Hedervari, Peter; "The Possible Correlation between Crustal Deformations Prior to Earthquakes and Earthquake Lights," Seismological Society of America, Bulletin, 71:371, 1981. (X50)

R54. Lewis, Richard S.; "Is the Earth a Giant Methane Store?" New Scientist, 78:277, 1978. (X51)

R55. Wagner, W.S., and Visvanathan, T.R.; EOS, 59:329, 1978. (X52)

R56. Nelson, Edward M.; "Earthquakes and Electricity," English Mechanics, 119:66, 1924. (X53)

R57. Lake, J.J.; "Earthquakes and Their Causes," English Mechanic, 21:51, 1875. (X56)

R58. Mallett, Robert; "On the Facts of Earthquake Phenomena," Report of the British Association, 1854, p. 130. (X57)

# GLD9 Volcano Lights

Description. Unusual exhibitions of lightning, expanding luminous arcs, bolts of light emanating from craters, and aurora-like beams associated with volcanic eruptions.

Background. The clouds of ash and water vapor emitted by active volcanos carry sufficient electric charge to guarantee ordinary lightning in the environs of many volcano eruptions. The phenomena reported here seem to transcend ordinary lightning.

Data Evaluation. Ordinary volcano lightning is rather common, but the more bizarre luminous phenomena described below are not seen often. Rating: 3.

Anomaly Evaluation. With so much charge-laden dust and steam accompanying volcanic eruptions, conditions exist for all manner of electrical discharges and associated luminous phenomena. Luminous structures may also originate in the powerful shock waves emitted from volcanic craters. In other words, most volcano lights are probably compatible with the extreme physical conditions generated. Rating: 3.

Possible Explanations. Electrical discharges between clouds of electrically charged dust and steam. Highly compressed shock waves.

Similar and Related Phenomena. Earthquake lights (GLD8), tornado lights (GLD10), low-level auroras (GLA4), visible sound waves (GE ).

## Examples of Volcano Lights

X1. No date given. Mount Vesuvius, Italy. "M. L. Palmeri first observed the flashes of volcanic lightning at a distance of a few hundred yards from the new crater at Torre del Greco. They always originated in large 'globes of smoke,' and were followed by explosions like pistol-discharges. Afterward from the Observatory he noticed similar flashes between the smoke and cinder masses and aqueous vapor above them, but very seldom between the 'globe of smoke' and the earth beneath it." (R1)

X2. April 7, 1906. Mount Vesuvius, Italy. During a vigorous eruption. "The frequency of the explosions varied from approximately one every three or four seconds to at least three per second. Although powerful, they were very sharp and sudden in their nature, and at the instant of each---but before it could be sensed by the eye or ear---a thin, luminous arc flashed upward and outward from the crater and disappeared in space. Then came the sound of the explosion and the projection of gas and detritus above the lip of the crater. The motion of translation of the arcs, while very rapid in comparison with that of the detritus, was not above the limits of easy observation and there could be no doubt as to the reality of the phenomenon, which was repeated some hundreds of times." (R2)

X5. 1902. Mont Pelee, Martinique. "At 11 p.m., lightning shot out from the mountain in all directions, zigzagging and flickering flashes alternating with, or being accompanied by, reddish globes, which ascended and exploded, and shot out stars and long rays. Away towards the south-west was another large focus of electric energy, which appeared to me to have a distinct relation to the volcanic electric discharges from Mont Pelee. This spot was, I reckoned, at least forty miles from the volcano, from which it bore almost west two points north. This latter electrical display was similar, but less extensive than that from Mont Pelee, and it was accompanied by curious glowing globes, which burst and shot out tongues of lightning. The most curious part of the magnificent sight, however, was that occasionally long rays of light, very like to the rays of a searchlight, shot out from the direction of Mont Pelee downwards to the secondary and distant electrical display, and on this broad ray reaching the western focus, the lightning there became more vivid, intense and extensive." (R5) In other words there seems to have been an atmospheric "connection" between the two active centers. (WRC)

*Expanding volcanic arcs over Vesuvius, 1906.*

X3. August 4, 1979. Mount Etna, Italy. "Scientists at Catania's Institute of Volcanology said the eruption was accompanied by strange bolts of bright light shooting out of the craters that appeared to be lightning---a phenomenon never seen before on Etna. The scientists said the bolts of light, followed by loud, explosive roars, were apparently caused by electrical activity coinciding with the eruption, an occurrence they described as unprecedented. (R3)

X4. 1902. Mount Misery, Leeward Islands. "A fortnight after the destruction of St. pierre, however, a loud explosion was heard by labourers working on the side of Mount Misery; flames seemed to leap out of the ground, and a strong wind swept by, overturning two small houses. At the same time a heavy thunderstorm occurred, with vivid lightning flashes." (R4)

## References

R1. "Electrical Phenomena of Vesuvius," Eclectic Magazine, 56:570, 1862. (X1)

R2. Perret, Frank A.; "The Flashing Arcs: A Volcanic Phenomenon," American Journal of Science, 4:34:329, 1912. (X2)

R3. "Mount Etna Explodes in Volcanic Fury," Washington Star, August 5, 1979, p. A-1. (X3)

R4. Nature, 66:378, 1902. (X4)
R5. Nicholls, H.A. Alford; "Notes on the Recent Eruptions of Mont Pelee," Nature, 66:638, 1902. (X5)

# GLD10 Tornado Lights

Description. Glowing columns inside tornado funnels, blinking luminous patches in tornado clouds, searchlight beams, ground-level flashes of light, and general surface luminosity in the region near funnels. Waterspouts, too, may exhibit similar luminous phenomena but usually to a much lesser degree. Tornadoes and waterspouts are frequently accompanied by ordinary lightning, ball lightning (GLB1), and vivid, almost continuous flickers of electrical discharge; these are not considered unusual enough to include here.

Background. Like earthquakes, tornadoes and waterspouts were widely thought to be manifestations of electricity two centuries ago. Today, the role of electricity in generating tornadoes is generally denied. Even the existence of tornado lights, which seem rather reasonable phenomena in the vicinity of powerful tornadoes, are not recognized as legitimate phenomena by many scientists.

Data Evaluation. Several modern reports by experienced observers plus some provocative photographs lend credence to tornado lights. Rating: 1.

Anomaly Evaluation. Although ordinary lightning is normally associated with tornadoes, the glowing columns and other luminous structures---all suggestive of strong electrical discharges---are not part of prevailing tornado theory. Even less likely, according to current thinking, is the possibility that tornadoes are actually caused by electricity. Rating: 2.

Possible Explanation. The intense electrical fields generated in the neighborhood of the tornado funnel are strong enough to cause glow discharge.

Similar and Related Phenomena. Earthquake lights (GLD8), volcano lights (GLD9), the burning and dehydration effects of tornadoes (GWT), and low-level auroras (GLA4).

## Examples of Tornado Lights

X1. December 12, 1846. St. Just, England. During the passage of four powerful whirlwinds. "At the distance of about a mile SW. of the mine, in the direction of Cape Cornwall, there suddenly rose from the sea, or the land near to it, to a vast height, a pillar of fire, exceedingly vivid, and apparently of the thickness of a man's arm. On reaching its highest elevation, it spread itself from the top in all directions, with splendid coruscations, followed by a terrific peal of thunder. This form of the luminous appearance renders it probable that the electric fluid passed at that moment from the earth into the clouds, along the axis of the fourth or most western whirlwind, which must at the time have been near the Cape." (R1)

X2. June 22, 1928. Greensburg, Kansas. "Steadily the tornado came on, the end gradually rising above the ground. I could have stood there only a few seconds but so impressed was I with what was going on that it seemed a long time. At last the great shaggy end of the funnel hung directly overhead. Everything was as still as death. There was a strong gassy odor and it seemed that I could not breathe. There was a screaming hissing sound coming directly from the end of the funnel. I looked up and to my astonishment I saw right up into the heart of the tornado. There was a circular opening in the center of the funnel, about 50 or 100 feet in diameter, and extending straight upward for a distance of at least one half mile, as best I could judge under the circumstances. The walls of this opening were of rotating clouds and the whole was made brilliantly visible by constant flashes of lightning which zigzagged from side to side." (R2)

X3. Summer 1942. Champaign, Illinois. "I was looking....up at the clouds when I saw something that looked like a searchlight beam extend out of the cloud and reach to the skyline. It seemed a bit brighter than the cloud background. Edges were sharp, overall intensity even, sides parallel. Width about a degree of arc. No movement or turbulence evident..... abruptly the ray was instantly replaced by a normal tornado funnel. No transition stage was noted. The funnel did not descend from the cloud layer. It appeared over all, in situ." (R3)

X4. May 25, 1955. Blackwell, Oklahoma. "Eyewitnesses to the tornado in Blackwell also have reported evidence of intense electrical activity. Montgomery, who viewed the tornado from a distance of about 3000 ft, reports: 'As the storm was directly east of me, the fire up near the top of the funnel looked like a child's Fourth of July pin wheel.' 'There were rapidly rotating clouds passing in front of the top of the funnel. These clouds were illuminated only by the luminous band of light. The light would grow dim when these clouds were in front, and then it would grow bright again as I could see between the clouds. As near as I can explain, I would say that the light was the same as an electric arc welder but very much brighter. The light was so intense that I had to look away when there were no clouds in front of it. The light and the clouds seemed to be turning to the right like a beacon in a lighthouse. Montgomery also tells of other eyewitnesses who had

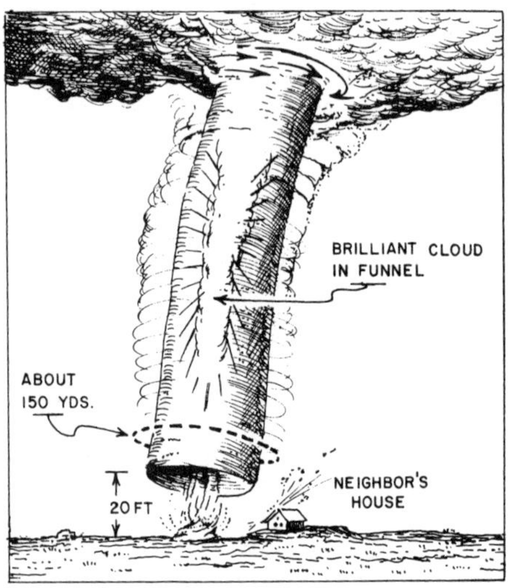


*Glow inside tornado funnel at Blackwell in 1955*

different views of the phenomenon, such as Mrs. Carl Sjoberg whose house in the direct path was completely demolished: 'She saw lightning coming up from the ground two or three feet high and about half as wide as adding machine tape. It was a deep blue and forked on the end like a Y or like a snake's tongue.' According to Montgomery, Lee Hunter, who was 4 mi north of Blackwell, described the tornado as follows: 'The funnel from the cloud to the ground was lit up. It was a steady, deep blue light ---very bright. It had an orange color fire in the center from the cloud to the ground. As it came along my field, it took a swath about 100 yards wide. As it swung from

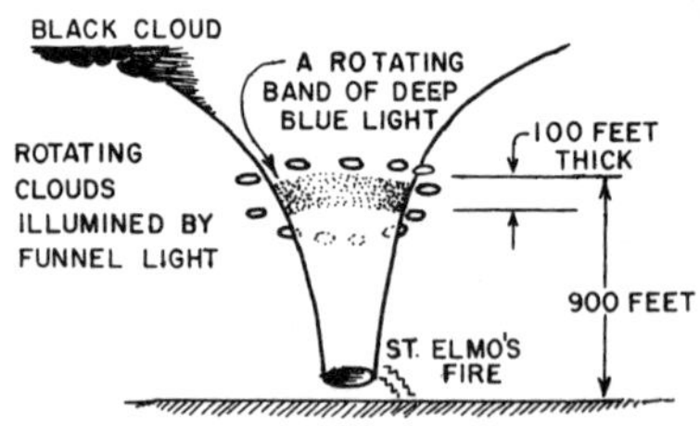


*Other luminous phenomena observed during the 1955 Blackwell tornado*

left to right, it looked like a giant neon tube in the air, or a flagman at a railroad crossing. As it swung along the ground level, the orange fire or electricity would gush out from the bottom of the funnel, and the updraft would take it up in the air causing a terrific light---and it was gone! As it swung to the other side, the orange fire would flare up and do the same.'" (R4-R5)

X5. May 31, 1956. No place given. Double tornado pulse generator in a thundercloud. "... there was no apparent electrical activity whatsoever except for an odd flashing patch of illumination in the top section of the thundercloud, as shown in figure 1, a flashing that appeared as a circular patch of light on the side of the cloud structure between the observer and the center of action where the electrical activity appeared to be. At that time these patches of light appeared as circular patches of pale blue illumination for an estimated time of two seconds; this circular area would then become dark for another estimated period of two seconds, and again would reappear. The on-and-off flashing continued for the entire time of visual observation. It was definitely obvious that there was absolutely no cloud-to-ground or cloud-to-cloud lightning visible or even any lighting of the sky in that particular region." (R6)

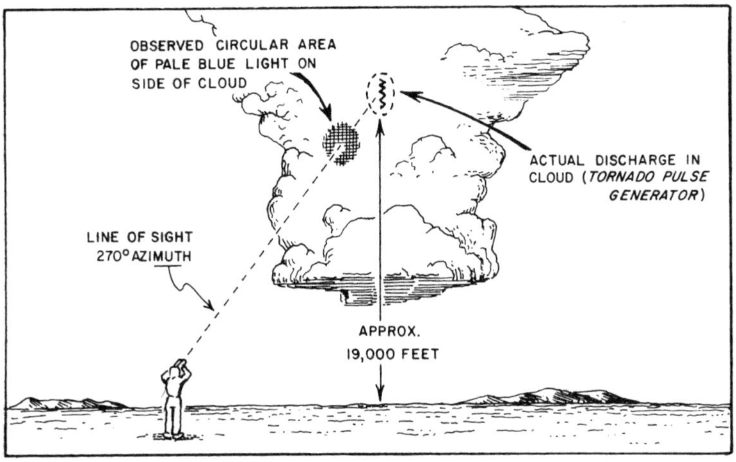


*Tornado pulse generator in tornado-spawning clouds near Blackwell, 1955*

X6. June 17, 1959. Miami, Florida. The "lower extremity was continuously illuminated with a blue-green light flashing like an electric welding torch. A part of this was no doubt produced by the tornado disrupting...power lines...However, the tornado was associated with a thunderstorm which had an extreme amount of electrical activity with almost continuous cloud to cloud and cloud to ground lightning strokes." (R3)

X7. April 11, 1965. Toledo, Ohio. "The beautiful electric blue light that was around the tornado was something to see, and balls of orange and lightning came from the cone point of the tornado." "The most interesting thing I remember is a surface glow---some three or four feet deep---rolling noise etc.... there was sort of a general brightness for a few seconds..." "We thought we saw searchlights all around us, but there were no light beams shooting up to the clouds from the ground. The lights darting around in the clouds were sort of luminous and appeared to be more round in shape than anything else, also they were quite large. The lights were not as bright as a stroke of lightning, but they were above a dense layer of clouds, and bluish white in color...they were shooting around." Photographs accompanying this article show vertical beams of light associated with the funnel. (R3, R7)

X8. April 3, 1974. Huntsville, Alabama. "What we first saw was a green spike of light...The spike would move along and there would be a burst of light and it looked like some bombing taking place...The green light looked very much like the kind of light you would get from a laser made to lase in the green, very fluorescent in nature." "...I noticed...a brilliant blue flash like an electric arc, off on the horizon...many miles away. It appeared to be a gigantic electric arc, and I thought at the time it was a power transmission line shorting out, but I've never found where the line was, if that's the case. That was followed about 5 to 10 seconds later by another gigantic flash of bright blue light in essentially the same location. There were just two flashes, nothing in between. At that time and after that, the entire sky in this area was just constant blue lightning flashes." "It appeared.... that in the lower impact areas there was a light blue discharge present---roughly that associated with the ionization of molecular oxygen. In the high impact area.... we saw a slightly green glow, which would be very typical of the ozone discharges I have seen in plasma physics. These were very faint glows from my point of observation. They were not a strong or intense glow. After the tornado had passed through.... there remained, for a few minutes, a purplish glow which would be very typical of ionized air.... This glow was more intense up near the clouds and tended to fade out as it worked toward the surface." (R8)

X9. General observations, no times or places given. Speaking of a tornado funnel. "...suddenly it turned white outside. This whiteness definitely was not fog. I would say it appeared to be giving off a light of its own." "The beautiful electric blue light that was around the tornado was something to see, and the balls of orange and lightning came from the cone point of the tornado." "The most interesting thing I remember is a surface glow---some three or four feet deep---rolling noise, etc." (R9) Obviously some of these observations are derived from the preceeding examples. (WRC)

X10. 1885. Greene County, Alabama. Tornado preceded by a phosphorescent glow resembling a sunrise. (R10)
X11. October 17, 1909. Mediterranean Sea. A streak of light was seen along the axis of a waterspout accompanied by vivid lightning. (R11)
X12. March 23, 1913. Omaha, Nebraska. The lower end of the tornado was a lurid, fiery mass. (R12)
X13. August 21, 1933. the Orkneys, England. Strange bright object seen in the sky after the passage of a tornado. (R13)
X14. February 10, 1959. St. Louis, Missouri. Blue flash of light, a big ball of fire in the sky, strong glow of light. (R14)
X15. March 26, 1968. Monroe, Michigan. A beautiful blue light associated with the tornado. (R15)
X16. May 15, 1968. Tuckerman, Arkansas. Pale-green, neon-like light inside the funnel. (R15)

References

R1. "An Account of Four Whirlwinds Which Passed through St. Just on the 12th of December 1846," Edinburgh New Philosophical Journal, 45:111, 1848. (X1)
R2. Justice, Alonzo A.; "Seeing the Inside of a Tornado," Monthly Weather Review, 58:205, 1930. (X2)
R3. Vonnegut, B., and Weyer, James R.; "Luminous Phenomena in Nocturnal Tornadoes," Science, 153:1213, 1966. (X3, X6, X7)
R4. Montgomery, Floyd C.; "Tornadoes at Blackwell, Okla., May 25, 1955," Monthly Weather Review, 83:109, 1955. (X4)
R5. Vonnegut, B., and Moore, C.B.; "Electrical Activity Associated with the Blackwell-Udall Tornado," Journal of Meteorology, 14:284, 1957. (X4)
R6. Jones, Herbert L.; "The Tornado Pulse Generator," Weatherwise, 18:78, 1965. (X4, X5)
R7. Shepard, Roger N., et al; "Tornadoes, Puzzling Phenomena and Photographs," Science, 155:27, 1967, and 155:1037, 1967. (X8)
R8. Vaughan, Otha H., Jr., and Vonnegut, Bernard; "Luminous Electrical Phenomena Associaed with Noctural Tornadoes in Huntsville, Ala., 3 April 1974," American Meteorological Society, Bulletin, 57:1220, 1976. (X8)
R9. Condon, Edward U.; Scientific Study of Unidentified Flying Objects, 1969, p. 728, (X9)
R10. "Middle Alabama Tornado," American Meteorological Journal, 2:44, 1885. (X10)
R11. Gandy, G.K.; "Notes on a Water Spout," Symons's Meteorological Magazine, 45:10, 1910. (X11)
R12. S., S.; "A Typical Tornado," American Meteorological Society, Bulletin, 12:131, 1931. (X12)
R13. "Waterspout in Kirkwall Bay," Meteorological Magazine, 68:210, 1933. (X13)
R14. Vonnegut, Bernard; "Luminosity Accompanying St. Louis Tornado---February 10, 1959," Monthly Weather Review, 87:64, 1959. (X14)
R15. Ryan, R.T., and Vonnegut, B.; "Eyewitness Tornado Observations Obtained with Telephone and Tape Recorder," Weatherwise, 23:126, 1970. (X15, X16)

# GLD11 Whirlwinds of Fire and Smoke

Description. Revolving columns or wheels of fire and smoke that sweep destructively along the surface of the ground or water. Much smaller than tornadoes and considerably larger than ball lightning, these fiery pillars often leave behind sulfurous odors and burnt vegetation.

Background. This phenomenon is difficult-to-classify. It resembles ball lightning in some respects, the tornado in others, and even some low-level meteors. The characteristic explosive onset and sulfurous odor suggest that it be placed here along with electrical discharge phenomena.

Data Evaluation. The observations of whirlwinds of fire and smoke are generally old and the product of untrained observers; but they are rather consistent. Rating: 3.

Anomaly Evaluation. The flames, smoke, odor, and burning effects make this a most puzzling phenomenon. If electricity is the basis of these strange whirlwinds, current theories are deficient. Rating: 2.

Possible Explanations. Some whirlwinds are so powerful that they can account for the physical destruction noted in the examples below. The fire and smoke might result if the whirlwinds picked up materials already ignited. Electrical heating is another possible mechanism.

Similar and Related Phenomena. Ball lightning (GLB), tornado lights (GLD10), the burning and dehydration effects of tornadoes (GWT), the explosive onsets of whirlwinds (GWW). Low-level meteors (GLM1).

Examples of Whirlwinds of Fire

X1. August 26, 1823. Boncourt, France. "A waterspout was seen not far from the village of Boncourt, having its broad base resting on the ground, and its summit lost in the clouds. It consisted of a thick and blackish vapour, in the middle of which were often seen flames in several directions. Advancing along with the storm, it broke or tore up by the roots, in the space of a league, seven or eight hundred trees of different sizes, and at last burst with great violence in the village of Marchepoy, one half of the houses of which were instantly destroyed." (R1)

X2. August 26, 1826. Carcassone, France. "Fiery whirl.... great heat in the morning. At noon massing clouds in the west. Then a thick black cloud suspended over a field. Crackling and hissing and air rushing toward the cloud. A loud detonation and an enormous column of fire hovered over the field---throwing down walls and picking up and killing sheep. A strong sulfurous odor. Nobody said it burned anything." (R2)

X3. Circa 1868. Lancashire, England. "I myself, about fifteen years ago, whilst driving one evening in the country district a few miles north of Bolton, in Lancashire, saw three pillars of fire rise out of the earth within a few yards from my phaeton. The first pillar rose on my left-hand side, then, after I had proceeded about a quarter of a mile, a second pillar rose on the right-hand side; and the third one appeared at the junction of two roads a few hundred yards further on. These fiery pillars seemed to be, as far as I can recollect, about twelve feet high, and in circumference about the same as the funnel of a large steamer." (R3)

X4. 1869. Ashland, Tennessee. "...a remarkably hot day.... a sort of whirlwind came along over the neighboring woods, taking up small branches and leaves of trees and burning them in a sort of flaming cylinder that travelled at the rate of about five miles an hour, developing size as it travelled. It passed directly over the spot where a team of horses were feeding and singed their manes and tails up to the the roots; it then swept towards the house, taking a stack of hay in its course. It seemed to increase in heat as it went, and by the time it reached the house it immediately fired the shingles from end to end of the building, so that in ten minutes the whole dwelling was wrapped in flames. The tall column of travelling caloric then continued its course over a wheat field that had been recently cradled, setting fire to all the stacks that happened to be in its course. Passing from the field, its path lay over a stretch of woods which reached the river. The green leaves on the trees were crisped to a cinder for a breadth of 20 yards, in a straight line to the Cumberland. When the 'pillar of fire' reached the water, it suddenly changed its route down the river, raising a column of steam which went up to the clouds for about half-a-mile, when it finally died out. Not less than 200 people witnessed this strangest of phenomena, and all of them tell substantially the same story about it." (R4)

X5. April 1869. Carlisle, England. "On Saturday morning, between four and five o'clock, a ball---or more properly speaking, a pillar of fire---passed from east to west over the City of Carlisle, and was plainly visible for fully a score of miles around. It resembled an ordinary gate-post in size and shape, and seemed as though it were prevented from falling by some invisible connecting cord. It travelled in a westerly direction, and was plainly visible at Cummerdale, and at Glasson, 10 miles distant from Carlisle. At Glasson, a respectable yeoman watched the pillar intently, and he says that it caused great heat, while another person in the same locality says that the fiery substance seemed to pass within a score of yards of him, and that the heat was almost overpowering. It exploded in the air, and immediately after the report, resembling the sound of the discharge of cannon, a singular brightness lit up the heavens." (R5, R6)

X6. November 30, 1872. Banbury, England. "About 12 o'clock we had a heavy storm of rain and hail, in the middle of which there was a very vivid flash of lightning, with almost instantaneous thunder of a very peculiar rattling sound. About five minutes afterwards, as I was leaving the house, my gardner called me to come quickly and see the ball of fire. I was unfortunately half a minute too late, but I have seen four persons who saw it from different points, and who all agree they heard a whizzing, roaring sound like a passing train, which attracted their attention, and then saw a huge revolving ball of fire travelling from six to ten feet off the ground. The smoke was whizzing round and rising high in the air, and a blast of wind accompanied it, carrying a cloud of branches along and destroying everything in its way....Where it first be gan the breadth of ground travelled over was very narrow, but increased as it proceeded, till in the last field the debris covered a space quite 150 yards wide, and here it seems to have exhausted itself, as all witnesses agree that the ball of fire seemed to vanish at this spot without any explosion. Here the ground had been cut in places as if by a cannon ball, but I could find no cause for this, and I saw no signs of fire on its route."....."William Marshall, gardner at Newbottle Manor, was returning from the stables to the house. He heard a noise like a long railway train crossing a bridge, and saw leaves and branches whirled into the air above the Spinney, and immediately afterwards 'a dark ball, as big as a carriage, and sending up 'a cloud of smoke,' come out of the trees with a shower of branches, and roll 'over and over', down the hill in the direction of the bridle road; the cloud of smoke at the same time whirling 'round and round' with a 'buzzing noise.' He distinctly saw sparks of a red colour emitted from the ball about six feet from the ground, and this is confirmed by another man, William Jilson, of Astrop, who, from a field on the west, saw fire and ran away affrighted." (R7-R9, R13)

X7. July 18, 1881. Americus, Georgia. "...at some distance from the town a small whirlwind, about 5 feet in diameter and sometimes 100 feet high, formed over a cornfield where it tore up the stalks by the roots and carried them with sand and other loose materials high into the air. The body of the whirling mass was of vapourous formation and perfectly black, the centre apparently illuminated by fire and emitting a strange 'sulphurous vapor' that could be distinguished a distance of about 300 yards, burning and sickening all who approached close enough to breath it. Occasionally the cloud would divide into three minor ones, and as they came together again there would be a loud crash, accompanied by crackling sounds, when the whole mass would shoot upwards into the heavens." (R10)

X8. June 1928. Travancore, India. "The account describes the passage of 'a cylindrical column of water 20 to 30 feet high' across a paddy-field, the column 'emitting fire and making a dreadful noise,' and the ground is said to have been torn up so as to leave a deep well-like hole." (R11)

X9. February 22, 1934. La Rinconada, Spain. "Suddenly I heard a strange noise. My son thought it was an airplane flying low and went out of the hut. Almost immediately I heard him say 'Papa, look!' I went out of the hut, and found myself within a black cloud. I was overcome and fell to the ground. My cap was violently torn from me and lost. The dust cloud made a black column which extended, strange to say, in a direction against the strong wind which was then blowing. It seemed to rapidly take fire. Of the hut only the posts remained. The furnishings were totally destroyed by the fire. The cup of a charcoal brazier, recently bought, was melted into several pieces and mixed with other material. The beam of a plow fifty meters from the hut was scorched, as were also some olive trees around the hut." (R12)

## References


R1. "Remarkable Water-Spout in France, in 1823," American Journal of Science, 1:10:183, 1826. (X1)

R2. Fortean Society Magazine, 1:12, Spring 1944. (X2)

R3. M., G.; "A Curious Phenomenon," Knowledge, 4:293, 1883. (X3)

R4. "A Fiery Wind," Symons's Monthly Meteorological Magazine, 4:123, 1869. (X4)

R5. "A Pillar of Fire," Symons's Monthly Meteorological Magazine, 4:37, 1869. (X5)

R6. "Singular Atmospheric Phenomenon," Eclectic Magazine, 9:765, 1869. (X5)

R7. Nature, 7:112, 1872. (X6)

R8. Cartwright, T. L. M.; "An Extraordinary Phenomenon," English Mechanic, 16:309, 1872. (X6)

R9. Beesley, T.; "The Newbottle Whirlwind of Nov. 30th, 1872," Symons's Monthly Meteorological Magazine, 8: 149, 1873. (X6)

R10. Monthly Weather Review, 9:19, 1881. (X7)

R11. Nature, 122:214, 1928. (X8)

R12. Wylie, C.C.; "Meteorite or Whirlwind in Spain," Popular Astronomy, 42: 277, 1934. (X9)

R13. Preece, W. H.; "Lightning and Lightning Conductors," Society of Telegraph Engineers, Journal, 1:362, 1873. (X6)

# GLD12 Anomalous Flashes Detected by Satellite

Description. Satellite-observed light flashes of very short duration distributed randomly over the earth's surface. See details in X1.

Data Evaluation. The data are from satellite photometers. Although the experimenters believe they have eliminated instrumental errors, that possibility always exists. Rating: 3.

Anomaly Evaluation. No known natural process generates these flashes. Rating: 2.

Possible Explanations. None, assuming instrument errors are eliminated.

Similar and Related Phenomena. Many short-duration light flashes illuminating much of the visible hemisphere have been observed from the ground (GLA14).

## Examples of Satellite-Detected Flashes

X1. General observation. From the OSO-2 and OSO-5 satellite observations. "On some occasions apparent 'lightning strokes' were detected that could not be assigned as such because at the time the fields of view of the telescopes were well above the earth. Some of these were detected by as many as three photometers simultaneously when the photometers were looking above the satellite.... The distribution of these signals over the earth's surface is approximately random, suggesting that the source is not lightning. Single occurrences are seen about once every ten orbits. No acceptable mechanism has yet been found to explain these anomalous spikes. These increases, which seem to be due to short duration (less than 0.1 sec) 'light' flashes, arise from processes other than lightning and must be eliminated from the data before any discussion of lightning distribution can be entered into." (R1

## References

R1. Sparrow, J.G., and Ney, E.P.; "Lightning Observations by Satellite," Nature, 232:540, 1971. (X1)

# GLD13 Enhanced Luminosity of Rocks and Other Solids

Description. The enhanced luminosity of rocks (especially chalk) and other solids during earthquakes, electrical storms, and other natural stimuli. The effect is reputed to be the strongest over buried ore bodies.

Background. The few simple experiments conducted on this phenomenon aimed at finding the cause of earthquake lights. This phenomenon is also probably closely related to some nocturnal lights and the so-called "mineral lights." In general, the intriguing possibility of enhanced phosphorescence has been greatly neglected, and we have little to go on.

Data Evaluation. The observational and experimental foundation for this phenomenon is rather weak. Rating: 3.

Anomaly Evaluation. The enhanced luminosity of solids due to natural causes, assuming the effect exists, probably has simple electrochemical explanations. Rating: 3.

Possible Explanation. Earth currents, often strong near ore bodies, and chemical reactions may stimulate luminosity during quakes, sharp atmospheric pressure changes, and other natural phenomena. A biological origin is also possible, perhaps enhanced by earth currents.

Similar and Related Phenomena. Earthquake lights (GLD8), nocturnal lights (GLN), especially the Brown Mountain Lights, and marine phosphorescent displays (GLW).

Examples of Luminous Rocks and Other Solid Materials

X1. General observations. "The fact that a luminous emanation of variable shape will appear in the dark at such points on the surface of the earth below which there are extensive ore deposits at more or less considerable depth, was recorded in Germany as far back as 1747. Immediately before or during a thunderstorm these phenomena are said to be especially striking. Similar observations have more recently been made in North America in the neighborhood of ore deposits. Though much should be ascribed to superstition and to errors of observation, the fact nevertheless has been confirmed by recent investigation. The electric emanation given off from the surface of the earth (see Prometheus No. 891) has in fact been repeatedly ascertained photographically by Mr. K. Zenger. Plates coated with fluorescent substances were used. It may thus be taken for granted that the emanations in question occur with an especially high intensity at those points of the ground where good conductors of electricity are found in large amounts in the neighborhood of the surface of the earth, in other words, above ore deposits, which are very good conductors of the electric current." (R1, R2)

X2. General observations. Isle of Wight. Photographic experiments on rock luminosity. "The object was to determine whether there was or not at the time of a large earthquake a practically instantaneous transmission of energy to distant regions other than that recorded by seismographs. It was observed and still is observed, by many persons that the face of a very large chalk pit at Shide exhibits, after dull damp days, a flaring luminosity. In a chamber at the end of a tunnel in this pit, a cylinder carrying photographic paper was installed. This cylinder was enclosed in a box, one end of which was a metal plate containing three holes. The plate touched a flat chalk surface. The cylinder took one week to turn; therefore parts of the paper before the holes were very slowly exposed to a chalk surface about 3/16th of an inch distant. On certain weeks the results were nil. Other weeks, after the development of the paper, there were three dark bands corresponding to the position of the holes, suggesting that the chalk had acted like an extremely feeble light....The conclusion was that the photographic effects were in no way connected with radio-activity, but they were probably electrical." (R3, R4)

X3. July 7, 1875. Switzerland. During a violent electrical storm. "Electrical phosphorescence was remarkably intense before and during the hail. The ground, animals, prominent objects, as well as the hailstones, were strongly phosphorescent. Immediately after the hail, ozone was greatly developed, the smell being so pronounced as to be compared by nearly all observers to garlic. The incessant electrical discharges passed from cloud to cloud over a central point from which the hail fell, but thunder was very rarely heard." (R5)

X4. No date or place given. "A peculiar light effect was observed by a party of German ski runners who had passed the day in fog and lightly falling snow at an elevation of 4,750 feet above sea level, and who had begun to experience slight symptoms of snow blindness. On entering a hut at dusk and opening the window they saw narrow, bright blue bands surrounding the lumps of snow that adhered to their gloves and shoes, and marking the boundaries between light and shade on all parts of their persons and clothing. The effect soon vanished with the fading twilight and not a trace of it was obser- on the following day." (R6)

X5. General observation. Phosphorus becomes luminous during and on the approach of storms. The incidence of some diseases increases markedly during the same period. (R7)

References

R1. Scientific American, 96:90, 1907. (X1)
R2. English Mechanic, 85:10, 1907. (X1)
R3. Milne, John, "Earthquakes and Luminous Phenomena," Nature, 87:16, 1911. (X2)
R4. Turner, H.H., et al; "On Seismological Investigations," Report of the British Association, 1907, p. 87. (X2)
R5. Nature, 12:447, 1875. (X3)
R6. "A Singular Effect," Scientific American, 104:103, 1911. (X4)
R7. "Phosphorescence in Connection with Storms and Disease," English Mechanic, 2:21, 1865. (X5)

# GLD14 Luminous Phenomena in Water and Ice

Description. Phosphorescence, sparks, and flashes of light in waterfalls, about reefs and rocky shores, and in breaking or deforming ice. This type of light production is separate and distinct from marine bioluminescence (GLW).

Background. There may actually be several different phenomena lumped together here. Only further study can separate them.

Data Evaluation. Generally poor. No systematic observation of the phenomenon. Rating: 3.

Anomaly Evaluation. Since several reasonable mechanisms of light generation have been proposed to explain this phenomenon, it has low anomaly content. No one, however, has gone much beyond merely suggesting ideas. Rating: 3.

Possible Explanations. (1) Charge separation followed by luminous electrical discharges; (2) Triboluminescence (as in crushing sugar cubes); (3) Sonoluminescence, in which trapped gases become incandescent due to sudden, violent pressure changes; (4) The release and ignition of contained gases (especially in glacier ice).

Similar and Related Phenomena. Marine phosphorescence (GLW), earthquake lights (GLD8).

Examples of Luminous Phenomena in Water and Ice

X1. General observation. Giant rollers breaking over obstructions are often accompanied by blinding white flashes. (R1, R2) Given a bright moon, such flashes might be produced by simple reflection by white spray. (WRC)

X2. General observation. Strange flashes of light are sometimes seen at the bases of waterfalls. (R3, R4)

X3. General observation. Phosphorescence, sparks, and light flashes are occasionally observed in sea ice when it is broken up by ships. Cracking and deforming glacier ice will sometimes seem to give off sparks. (R5, R6)

References

R1. "The White Flash," English Mechanic, 99:14, 1914. (X1)
R2. "The White Flash," English Mechanic, 99:37, 1914. (X1)
R3. "Electricity of Waterfalls,: American Journal of Science, 3:44:423, 1892. (X2)
R4. "An Explanation for Those Flashes of Light from Waterfalls,: New Scientist, 19:399, 1963. (X2)
R5. Allen, J. Allen; "Luminous Phenomena on Rupture of Sea-Ice," Nature, 24:459, 1881. (X3)
R6. "Sparks in Glacier Ice?" New Scientist, 18:738, 1963. (X3)

# GLD15 Dazzling Lights in and on Clouds

Description. Brilliant, stationary, greenish-white lights seen on and near cloud surfaces. The light source is usually spherical and blindingly intense. Lifetime: a few minutes.

Data Evaluation. Very few observations exist. Rating: 3.

Anomaly Evaluation. Observers of this phenomenon are quite sure it is electrical in nature, perhaps a form of St. Elmo's Fire. But St. Elmo's Fire is not nearly as intense as the data indicate for this phenomenon. We must assume that these brilliant spheres are some sort of unrecognized electrical discharge. Rating: 2.

Possible Explanations. A particularly energetic form of ball lightning; some rare and undescribed form of electrical discharge from cloud to atmosphere.

Similar and Related Phenomena. Ball lightning (GLB), the tornado pulse generator (GLD10).

## Examples of Dazzling Lights on Clouds

X1. February 29, 1936. New Zealand. While observing a distant electrical storm, a faintly glowing light was seen among the clouds. The light was stationary for about 5 minutes. "At 10.15, as stated, I realised the glowing light was proceeding from a cloud. Before I had time to conjecture what that could mean, I witnessed one of the most weird and uncanny sights I have ever seen. It suddenly seemed to pulsate, it took definite shape as a molten ball of soft light, but although in itself not dazzling to the eyes, threw off an indescribably bright, greenish white light, or rather radiance. This radiance lit up the whole of the upper surface of the cloud bank and showed the ball of light balanced on a finger of cloud. On either side of this finger were ugly looking black peaks, and all these

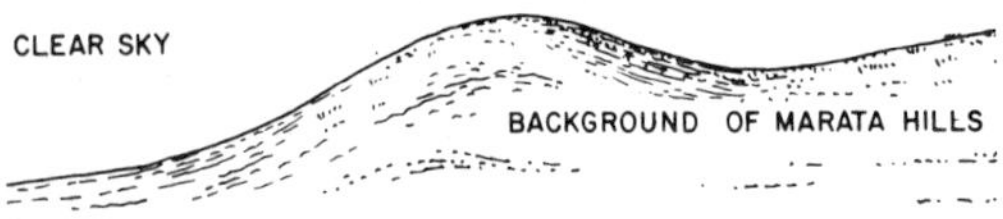


*Brilliant sphere of light atop a cloud deck over New Zealand*

were silhouetted against the radiance. From being a small ball of light, it instantly became larger, until it was the size of a half crown seen from a distance of 18 ft. The radiance became brighter and lit up all the landscape and countryside. The ball itself although molten in appearance, did not hurt or dazzle the eyes. It held this size for about 15 seconds (I was far too interested in watching it to take the time). Then it pulsated again (seeming to slightly contract and expand once or twice) and almost immediately became much enlarged.... This time the radiance was terrific. Thousands of searchlights would not have equalled the intensity. But still the focal point or ball was not blinding." In about 15 seconds, the ball contracted and later disappeared. (R1)

X2. September 5, 1936. Leiden, Holland. Cloudy sky. "In the west-north-west at a height of about 30° an intense light was seen. At first I thought it was a mock sun (there was actually a slight part of a halo visible above the sun), but it was too high for it and the light was of one colour and of great intensity. As far as I can state it, the colour was greenish white. The light did not pulsate, but I did not see its development.... There remained no question about its being an electrical phenomenon. The light was so intense that it was unpleasant for the eye." (R2)

## References

R1. Laurenson, M.D.; "Phenomenon Witnessed in New Zealand on February 29th, 1936," Meteorological Magazine, 71:134, 1936. (X1)

R2. van der Heyden, K.; "Again: St. Elmo's Fire on a Cloud," Meteorological Magazine, 71:231, 1936. (X2)

# GLD16 Luminous Patches Moving on Cloud Surfaces

Description. Patches of light on the bottoms of clouds that move as if they were caused by searchlight beams playing on the clouds. The phenomenon is silent.

Data Evaluation. Only a single observation of good quality. Rating: 3.

Anomaly Evaluation. If the moving patches of light are simply manifestations of electrical discharges within the clouds, the major problem seems to be in accounting for the curious motion of the patches. In any event, this phenomenon presents only minor problems to the meteorologist. Rating: 3.

Possible Explanation. Intra- and inter-cloud electrical discharges, or possibly cloud-level ball lightning in motion.

Similar and Related Phenomena. Heat lightning is also silent and confined to the clouds, but it is characterized by rapid flickering on a large scale. The moving surface patches of luminosity described in GLD4 may be related.

Examples of Luminous Patches Moving on Clouds

X1. April 12, 1939. Wellington, New Zealand. "...the writer and two friends came out of a house on Kelburn Parade, Wellington, New Zealand, altitude about 130 metres, and were immediately arrested by what appeared to be a searchlight laying on the undersurface of broken clouds lying over the Wellington Harbour. The clouds were very low fracto-cumulus, at about 200 metres, as judged by their level on the hills. As no searchlight beam was visible, one observer said that he thought that the phenomenon was due to 'lightning'....It was seen that the clouds were moving very slowly from the south, were about six-tenths in amount (mostly overhead and towards the east) and were illuminated almost continuously by silent discharges which passed comparatively slowly, like 'blushes' (though the light was whitish) over their surfaces. The slower discharges were estimated to take as long as half a second to pass over the surface of a cloud. Others were much quicker and seemed to be brighter. Occasionally a bright but silent flash passed very rapidly from one cloud to the next. Although now and then several clouds discharged simultaneously, for the most part the 'blushes' affected first one cloud and then another." (R1)

References

R1. Palmer, C.E.; "An Unusual Electrical Discharge at Wellington," Meteorological Magazine, 74:185, 1939. (X1)

# GLD17 Ground-Level Light Flashes

Description. Quick, bright flashes of blue-green light occurring in midair a few feet from the ground. The phenomenon is silent and narrowly localized.

Data Evaluation. A single, unpublished, but careful and thoughtful account of four separate flashes. Rating: 3.

Anomaly Evaluation. This phenomenon is more than usually significant because no electrical storms were in the area and there were so other pertinent phenomena, such as earthquakes. There seems to be no mechanism for charge separation followed by a discharge emitting light. Rating: 2.

Possible Explanation. None.

Similar and Related Phenomena. Many cases of general sky flashes have been reported (GLA). Similar flashes sometimes accompany earthquakes (GLD8) and dust and snowstorms. (GLD5)

Examples of Ground-Level Flashes

X1. October 21, 1953. Alliance, Ohio. "On the evening of October 21, 1953, my brother David and I were returning home after a twilight jaunt. The full moon had risen about two hours before and was brilliantly illuminating a clear, star-filled sky. As we walked north down the east side of our street (South Seneca Avenue), we suddenly saw an instantaneous green-blue flash of light seven or eight houses down the opposite side of the street. Although most of the flash appeared to be hidden between two houses, the top of the flash and the brightly illuminated side of one of the dwellings could be seen. Assuming the light was perhaps a television tube exploding or a flashgun popping, we continued walking. After a short interval, we approached the house owned by a Mr. and Mrs. Mather. All at once another huge silent green-blue flash occurred almost opposite us to our right between the Mather house and a tall apple tree. Lasting but a split second, the flash seemed to have its center perhaps one to two meters (3 to 6 feet) above the yard. There were no wires or other objects from which the flash could have originated. The distance separating us from this second flash was, to the best of my recollection, no more than nine or 12 meters (30 or 40 feet). Needless to say, we stood there completely astonished. Seconds later a third flash exactly like the first two appeared across the street, again almost opposite our position. It occurred in mid-air on the north side of a house. This was followed almost immediately by the fourth and final flash which took place in mid-air on the south side of the same house. Again, there was no sound. We were the only ones on the street, and no one came out to investigate." Weather: very dry and unseasonably warm. (R1)

References

R1. Webb, Walter N.; Personal communication, February 12, 1979. (X1)

# GLL LIGHTNING ANOMALIES

### *Key to Phenomena*



## GLL0 Introduction

Ordinary lightning is the rapid, concentrated discharge of electricity through the atmosphere. The adjectives "rapid" and "concentrated" must be specified to distinguish ordinary lightning from the "slow" and "diffuse" discharges of electricity that occur in aurora-like and electric discharge phenomena (GLA, GLD). Actually, some lightning phenomena cataloged below, such as thunderless lightning, may be forms of diffuse glow discharge rather than conventional lightning.

The overwhelming majority of lightning strokes occur between clouds and between clouds and the earth. A very small fraction of these discharges, however, reach upwards from the clouds towards the ionosphere and outer space. Somewhere high above the earth there seems to exist an electrical terminal for this upward-directed "rocket lightning." (Could auroras and mountain-top glows use the same unseen terminal?)

A singular and most impressive form of lightning is pinched or bead lightning. Terrestrial experiments with strong electrical currents in gases demonstrate that an electrical current can magnetically constrict itself. Quite possibly bead lightning begins as ordinary lightning and ends up strangling itself with its own magnetic field, breaking the stroke up into many brilliant beads.

Long horizontal discharges of lightning, some of which meander close to the ground, pose a serious problem in explanation, for there seems no obvious reason for lightning to digress five miles parallel to the earth's surface when the ground is only a few hundred feet straight down, or in some instances just inches away. This apparent aimlessness of some lightning strokes is counterbalanced by lightning's strange preference for oak trees and certain areas of the terrestrial surface. Do some specific regions, perhaps because of geological conditions, have a special attraction for lightning and thus entice it from afar?

Rarely, lightning will strike without warning and without a cloud in sight. In such cases, electrical charge must accumulate in the atmosphere without the usual accompaniment of water vapor in the form of visible clouds. Little is known about these invisible concentrations of electrical charge. All we know is that "bolts out of the blue" really do occur.

Thunder is usually thought to be inseparable from lightning. This is not necessarily so! The heat lightning so common on sultry summer nights is often silent, flickering eerily for hours like a defective fluorescent bulb.. This lack of sound seems to suggest that heat lightning is really a diffuse discharge of electricity rather than distant, concentrated thunderbolts.

Sounds other than thunder sometimes precede nearby lightning strikes. The rapid buildup of electrostatic forces in the lightning's target area creates brush discharge from twigs and other sharp objects, which buzzes and hums like St. Elmo's Fire, but only for an instant until the electrical stresses are relieved suddenly by a lightning flash.

Lightning shadowgraphs have been seen on surfaces exposed to intense light. The blinding light from a lightning flash may create some temporary chemical reaction on the surface that freezes silhouettes for a few minutes. Another graphic (and rather gruesome) effect of the thunderbolt is the "lightning figure", a vein-like pattern that sometimes appears on the bodies of lightning victims. In less sophisticated times, lightning figures were said to be the images of nearby trees on the skin. Today's thinking is that they are simply the branching flow of electrical currents on the body's surface. Nevertheless, the literature contains many strange stories in this area.

Finally, lightning is an unpredictable prankster; smashing a house to bits one moment and then delicately cracking eggshells but leaving the thin membrane underneath intact. It may carefully avoid a child playing on the floor and next melt all the wiring in the house. Some of these mischievous acts try the laws of science as well as one's credulity.

# GLL1 Rocket Lightning

Description. Ascending strokes of lightning, usually originating in clouds and terminating in clear sky. The bolt often shoots up in a single column "like a rocket" and bursts like fireworks into innumerable fingers. Occasionally strokes rise directly from the surface of the ground or sea. Rocket lightning strokes frequently seem thicker and slower than those of ordinary lightning.

Background. Ordinary lightning is a cloud-to-earth phenomenon, although individual lightning bolts may travel long distances horizontally.

Data Evaluation. Many excellent reports exist, mostly from ocean-going vessels. Rating: 1.

Anomaly Evaluation. Lightning is classically viewed as an electrical discharge between two diffuse terminals, earth and cloud, separated by a high potential difference. It is difficult to visualize how a volume of clear sky could act as a highly charged terminal. Rating: 2.

Possible Explanations. The charge-separating mechanism of the thunderstorm may extend well above the visible clouds. Intense showers of cosmic rays may ionize the regions above thunderclouds (GLL27).

Similar and Related Phenomena. Lightning from a clear sky (GLL6), thunderstorms correlated with space radiation (GLL27), low-level auroras (GLA4), and mountain-top glows (GLD1).

Examples of Rocket Lightning

X1. August 18, 1927. West Indies. "...on running out of a heavy shower of rain, the horizon to the S.W. having been ablaze with sheet lightning all evening, again became visible and the lightning with greater play and intensity. Out of the centre of this silent confusion of light one thin streak of lightning was seen to emerge, shooting up the sky in a perfectly straight line until apparently right overhead, and then scattering at the extremity like a rocket. Unlike the usual fork lightning, it was not blinding, and gave the impression of being decidedly sluggish in its upward path." (R1, R2)

X2. February 16, 1950. Near Fiji, South Pacific Ocean. "Three or four times a remarkable phenomenon occurred. Out from the top of the cloud shot a burst of light like a firework display. The flash took several seconds to reach its maximum height at about 3,000 ft above the cloud top; it was not just a burst of light but rather a series of streamers extending from a single point at the centre of the anvil and spreading out like a water fountain. The light was so bright and so distinct that passengers and crew were entertained by the sight for several minutes." (R3)

X3. July 30, 1951. Central American Waters. "Flashes of light were observed behind a mass of Cu just above the horizon. A small sector of the sky was brilliantly illuminated and, on an average, immediately after every second flash, lines of orange-coloured flashes shot up to an altitude of $15^{\circ}$, terminating at a layer of St. On two occasions these appeared to mushroom out at the top. The phenomenon persisted fairly regularly until 2040 A.T.S." (R4)

X4. November 19, 1951. Arabian Sea. "A brilliant greenish explosion, bearing $015^{\circ}$ (T), seemed to occur behind a large Cu, lighting it up and in particular illuminating its outer edges for about two seconds; at the same time a fiery shaft, similar to a brilliant rocket, shot up to a height of about $18^{\circ}$. This was visible for about one second and then seemed to extinguish itself in a downward direction." (R5)

*Rocket lightning over the Arabian Sea, 1951*

X5. February 16, 1964. South Atlantic Ocean. "Between 0030 and 0130 GMT very vivid lightning was observed to the westward, shooting upwards from a bank of Cb cloud, as shown in the sketch." (R6)

X6. October 13, 1967. North Atlantic Ocean. "At 0215 GMT lightning of unusual character was seen in the south-east. Instead of travelling downwards towards the sea it began at an elevation of about $35^{\circ}$ and travelled upwards to an altitude of $45^{\circ}$ to $50^{\circ}$. In some respects it resembled a firework display. No thunder was heard, nor was there any rain in the vicinity of the ship, but the radar indicated heavy rain clouds about 15 miles away and moving in a NWly direction. This unusual type of lightning was seen over a period of 3 hours." (R7)

Rocket lightning over the South Atlantic, 1964

X7. July 16, 1850. Manchester, England. Lightning travelled 6 or 7 miles, terminating in midair in a digital pattern. (R8)
X8. Circa 1876. Wing, England. Several simultaneous flashes of rocket lightning. (R9)
X9. July 5, 1884. Duncannon, Pennsylvania. Lightning darted out into midair with digital terminations. (R10)
X10. November 23, 1885. Kingston, Jamaica. Ascending darts of lightning. (R11, R12)
X11. July 22, 1903. Calcutta, India. Rocket lightning without digital terminations. (R13, R14)
X12. General observation. Rocket lightning definitely exists. It may result when a normal bolt of lightning fails to sustain itself. (R27) This explanation would not account or rocket lightning's upward direction and usual digital termination. (WRC)
X13. August 31, 1903. Lake Superior, Quebec, Canada. (R15)
X14. June 15, 1908. North Atlantic Ocean. The bolts burst like rockets. (R16)
X15. September 1917. Kent, England. Many reports of rocket lightning display, some from over 150 miles away. (R17)
X16. April 21, 1933. No place given. Purple flashes ending in clear sky. (R18)
X17. May 11, 1956. South Pacific Ocean. (R19)
X18. May 24, 1957. South Pacific Ocean. Up to 36 flashes per minute. (R20)
X19. May 10, 1961. Malayan waters. Thick upward flashes exploding in plumes. (R21)
X20. September 25, 1962. Mediterranean. Lightning rose from the horizon and burst into showers of sparks. (R22)
X21. March 27, 1971. Ionian Sea. Several simultaneous displays of rocket lightning. (R23)
X22. April 24, 1971. Indian Ocean. (R24)
X23. July 29, 1978. Sussex, England. The lightning branched upwards in tree-like fashion. (R25)
X24. February 7, 1945. Broome, Australia. Purple lightning like the roots of an inverted tree. (R26)
X25. 1933. Germany. (R28)
X26. No date given. Gulf of Tonkin. Pilot sees bolt of lightning shoot straight upward, like a beam of light. (R28)
X27. No date given. Amarillo, Texas. A pilot near a cold front with considerable

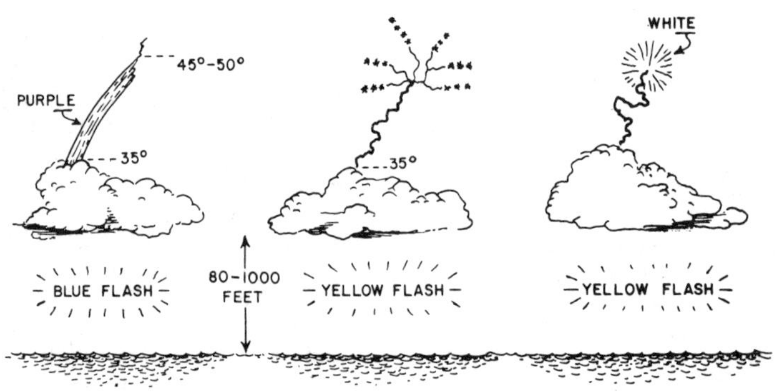


*Rocket lightning ascends from cloud over the North Atlantic. Simultaneous light flashes appeared beneath the cloud.*

lightning estimates that 1-2% of the bolts rose upwards toward the ionosphere. (R28)

References

R1. Meethan, J.T.; "Unusual Form of Lightning," Marine Observer, 5:157, 1928. (X1)
R2. "Strange Forms of Lightning, " Literary Digest, 111:22, October 10, 1931. (X1)
R3. Wright, J.B.; "A Thunderstorm in the Tropics," Weather, 5:230, 1950. (X2)
R4. Fraser, N.; "Rare Form of Lightning," Marine Observer, 22:123, 1952. (X3)
R5. Wood, H.S.; "Rare Form of Lightning," Marine Observer, 22:194, 1952. (X4)
R6. McKay, G.A.; "Lightning," Marine Observer, 35:9, 1965. (X5)
R7. Powell, G.; "Lightning,: Marine Observer, 38:173, 1968. (X6)
R8. Joule, J.P.; "On a Remarkable Appearance of Lightning," Philosophical Magazine, 3:37:127, 1850. (X7)
R9. Boys, C.V.; "Progressive Lightning," Nature, 118:749, 1926. (X8)
R10. Brown, W.; American Meteorological Journal, 1:198, 1884. (X9)
R11. MacKenzie, T.; "Meteorological Phenomenon," Nature, 33:245, 1885. (X10)
R12. MacKenzie, T.; "A Meteorological Phenomenon," Royal Meteorological Society, Quarterly Journal, 12:62, 1886. (X10)
R13. Everett, W.H.; "Rocket Lightning," Nature, 68:599, 1903. (X11)
R14. Lee, W.A.; "Rocket Lightning," Nature, 69:224, 1904. (X12)
R15. Nature, 114:513, 1924. (X13)
R16. "Lightning and Compasses," Royal Meteorological Society, Quarterly Journal, 34:269, 1908. (X14)
R17. Ashmore, S.E.; "Unusual Lightning," Weather, 5:331, 1950. (X15)
R18. Shipley, John P.; "Upward Discharge of Lightning from a Cloud," Meteorological Magazine, 75:49, 1940. (X16)
R19. Lavers, F.J.; "Abnormal Lightning," Marine Observer, 27:80, 1957. (X17)
R20. Fulcher, L.W.; "Rare Form of Lightning," Marine Observer, 28:74, 1958. (X18)
R21. Martin, H.K.; "Lightning," Marine Observer, 32:114, 1962. (X19)
R22. King, G.T.; "Unusual Lightning Flashes," Marine Observer, 33:116, 1963. (X20)
R23. Taylor, T.J.; "Lightning," Marine Observer, 42:12, 1972. (X21)
R24. Field, A.J.; "Lightning," Marine Observer, 42:52, 1972. (X22)
R25. Clark, C.A.; "Thunderstorm in Sussex, Night of 29/30 July 1978," Weather, 33:442, 1978. (X23)
R26. Wood, C.A.; "Unusual Lightning," Weather, 6:64, 1951. (X24)
R27. Humphreys, W.J.; "Various Appearances of Lightning," English Mechanic, 108:84, 1918. (X12)
R28. Sullivan, Walter; "High-Flying Pilots Report Sighting Lightning Bolts That Shoot Upward," New York Times, May 18, 1982. (X25-X27)

# GLL2 Bead Lightning

Description. Lightning strokes that appear to dissolve into many luminous segments, usually spherical but sometimes rectangular, as the stroke fades away. In some cases, the long lines of bright beads seem to appear without any precursor normal lightning stroke. In both cases, the beads may persist for 1 to 2 seconds. Also called pearl lightning, chain lightning, segmented lightning, punctuated lightning, pinched lightning, perlschnurblitz, eclair en chapelet, and other names.

Background. Like ball lightning, bead lightning was long doubted as a real physical phenomenon. Some scientists claimed it was simply a form of afterimage resulting from the intensity of the lightning flash.

Data Evaluation. Abundant visual observations and photographs exist. There are, in fact, so many accounts that no special effort has been made to enlarge the bibliography below, which must be considered as only representative. Rating: 1.

Anomaly Evaluation. While spherical segments may yield to explanations from plasma physics, the sharply defined rectangular links do not seem to have ready explanations. Rating: 2.

Possible Explanations. Spherically segmented lightning or bead lightning may result from longitudinal waves set up in an ordinary lightning stroke by some perturbation. (R16) The so-called pinch effect, well known from plasma physics, may cause some constrictions in lightning strokes. In the past, observers maintained that the string of glowing beads was formed of burning bits of matter remaining after the lightning stroke.

Similar and Related Phenomena. Ball lightning (GLB1).

Examples of Bead Lightning

X1. August 18, 1876. Paris, France. During a violent thunderstorm with a great variety of lightning strokes. "Some of the most remarkable passed from the cloud to the ground in the form of an elongated S, lasting for an appreciable instant, like a string of brilliant grains distributed along a narrow luminous string. Observed from the heights of Meudon, this lightning appeared to strike Paris in the direction of Vaugirard, where it fell. This formation of luminous beads alternating with streaks of fire---a consequence of the electric fluid passing through a ponderable medium---is altogether analogous to a string of incandescent globules presented by a long metallic wire melted by a voltaic current, the extremities of which remain suspended an instant in a state of fusion at the poles of a pile or to the swellings and knots resulting from the flowing of every liquid vein. Such agglomerations of electrised and luminous matter should naturally dissipate themselves more slowly than the bolt that develops them, which explains their persistence in this case." (R1, R8)

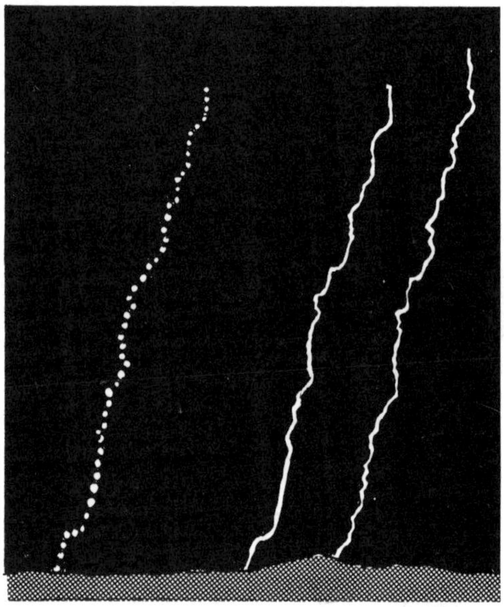

*One of two parallel strokes of lightning breaks up into beads, as shown on left*

X2. Circa 1894. Denver, Colorado. "To-day while driving over the country midst a terrific thunderstorm, I observed many fine flashes, one instance of which deserves particular mention. Two substantially parallel simultaneous flashes, somewhat wavy, seemed to pass toward the Earth. Length about $40^{o}$, distance apart about $7^{o}$, both brilliant, but not blindingly so. Both, as lines of light, vanished instantly, but the one to the right broke into a multiplicity of what seemed to be spheres. This string of beads, as it seemed, appeared to be formed of material furnished by the original flash." (R2)

X3. July 26, 1929. Balboa Channel, Central America. "At 2.15 (75th Mer. Time., approx.) a flash of lightning on the port hand appeared to enter the sea about 100 yards from the vessel. It appeared as a stream of molten metal being poured from a height into the sea and of about one to two feet wide. It first made a peculiar shattering noise (as a china plate would make on a stone floor), then, as it broke up the noise was a sizzling sound (as fat being thrown on a fire), On breaking up it appeared as a disjointed spinal column and the sea as if water was being poured into it." (R3, R4)

X4. December 5, 1935. Johannesburg, South Africa. "During the height of the storm. a particularly bright flash struck the ground about a hundred yards away; the flash appeared to be approximately a foot wide and to last for at least a second. After the flash had died away, there remained a string of bright luminous beads in the path of the flash. The beads, of which there were twenty or thirty, appeared to be about a quarter of the width of the flash, that is, say, three inches in diameter. The distance between the beads,

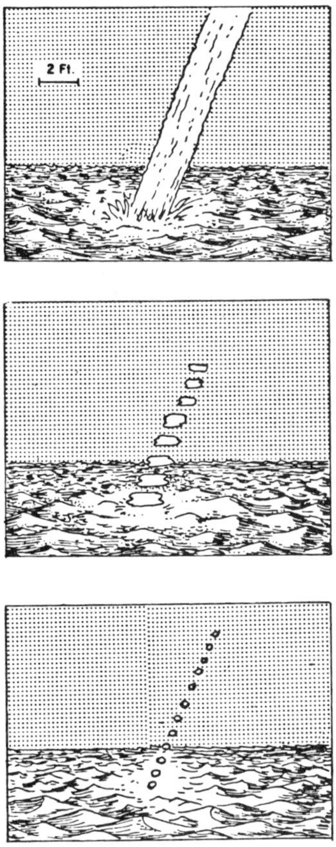


*"Stream" of lightning transforms into brilliant segments.*

which appeared to be nearly constant, seemed about two feet. They remained visible for approximately half a second; during this time they gave no indication of any movement." (R5)

X5. General observations. Berchetesgaden, Germany; Lake Lucerne, Switzerland, and Jackson, New Hampshire. "Each of these three storms was exceptionally violent, among the most violent I have ever witnessed. The phenomenon was observed only with flashes which were comparatively close, within perhaps 2,000 feet. In each storm several flashes left beaded trails, but not every flash which struck near by exhibited that peculiar appearance. I would estimate the time during which the beads remained visible as at least one second, a time amply sufficient to observe distinctly. It appeared to me that the entire course of the flash remained luminous, with a dull red glow, but that at intervals along the path bright points like sparks appeared to remain suspended in the air. The sparks appeared to be moving horizontally as though blown along by the wind." (R6)

X6. Circa 1838. Suffolk, England. Punctuated lightning. (R7, R8)

X7. June 19, 1857. London, England. Flashes persisted, melted into granular light. (R1)

X8. Circa 1871. London, England. (R9)

X9. August 24, 1873. Hampstead Heath, England. Chain lightning, links faded away gradually. (R8, R10)

X10. August 16, 1877. Southport, England. Stroke composed of small detached fragments. (R8, R11)

X11. September 5, 1880. Vegesack, Germany. (R8)

X12. July 5, 1888. Celle, Germany. (R8)

X13. October 1, 1892. Gottingen, Germany. (R8)

X14. September 1894. Schwanden, Switzerland. Small, regular pieces of light. (R8)

X15. May 27, 1907. Calexico, California. (R6)

X16. August 15, 1907. Grenoble, France. (R6)

X17. Spring 1914. Lincoln, Nebraska. (R12)

X18. October 3, 1927. Paraguay. (R13, R14)

X19. August 1961. Los Alamos, New Mexico. Photo taken of pinched lightning. (R15)

X20. July 17, 1908. Chicago, Illinois. Analysis of a photograph of bead lightning revealed the average bead length to be 36.5 feet, with an average separation of 14.6 feet. (R17)

References

R1. Lake, J.J.; "Beaded Lightning," English Mechanic, 24:234, 1876. (X1, X7)

R2. Steward, J.J.; Popular Astronomy, 2:89, 1894. (X2)

R3. Morgan, W.H.; "Thunderstorm," Marine Observer, 7:146, 1930. (X3)

R4. Barlow, E.W.; "Thunderstorms at Sea," Marine Observer, 12:49, 1935. (X3)

R5. Beadle, D.G.; "A Curious After-Effect of Lightning," Nature, 137:112, 1936. (X4)

R6. Lockyer, William J.S.; "Beaded or Pearl Lightning Flashes," Knowledge, 5:53, 1908. (X5, X15, X16)

R7. Lawrence, E.J.; "Remarkable Form of Lightning," Nature, 18:278, 1878. (X6)

R8. Lockyer, William J.S.; "Beaded or Pearl Lightning Flashes," Knowledge, 4:121 and 4:145, 1907. (X6, X9-X14)

R9. "Beaded Lightning," English Mechanic, 24:291, 1876. (X8)

R10. Smith, B. Woodd; "Remarkable Form of Lightning," Nature, 18:302, 1878. (X9)
R11. Joule, B. St. J. B.; "On a Remarkable Flash of Lightning," Nature, 18: 260, 1878. (X10)
R12. Humphreys, W. J.; "Various Appearances of Lightning," English Mechanic, 108:84, 1918. (X17)
R13. Talman, C. F.; "The Strangest Lightning," American Meteorological Society, Bulletin, 12:130, 1931. (X18)
R14. "Strange Kinds of Lightning," Literary Digest, 111:22, October 10, 1931. (X18)
R15. Matthias, B. T., and Buchsbaum, S. J.; "Pinched Lightning," Nature, 194:327, 1962. (X19)
R16. Barry, James Dale; Ball Lightning and Bead Lightning, New York, 1980.
R17. Abbot, C. G.; "Remarkable Lightning Photographs," Smithsonian Miscellaneous Collections, 92:1, no. 12, 1934. (X20)

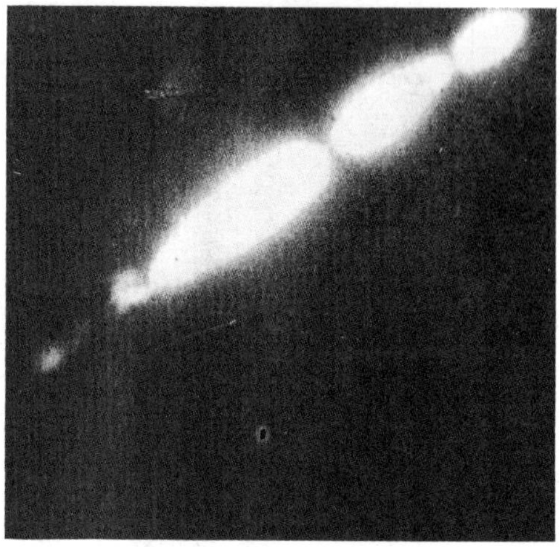

*Still camera photograph of pinched lightning (R15. Courtesy J. D. Barry)*

# GLL3 Colored Lightning with Unusual Features

Description. Yellow, red, blue, and green lightning possessing additional unusual characteristics, such as single strokes with more than one color. Red and yellow lightning are rather common, green is rare; but no colored lightning can be considered anomalous unless it possesses other puzzling features.

Data Evaluation. The unusual cases of colored lightning singled out below are one-of-a-kind. Rating: 3.

Anomaly Evaluation. The examples presented were selected for curiosity value. Rating: 3.

Possible Explanations. The color of lightning depends upon the composition of the air and the temperature of the lightning channel. Red lightning, for example, occurs when hydrogen derived from water vapor is excited electrically and emits red light. The degree of excitation along a lightning stroke may vary, creating the possibility of different colors in the same bolt of lightning.

Examples of Colored Lightning with Unusual Features

X1. August 1890. England. Red, yellow, and blue lightning in the same part of the sky, indicating local changes in atmospheric composition or lightning temperatures. One flash was yellow with a blue center. (R1)

X2. May 13, 1906. Somerset, England. Flashes of dull, "nonluminous', red lightning which was not synchronized with any of the thunder. Normal lightning occurred at the same time. (R2)

X3. October 6, 1927. Brantford, Ontario. Lightning flash began as white at the top and turned vivid green at the bottom. (R3)

X4. General observation. Red lightning seems to set very few forest fires. Since the red color is due to hydrogen from water vapor, the author concluded that red discharges occur in precipitation which would naturally suppress any fires that were started. (R4)

References

R1. Pincott, Frederic; "Red, Yellow and

Blue Lightning," Knowledge, 13:237, 1890. (X1)
R2. Boys, H.A.; "Red Lightning," Meteorological Magazine, 60:191, 1925. (X2)
R3. Stephenson, H.H.; "Green Lightning," Nature, 120:695, 1927. (X3)
R4. Humphreys, W.J.; "White Lightning Versus Red as a Fire Hazard," Monthly Weather Review, 59:481, 1931. (X4)

# GLL4 Silent Lightning

Description. Silent, rapidly flickering, diffuse lightning, usually occurring from cloud to cloud. Silent lightning, which is also called summer or sheet lightning, resembles a large area, slow discharge of electricity as opposed to the usual highly concentrated lightning stroke. Subdued rumbling sometimes accompanies this variety of lightning, but there are no loud peals of thunder.

Data Evaluation. Silent lightning is very common. The examples offerred below have been selected because they emphasize the slow discharges of electricity on massive scales, which is evident only from good vantage points, such as mountain tops. Rating: 1.

Anomaly Evaluation. Although silent, sheet lighting is a recognized form of lightning, it is not generally appreciated that it has the characteristics of slow, large-area discharges and is not accounted for by conventional theories of lightning discharges. Rating: 2.

Possible Explanations. Slow, large-area discharges of electricity.

Similar and Related Phenomena. Mountain-top glows (GLD1), low-level auroras (GLA4), earthquake lights (GLD8), tornado lights, especially the neon-tube-like glows in funnels (GLD10). Luminous patches seen on the horizon, sometimes called weather lights (GLA16), may derive from distant sheet lightning.

## Examples of Silent Lightning

X1. July 19, 1899 and August 10, 1899. Shanghai, China. The July 19 event. "The northern sky was in an almost constant blaze of light. Flashes came sometimes from two centres, as though there were an elliptical area of disturbance from whose foci were sent forth the shafts of lightning. At times these flashes would take the opposite course, and starting from the circumference make their way to the foci. Though the lightning flashes reached within twenty-five degrees of the zenith, and were vigorous enough in all conscience, yet nothing but the faintest rumbling could be heard. On August 10, the reflection of lightning was seen from the S.W. and gradually increased in brightness until at about 7.50 it had reached the zenith... the lightning played over nearly the whole of the exposed sky, sometimes six or seven streamers at a time lighting up the sky. They were different in appearance from ordinary forked lightning, having rather the appearance of a network of ribbons crossing the exposed sky in all directions, like the discharges in a vacuum tube. The most unusual circumstance was that these discharges, though most vivid, were almost noiseless, and could scarcely be heard above the ordinary jinrickshaw traffic of the street, the only accompanying sound to the brightest display even in the zenith being a low rumbling, as of ordinary very distant thunder." (R1)

X2. August 26, 1900. Gapland, Maryland. "I saw the most beautiful display of lightning without thunder; the flashes appeared in the southwest corner of the valley known as Middletown Valley, followed the Potomac River and mountains on the Virginia side, then passed to the Blue Ridge at Weverton and followed the mountain top, making a circuit of at least 60 miles, this appeared to occur twice, when gradually the flashes spread, as it were, to the valley, in appearance resembling the discharge of a roman candle. This most beautiful phenomenon lasted from about 7 to 10:30 p.m., and when near the house the light was so vivid that at times one could easily have read a book." (R2)

X3. October 3, 1927. Paraguay. After describing an immense display of streak and bead lightning. "Dr. Knoche tells also of various anomalous phenomena seen during this display, including curious, rapidly-moving orange-colored discharges, which he says resembled cylindrical masses of glowing gas; flashes that revolved like pinwheels; and, at one period of the storm, hundreds of luminous arcs crowded together near the zenith, so dazzling in their brilliancy that he had to close his eyes. Strange to say, this spectacular display went on for hours without thunder. A ghastly quiet prevailed; not a breath of air stirred." (R3)

X4. No date given. Shanghai and Madagascar. "But the most remarkable instance occurred in May, off the south coast of Madagascar, when, from 7 to 11 p.m., I watched a glare, which left the sky dark for only a second once in an hour or so, the central point of which seemed elevated 10° or 15° above the horizon. As the nearer clouds cleared away, I watched for hours the unceasing flashes, tongues of fire darting out and around the distant clouds, radiating in five or six distinct streams of flame from a given point, like the thunderbolt in the hand of Jove, coursing along the sky, or dashing down into the sea at the horizon like liquid fire; yet all the while not a sound was heard." (R4)

X5. No date given. Italy. Describing the passage of two formations of clouds. "When the clouds from the south passed those from the northwest the flashes passed between the strata of cloud so rapidly and yet so silently as to produce the appearance of majestic columns of light, lasting in some cases as much as 7 seconds. Prof. Mandoj supposes that the two clouds were charged with electricity of the same kind, but as they were moving in different directions they might be compared to an immense electrical machine, of which the upper cloud was the conductor, the lower the disc, and the wind the motive force. The moisture of the air served as a conductor for a continuous silent discharge of the electricity that was induced by the motion of the clouds." (R5)

X6. No date given. Montenegrin highlands. "On these nights, the lightning was so incessant and vivid that we were able to walk about, choosing our way among the stones and shrubs as readily as by daylight, the intervals between the flashes being, I should judge, never more than a minute, while much of the time they seemed absolutely continuous, the landscape being visible in all details under a diffused violet light. Looking overhead, the movements of the lightning were easily discernable, the locality of the discharges varying from one part of the vault to another in a manner which it was impossible to confound with the reflection of lightning from a distance. Like the storm of last night, those were followed by copious rain, but not a single peal of thunder was heard during the whole night." (R6)

X7. No date or place given. No thunder heard from an intense thunderstorm seen over a town 10-11 miles away. Later, the town's residents claimed normal thunder prevailed. (R7) Anomalous sound propagation might have put the observer in a silent zone. (WRC)

References

R1. Nature, 60:511, 1899. (X1)
R2. "Notable Lightning," Monthly Weather Review, 28:290, 1900. (X2)
R3. Talman, C. F.; "The Strangest Lightning," American Meteorological Society, Bulletin, 12:130, 1931. (X3)
R4. Collingwood, C.; "Silent Lightning," Knowledge, 4:381, 1883. (X4)
R5. "Columnar Lightning," Franklin Institute, Journal, 113:150, 1882. (X5)
R6. "Lightning without Audible Thunder," Popular Science Monthly, 24:575, 1884. (X6)
R7. "Silent Lightning," Knowledge, 4:308, 1883. (X7)

# GLL5 Horizontal Lightning

Description. Lightning that travels horizontally many times the cloud-to-earth distance before striking the earth. Many apparently inviting targets, such as church spires and tall chimneys may be bypassed in favor of distant low-level targets located miles from the cloud-of-origin and often in bright sunlight.

Data Evaluation. Only a handful of good cases have been found. Rating: 2.

Anomaly Evaluation. The curious aspect of this phenomenon is the avoidance of what would seem to be much more attractive targets. Doubtless no laws of physics are in danger here, but horizontal lightning does hint at unusual meteorological conditions. Rating: 3.

Possible Explanations. Paths of especially low electrical conductivity may be created by air circulation well away from the visible clouds. Ionizing radiation and even meteors might be involved. Something distorts the normal cloud-to-earth electrical circuit.

Similar and Related Phenomena. Lightning from a clear sky (GLL6), rocket lightning (GLL1), meandering lightning (GLL25).

Examples of Horizontal Lightning

X1. July 16, 1873. Hereford, England. "About 10 a large cumulus rose in W.S.W., and broke into rain about 5 miles from this place, its summit at the same time, as is usual in incipient storms, assuming the cirriform appearance. The first thunder-clap soon followed, and was succeeded by seven or eight others as the storm travelled to N.W. The sky overhead was quite clear, and also over Hereford, 3 miles east of this place, and the storm-cloud had a very isolated appearance, though there were other distant clouds on the horizon. I was standing in my garden watching the distant lightning, when a flash left the cloud 2 or 3 miles to W.N.W. of this place, passed almost directly overhead, but a little to N.E., and descended upon Hereford, traversing in a horizontal direction a space of about five miles of clear blue sky devoid of cloud.... Hereford lies comparatively low, and the electric fluid travelled near the earth over high ground covered with trees, buildings, &c.; avoided All Saints' and St. Peter's spires, very near which it must have passed, and singled out a house in the more eastern part of the city. The house struck is lower than the others in the same row, and the adjoining house on the west has much higher chimneys.....As observed here, the flash appeared straight, almost resembling a rocket fired horizontally, but Mr. Isbell informs me that as seen in Hereford it was very zigzag, and seemed to come along near the ground." (R1)

X2. July 23, 1926. Near Bloomington, Indiana. After a stroke that killed two children. "They had heard no thunder previous to this stroke and heard only two or three of distant thunder afterwards. At the time of this stroke we were having frequent strokes of lightning and thunder here at Bloomington and we had 0.23 inch of rain.... The lightning was severe. Witnesses near where the children were killed say the lightning travelled horizontally from north to south. It passed three buildings, missing them by about 100 feet, and struck this little house just above the top of the corner foundation post." (R2)

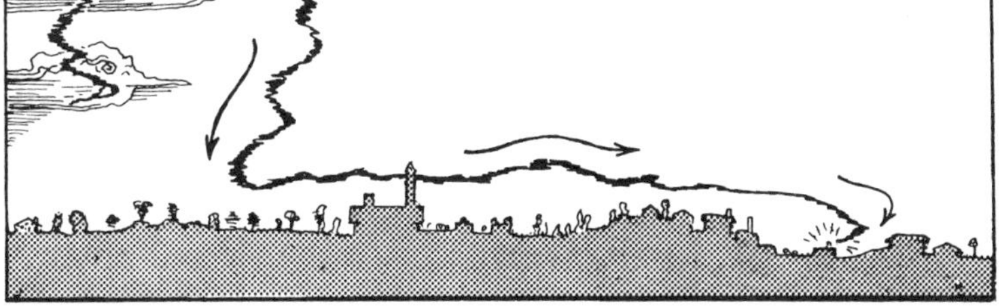

*Horizontal lightning strikes 5 miles away from cloud at Hereford, 1873. (Artist's interpretation)*

X3. May 14, 1965. Malacca Strait. "At 2040 GMT, a single jagged flash of lightning was seen coming from a well developed Cb cloud with anvil, several miles to the southwards. Instead of following the normal cloud-to-earth direction, this flash appeared to begin in a horizontal direction, curving to earth a considerable distance to the east and well clear of the cloud itself." (R3)

X4. July 7, 1969. Lugano, Switzerland. "Lightning strikes downward, as everybody knows, and sometimes up. But there is now at least one authenticated record of a bolt that struck sideways. Nor was this the only peculiar aspect of the freak flash. It took off upward, from a 92-meter-high television tower atop Mt. San Salvatore just south of Lugano, inscribed a complete loop in the sky and finally shot off to the west for about two kilometers. It also lasted an unusually long time---more than a tenth of a second, compared with most bolts which are measured in millionths of a second." (R4)

References

R1. Ley, W. Clement; "Extraordinary Flash of Lightning," Symons's Monthly Meteorological Magazine, 8:106, 1873. (X1)

R2. "A Lightning Stroke Far from the Thunderstorm Cloud," Monthly Weather Review, 54:344, 1926. (X2)

R3. Cook, J.W.; "Lightning," Marine Observer, 36:53, 1966. (X3)

R4. "A Strange, Looping, Sideways Lightning Bolt," Science News, 104:105, 1973. (X4)

# GLL6 Lightning from a Clear Sky

Description. Lightning discharges that appear to originate in clear portions of the sky. Thunder often follows such discharges, but not always. Most examples cited involve well-defined bolts, but other flashes are diffuse, like heat lightning.

Data Evaluation. Many observations. Rating: 1.

Anomaly Evaluation. Lightning usually originates in visible thunderclouds, where convection cells exist for separating electrical charge. Lightning from a clear sky implies that charge separation can occur without other visible phenomena. Rating: 3.

Possible Explanation. Invisible convection currents may be sufficiently strong so that strongly charged regions of the atmosphere are built up. Extraterrestrial radiation might conceivably contribute ionizing particles.

Similar and Related Phenomena. Horizontal lightning (GLL5), general sky flashes (GLA), rocket lightning (GLL1).

Examples of Lightning from a clear Sky

X1. September 10, 1909. Cyprus. "I was riding along the foothills to the south of Mount Troodos at an elevation of about 3000 feet about 10.30 a.m. on September 10, 1909, when straight in front of me I distinctly saw this flash. It was exactly similar to the ordinary flash usually seen against a dark thundercloud. It was followed by a very sharp crash of thunder, so that there can be no doubt it was an electric discharge. Heavy thunderclouds were then forming on the high mountains to the side, and low thunder was rumbling continuously, but the flash was quite away from these clouds and where the sky was entirely blue." (R1)

X2. July 2, 1927. Missoula National Forest, Montana. From a report of topographic mapmakers. "As they were making one of these observations toward a point southeast of Gold Peak both men saw a flash of lightning strike the ground almost on their line of alidade sight, and about 15 miles from them. This flash was followed by four others within the next few minutes. The first strike started a forest fire, the others did not. The phenomenon was most peculiar because all of these strikes descended almost vertically, apparently out of a blue sky, the nearest clouds being about 15 and

25 miles, respectively, from the area struck....These lightning bolts, all of which struck within a small area not over half a mile in diameter, appeared to descend vertically, and they were not between the two clouds, but in a northeasterly direction and over 15 miles from them. No thunder was heard from these flashes, and no further bolts were seen." (R2)

X3. July 21, 1952. Near the Canal Zone. Observer in a twin engine metal aircraft. "I was at 5,000 feet altitude and had 5,000 feet of very small steel cable trailing behind the aircraft with a cloth sleeve attached. I was flying on a heading of 270 degrees magnetic, and off in the distance (about 50 miles) I noticed a flash of lightning. About two minutes later the plane I was flying was struck by lightning. The plane lurched and I was blinded momentarily. The lightning struck the plane on the left side of the nose and traveled down the side to the rear. It passed on to the small cable at the rear of the plane and completely demolished it by burning it to a crisp. My ordnance man reported that we had lost all the cable. I looked back and there was a trail of smoke to the rear about 2,000 or 3,000 feet caused by the burning cable....It is to be noted that no precipitation was encountered for more than one hour prior to the discharge and there were no clouds of any type pressent at my location." (R3)

X4. October 17, 1956. North Atlantic Ocean. Two cloud layers: high cirrus (Ci) and low

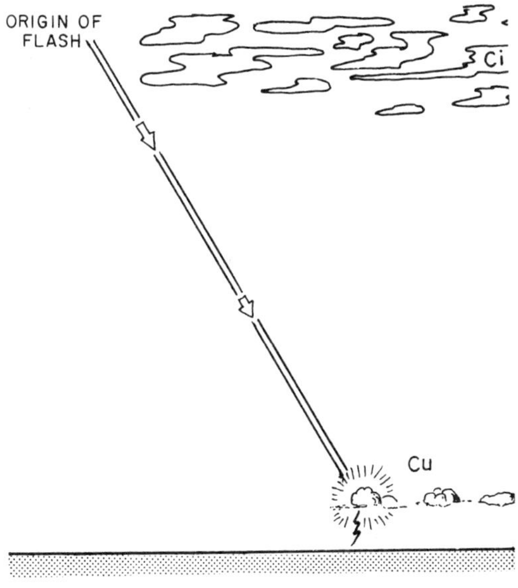


*Lightning from a clear region of the sky strikes a cloud, producing a normal cloud-to-earth flash.*

cumulus (Cu). "Lightning started to flash from a point slightly behind the Ci and travelled in a path about 30° from the vertical in an absolutely straight line, at what seemed to be a comparatively slow speed, towards the small Cu. It appeared to enter the Cu and then reappear, moving downwards in a different direction at normal speed. The lightning track from the Cu, which had a greenish glow when 'hit', did not seem to be so wide as the original one. Similar flashes occurred at regular intervals of between 2-3 min and each followed the same path to the Cu. The watch was relieved at 0400 but it is believed the lightning continued regularly for another 15-20 min. Neither of the observers had ever seen lightning before seeming to emerge from an almost cloudless part of the sky, and which travelled so straight and apparently slowly." (R4)

X5. General observation. Dominica. "The phenomenon of lightning from a cloudless sky seems to be regarded in the States as one of very rare occurrence, as it very likely is. I cannot remember ever observing it while in the States, but down here in the West Indies it is of very frequent occurrence, so frequent in fact that it is not regarded as remarkable by the people. ....The appearance of the flash is that of sheet lightning, generally, single flashes being seen at intervals of from two to five minutes, and again only two or three occasional flashes will be seen during an evening. They do not seem to be confined to any particular quarter of the sky for local reasons, as I have observed them in all quarters. I do not think flashes are due to falling meteors, but they may be reflected flashes of distant thunderstorms, although a clear sky certainly does not offer so good a reflecting surface as a clouded one." (R5)

X6. August 12, 1771. Portsmouth, New Hampshire. Clear sky outside. Flash came in an open window and burned a women. (R6)

X7. May 28, 1853. Weld, Maine. Many vivid flashes from various parts of a cloudless sky. (R7)

X8. Circa 1867. South Killingly, Connecticut. Three people prostrated by lightning from a clear sky. Heavy clap of thunder followed. (R8)

X9. April 27, 1886. North Atlantic Ocean. Clear sky but thin mist about ship. Vivid flash and violent thunder. (R9)

X10. August 3, 1899. Newburg, New York. (R10)

X11. August 7, 1899. Coldharbor, Virginia. Two men struck. (R10)
X12. October 4, 1900. Detroit, Michigan. (R11)
X13. Circa 1926. England. (R12)
X14. January 20, 1931. Tatoosh Island. Several flashes. (R13)
X15. June 17, 1932. Lincolnshire, England. (R14)
X16. August 18, 1940. Asheville, North Carolina. Several flashes from a clear sky observed from two locations. (R15)

References

R1. Williams, W.; "Lightning in Blue Sky," Royal Meteorological Society, Quarterly Journal, 36:126, 1910. (X1)
R2. Gisborne, H.F.; "Lightning from a Clear Sky," Monthly Weather Review, 56:108, 1928. (X2)
R3. Baskin, Donald; "Lightning without Clouds," American Meteorological Society, Bulletin, 33:348, 1952. (X3)
R4. Smith, N.W.; "Lightning," Marine Observer, 27:208, 1957. (X4)
R5. Ashcraft, Charles E., Jr.; "Lightning from a Cloudless Sky," Monthly Weather Review, 28:489, 1900. (X5)
R6. Massachusetts Spy, September 5, 1771. (X6)
R7. Masterman, Stillman; "Observations on Thunder and Lightning," Smithsonian Institution Annual Report, 1855, p. 265. (X7)
R8. Scientific American, 17:72, 1867. (X8)
R9. "Thunderstorm in a Clear Sky," Scientific American, 54:325, 1886. (X9)
R10. "Lightning from a Cloudless Sky," Monthly Weather Review, 28:292, 1900. (X10, X11)
R11. Pague, B.S.; "Lightning from a Cloudless Sky," Monthly Weather Review, 28: 429, 1900. (X12)
R12. Copus, H.H.; "Lightning," English Mechanics, 123:38, 1926. (X13)
R13. Myers, Fred; "Lightning from a Clear Sky, January 20, 1931," Monthly Weather Review, 59:39, 1931. (X14)
R14. Reid, N.C.; "Lightning Discharge during Sunshine," Meteorological Magazine, 68:139, 1933. (X15)
R15. Jemison, George M.; "Bolts from the Blue?" American Meteorological Society, Bulletin, 21:349, 1940. (X16)

# GLL7 Crown Flash

Description. The brightening of a thunderhead crown followed by the appearance of aurora-like streamers emanating into the clear atmosphere. Crown flash seems to be synchronized with lightning strokes from the cloud base to the ground.

Data Evaluation. Only a few observations have been discovered in the literature so far, but they are fairly convincing. Rating: 2.

Anomaly Evaluation. The apparent extension of the luminous display into the clear sky infers that atmospheric electric discharge phenomena may occur where no clouds are visible. The apparent cycling of the phenomenon through the cloud volume is in need of explanation. Rating: 2.

Possible Explanations. The original observers of this phenomenon speculated that it was some sort of ion discharge, perhaps occurring atop an ionized air column associated with the thundercloud. This dovetails with the occasional correlation of auroral phenomena with thunderstorms (GLA9).

Similar and Related Phenomena. Auroras correlated with thunderstorms (GLA9), tornado lights (GLD10), dazzling lights in clouds (GLD15), cyclic flashing in thunderstorms (GLL26), rocket lightning (GLL1), lightning from a clear sky (GLL6).

Examples of Crown Flash

X1. July 2, 1970. Ann Arbor, Michigan. While observing a thundercloud. "At and just above the peak of the storm cell the cloud mass seemed to be undergoing sudden changes in brightness lasting for several seconds at a time.... The phenomenon continued to occur repeatedly at intervals of

30-60 s during the next 15 or 20 min, providing the basis for the following description. The sudden brightening effect began concurrently with lightning strokes in the main cloud mass, but continued after the lightning flash was over. It had the appearance of a ripple-like upward and outward spread of radiance from the region just west of the peak of the cumulus cloud, resembling somewhat a fan-like display of aurora borealis. It lasted a substantial fraction of a second with each lightning stroke. On one or two occasions it had the appearance of a bright ring moving rapidly outward and upward above the cumulus peak. On these occasions it was clearly observed to extend beyond the cloud and into the blue sky. A linear shadow, apparently cast by one of the cumulus masses, appeared to shift its position suddenly up or down with each occurrence of the event." (R1)

X2. August 18, 1950. St. Albans, England. "At 12.25 GMT a cumulo-nimbus of moderate size was passing about four miles to the south, and giving rumbles of thunder about once every minute. While observing the cloud my attention was drawn to a bright streamer, apparently of cloud, projecting northwards from the anvil top for a distance of five degrees. As I watched, the streamer suddenly 'exploded' into a rapidly widening circle of light, fading as it did so. Immediately afterwards the streamer started to reform in the same place as before, only to repeat its 'disappearing' act again after a minute or so, presumably whenever an electrical discharge took place. On some occasions the streamer shifted violently to one side, changing its form greatly, but always returning to one spot. The spectacle lasted for half an hour, becoming fainter as the cloud moved away eastward." (R2)

X3. April 30, 1885. Denison, Texas. A glowing region repeatedly travels along the tops of thunderclouds that were arranged in a long bank. (R3)

References

R1. Gall, John C., jun., and Graves, Maurice E.; "Possible Newly Recognized Meteorological Phenomenon Called Crown Flash," Nature, 229:185, 1971. (X1)

R2. Hale, R. B.; "Unusual Lightning," Weather, 5:394, 1950. (X2)

R3. "Electrical Phenomena," Monthly Weather Review, 13:103, 1885. (X3)

# GLL8 The Preference of Lightning for Certain Trees

Description. The marked preference of lightning for some trees over others in a mixed forest after discounting soil moisture, tree prominence, and other contributing factors. Surveys are emphatic that oaks are strongly favored and beeches the least favored.

Background. Folklore has long maintained that oaks are the most dangerous trees in the forest to take shelter under during a thunderstorm, while beeches are least often struck. Scientific surveys have borne out these popular observations.

Data Evaluation. Several scientific, but rather old, surveys of forest damage by lightning. Rating: 2.

Anomaly Evaluation. This phenomenon is primarily of curiosity value because one would expect that trees with high electrical conductivity would be struck often. However, there exists no thorough study of the electrical properties of trees that would explain the statistical data. Rating: 3.

Possible Explanations. The surveys of lightning damage are probably explained in terms of the different electrical properties of trees.

Similar and Related Phenomena. Lightning districts (GLL13), the tendency of lightning to strike repeatedly certain rock outcroppings and buildings.

Examples of Lightning's Preference for Certain Trees

X1. General observation. "Investigations made by Dr. Carl Muller, and reported in Himmel und Erde, show that lightning prefers to strike certain kinds of trees. Under the direction of the Lippe-Detmold Department of Forestry, statistics were gathered showing that in eleven years lightning struck fifty-six oaks, three or four pines, twenty firs, but not a single beech tree, although seven-tenths of the trees were beech. It would seem, then, that one is safer in a storm under a beech tree than under any other kind." (R1)

X2. General observation. The danger of taking shelter under specific trees, taking beech as 1, is 6 for spruce, 37 for Scots pine, and 60 for oak. (R2)

X3. General observation. Oaks strongly favored over beech by lightning. (R3)

X4. General observation. Oaks and, to a lesser extent, pines are favored over other trees by lightning. (R4)

X5. General observation. Oak, pine, and poplar struck more often than other species. (R5)

X6. General observation. Relative frequency of lightning hits: oak, 48; beech, 1; spruce, 5; pine, 33. (R6)

References

R1. "Lightning and Trees," Symons's Monthly Meteorological Magazine, 31: 74, 1896. (X1)

R2. "Trees and Lightning," Royal Meteorological Society, Quarterly Journal, 33: 255, 1907. (X2)

R3. "Which Trees Attract Lightning?" Monthly Weather Review, 26:257, 1898. (X3)

R4. "Lightning and Trees," Royal Meteorological Society, Quarterly Journal, 20: 172, 1894. (X4)

R5. Nature, 68:351, 1903. (X5)

R6. Marriott, William; "Lightning and Trees," Knowledge, 11:372, 1914. (X6)

# GLL9 Lightning Figures

Description. Tree-like patterns or images of nearby objects appearing upon the bodies of people and animals struck or narrowly missed by lightning.

Background. The literature contains many examples of lightning figures, but only a handful have been recorded below. For many years, it was believed that the tree-like patterns on the skins of lightning victims were the true shadows of nearby foliage. Some investigators, however, could see no faithful resemblance to the surroundings; they claimed that the lightning figures were simply analogs of the dendritic patterns produced in the laboratory by electrical discharges on coated glass plates. This experimentally supported viewpoint became dominant and, probably justifiably, remains so today. Some instances of lightning figures, though, involve reproductions of nearby metallic objects on the skins of victims. There seems little room for mistaken identity here, and the laboratory dendritic patterns have no explanatory power. The evaluation that follows applies only to this type of lightning figure.

Data Evaluation. The cases of lightning figures involving nearby metallic objects are scarce, of poor quality, and rather suspect. Rating: 3.

Anomaly Evaluation. There is no known mechanism for transferring the shadow of a metal object to the skin of a lightning victim. If such a phenomenon exists, it possesses more than mere curiosity value. Rating: 2.

Possible Explanations. The most likely explanation is that the pertinent examples are just tall tales, and no amomaly exists. On the other hand, it is conceivable that the presence of a metallic object might somehow distort the lightning currents to form a rare, crude image. Lightning currents have frequently melted rings and other metal objects on people struck. For example, coins heated by lightning-induced currents might well leave circular burns on the victim's skin.

Similar and Related Phenomena. Lightning shadowgraphs (GLL22). See also GLL11 to gain an appreciation of the sometimes uncanny tricks played by lightning.

Examples of Lightning Figures

X1. September 1825. Bay of Armiro. "In September 1825, the brig Il Buon Servo, anchored in the Bay of Armiro, was struck by lightning, and a sailor who was sitting at the foot of the mizenmast was killed. Marks were found on his back, extending from the neck to the loins, including the impression of a horse-shoe, perfectly distinct, and of the same size as the one that was fixed to the mast." (R1)

X2. July 12, 1875. Americus, Georgia. After lightning had struck and severed a tree outside a house containing several people. "All these persons were rendered insensible for a time by the stroke which severed the tree, and on their recovery there were found impressed upon the bodies of them all more or less distinct images of this tree. It was most distinct in the case of the child near the center of the room. Mr. Simmons says, 'the child is impressed upon its back and exactly opposite upon its stomach. The entire tree is plain, and perfect in toto; every limb, branch, and leaf, and even the severed part, is plainly perceptible.'.... Where the leaves are, on the branch, the mark does not take the shape of the entire leaf, but only of the skeleton of the leaf." The marks on all faded away eventually.(R2)

X3. General observation. Results of a survey. "The phenomenon of lightning prints is one of which little is yet known, but which deserves attention. Prof. Poey mentions twenty-four cases of impressions like photographs made by lightning on the bodies of men and animals. Of these, eight were impressions of trees or parts of trees; one of a bird, and one of a cow; four of crosses; three of circles or of impressions of coins carried about the person; two of horseshoes; one of a nail; one of a metal comb; one of a number or numeral; one of the words of a sentence; one of the back of an arm-chair. Many other instances of similar impressions are recorded." (R3)

X4. 1595. Wells, England. Some of the worshippers in a church struck by lightning imprinted with crosses. (R4, R14)

X5. 1812. Bath, England. Skins of six sheep killed by lightning showed images of the surrounding countryside. (R4, R5)

X6. 1828. Cuba. Image of a passing railroad engine imprinted upon the body of a man killed by lightning. (R1)

X7. July 1851. Washington County, Maryland. Image of robin on branch appeared on sheep killed by lightning beneath tree. Feather edges in great detail. (R6)

X8. August 26, 1853. No place given. (R5)

X9. November 2, 1861. Lappion, France. Tree image. (R4)

X10. June 9, 1883. Berwickshire, England. Image of nearby yew tree imprinted upon boy hurt by lightning. Image faded later. (R7, R8)

X11. June 17, 1896. Pertuis, France. Tree image. Flammarion investigated and found little resemblance to actual tree. (R4)

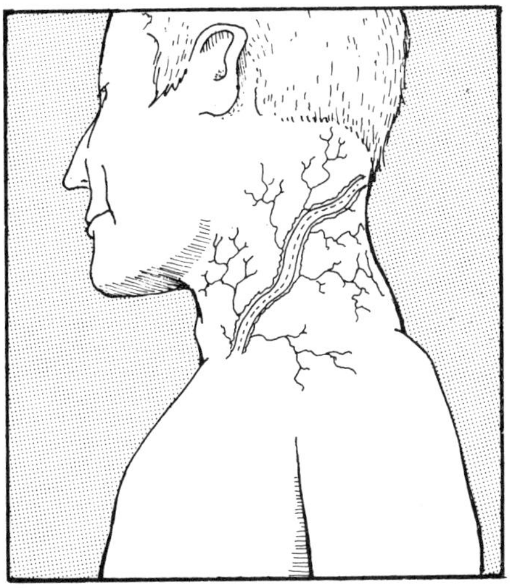

*Typical lightning figure*

X12. September 14, 1919. London, England. Leaves and branches imprinted. (R9,R10)

X13. General observation. Tree-like figures are easily produced by electrical discharges on glass plates with conducting films. (R11)

X14. General observation. Conditions need-for image formation; i.e., lens, light-sensitive surface; are missing in all examples of lightning figures. (R12)

X15. General observation. The tree-like images are not really like the surrounding foliage. Tree-like images have appeared on people struck in the open country. (R13)

X16. No date given. Zante, Greece. Image of the number 44 appeared on the body of a sailor struck by lightning. A metallic number 44 was attached to the rigging. (R1)

References

R1. Tomlinson, C.; "On Lightning Figures," Nature, 12:9, 1875. (X1, X6, X16)

R2. "Images Produced by Lightning," American Journal of Science, 3:10:317, 1875. (X2)
R3. "Curious Lightning Phenomena," Popular Science Monthly, 42:422, 1893. (X3)
R4. Talman, Charles Fitzhugh; "Lightning Prints," Scientific American, 108:576. 1913. (X4, X5, X7, X11)
R5. "A 'Lightning Figure' Photographed," Symons's Monthly Meteorological Magazine, 18:81, 1883. (X5, X8)
R6. Logan, Thos. J.; "Marks Produced by Lightning Stroke," Scientific American, 21:214, 1869. (X7)
R7. Williamson, John; "Thunderstorm at Dunse...," English Mechanic, 37:364, 1883. (X10)
R8. Williamson, John; "Thunderstorm at Dunse...," Symons's Monthly Meteorological Magazine, 18:82, 1883. (X10)
R9. Newell, A.G.; "Death by Lightning---An Unusual Phenomenon," Symons's Meteorological Magazine, 54:104, 1919. (X12)
R10. "Death by Lightning," Monthly Weather Review, 47:729, 1919. (X12)
R11. Tomlinson, Charles; "On Lightning Figures," Symons's Monthly Meteorological Magazine, 18:104, 1883. (X13)
R12. Bruce, George; "Lightning Prints on Boys," Symons's Monthly Meteorological Magazine, 18:83, 1883. (X14)
R13. Wood, James G.; "Lightning Figures," Symons's Meteorological Magazine, 54: 131, 1919. (X15)
R14. Bartlett, J. A.; "Effects of Lightning," English Mechanic, 37:504, 1883. (X4)

# GLL10 Lightning Sounds (Other than Thunder)

Description. "Vits," "clicks," or ripping sounds heard just prior to and simultaneous with nearby lightning.

Data Evaluation. A large number of observations exist, many by trained personnel at scientific observatories. Only of few of these examples are reported here. Rating: 1.

Anomaly Evaluation. If the lightning sounds are only electric discharge noises (brush discharge), we have only a poorly investigated curiosity. If the perceived sound is actually due to a burst of electromagnetic energy being translated into sound physiologically, we have something demanding more attention, but still a phenomenon that poses no serious challenge to prevailing theories. Rating: 3.

Possible Explanations. The sudden discharge of electricity from exposed conductors, trees, brush, etc. Electromagnetic waves perceived as sound (electrophonic sound).

Similar and Related Phenomena. The sound of the aurora (GSD), sounds associated with meteors (GSD).

Examples of Lightning Sounds

X1. April 24, 1925. Petersfield, England. "During the storm three men were working in a field; two of them were together close to a holly tree in a hedge; there was a very bright flash of lightning, with a just perceptible interval between the flash and the thunder. At the moment of the lightning there was quite a loud swishing sound in the holly tree, as though, they said, a sudden blast of air went through the tree; the sound occurred definitely before the thunder.... The flash must have been very close as they both smelt 'sulphur'---nitrogen peroxide; and they could scarcely see anything for some moments. The third man was about 230 metres away and was close to an oak tree to which he had his back; he says that when the flash came there was a noise in the tree as though it were 'on fire.'"(R1)

X2. August 29, 1930. No place given. "I did not see the actual flash, only the illumination of the garden through the open window; it was very brilliant and was followed instantly by a noise as though a shower of large water drops had been thrown onto a hot metal plate; this was followed almost instantly by the thunder. I think there was a slight interval between the swish and the thunder, but it must have been only a fraction of a second.... It was also heard by two people in a cottage about 50 yards from the house; one of them likened it to a red-hot poker being plunged into cold water, the other to the sound of the arc when two electric cables are short circuited." (R2)

X3. Circa 1915. Little Shelford, England. Like a sudden rending of calico, just before the thunder. (R3)
X4. July 30, 1921. No place given. Like canvas being ripped violently, just before the flash. (R4)
X5. July 9, 1923. Kent, England. Three vits heard just before the thunder. (R5)
X6. April 28, 1934. London, England. A swishing noise accompanying the flash. (R6)
X7. March 14, 1953. Off Sardinia. Lightning strikes the water accompanied by a loud hissing noise. Thunder followed. (R7)
X8. General observation. Vits or clicks preceding lightning are commonly heard wherever electrical conductors are exposed. (R8)
X9. General observation. Siam. Vits and clicks heard often preceding the flashes. In addition, animals sometimes are disturbed just prior to a flash and have been observed to look in the direction of the impending flash. (R9)
X10. General observation. The author of this item collected many cases of vits and clicks, noting that almost all occur simultaneously with the lightning, rarely preceding it. (R10)

References

R1. Cave, C.J.P.; "Lightning," Nature, 115:801, 1925. (X1)
R2. Cave, C.J.P.; "Noise Associated with Lightning," Nature, 126:401, 1930. (X2)
R3. Balls, W. Lawrence; "The Sound of Lightning," Nature, 115:912, 1925. (X3)
R4. Cave, C.J.P.; "The Sound of Lightning," Nature, 116:98, 1925. (X4)
R5. Best, Ronald L.; "Unusual Thunderstorm Phenomena," Meteorological Magazine, 63:135, 1928. (X5)
R6. Belasco, J.E.; "A Phenomenon Accompanying Lightning," Meteorological Magazine, 69:91, 1934. (X6)
R7. Cummins, K.A.H.; "Lightning," Marine Observer, 24:11, 1954. (X7)
R8. McAdie, Alexander; "Phenomena Preceding Lightning," Monthly Weather Review, 56:219, 1928. (X8)
R9. Breton, R.S.; "Unusual Thunderstorm Phenomena," Meteorological Magazine, 63:113, 1928. (X9)
R10. Ashmore, S.E.; "The Sounds Associated with Lightning," Weather, 11:269, 1956. (X10)

# GLL11 Lightning's Pranks

Description. Odd, almost incomprehensible actions of lightning on people and objects. Lightning pranks range from neatly excising circles of glass from window panes to breaking every other dish in a pile.

Background. The pranks of lightning are so varied that this must be a catch-all category. Excluded are such "reasonable" actions of lightning as melting glasses frames, moving furniture about, smashing holes in dwellings, plowing up long furrows in lawns, and the well-documented propensity for killing four-footed animals rather than humans.

Data Evaluation. Lightning pranks are mostly one-of-a-kind; and few have been investigated. Rating: 3.

Anomaly Evaluation. The vagaries of lightning seem to challenge no physical law. Rather, it would seem that detailed examinations of each prank would probably result in reasonable physical explanations. Rating: 3.

Possible Explanations. Strange though lightning's pranks may be, careful studies of electrical conductivities, moisture contents, presence of metal, air currents, etc. should lead to a good understanding of each event.

Similar and Related Phenomena. Tornado pranks (GWT), the peculiar actions of ball lightning (GLB).

Examples of Lightning's Pranks

X1. Circa 1872. Vernon, France. "A curious phenomenon has been observed on a property at Vernon, in France. Five or six years ago lightning struck a garden planted with gooseberry bushes and cherry trees, making a very deep hole, the orifice of which was not 10 ctm. in diameter. Subsequently everything died round about. The death circle enlarged year by year. It is now seven metres in diameter, and has just reached a cherry tree planted twelve years ago, which has died like the rest. The gooseberry bushes which we replanted died in two years." (R1)

X2. June 15, 1873. Kansas City, Missouri. A young couple was struck by lightning. "She was killed, and he was crippled; while the shock was felt for a hundred yards from the fatal point of impact, jarring buildings, and benumbing and prostrating human beings, so as to prevent spectators for a few moments from coming to the aid of the chief sufferers. When they did so, they found, to their horror, that the body of the young lady was actually undergoing combustion, making it necessary to extinguish the flames by pailsful of water." (R2) There may be a connection here with spontaneous human combustion (BHB). (WRC)

X3. June 8, 1878. Ashford, England. A workman had taken shelter under a tree. "His companions found the tree partly denuded of its bark, and the patient's boots standing at its foot. The patient himself was lying on his back two yards off, and though he was fully clothed previously, he was now naked, with absolutely nothing on except part of the left arm of his flannel vest. He was conscious, but much burnt, and his left leg was badly broken. The field around was strewn with fragments of the clothing; the clothes were split or torn from top to bottom, the edges of the fragments being often torn into shreds or fringes; they only showed evidences of fire where they came in contact with metal, such as his watch and the buckle of his waist belt. There were no laces in the boots. The left boot was torn and twisted into fantastic shapes, but the sole was uninjured, and there were no signs of fire upon it." (R3, R4)

X4. Circa 1886. Ribnitz, Germany. "At Ribnitz, in Mecklenburg, during a violent thunderstorm, with rain and hail, about 6 a.m., the lower pane of a window on the first floor was broken by lightning, and a jet of water was thrown upwards through the aperture to the ceiling, where it detached part of the ceiling, and this, falling with the water, broke a small cigar-table below. Three bucketfuls of water were afterwards taken from the room. The hole in the window was like that from a bullet, and there were radial cracks. The path of the lightning is not very clear, but that it passed through the glass could not be doubted. Some cigars on the table, it may be mentioned, were carbonised." (R5, R6)

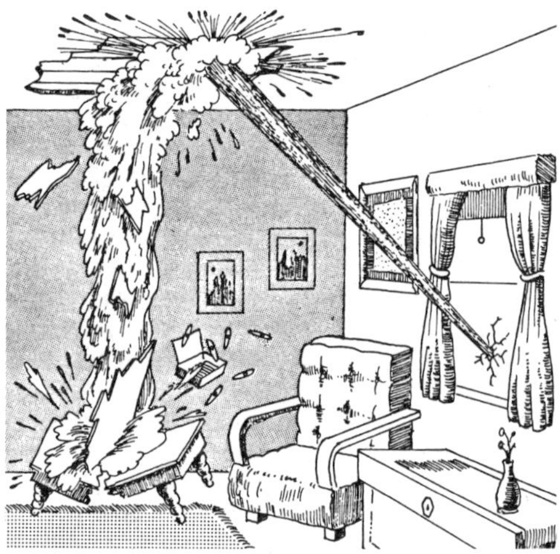

*Lightning breaks window at Ribnitz, Germany, and hurls stream of water at ceiling. (Artist's interpretation)*

X5. January 5, 1890. Co. Mayo, Ireland. Lightning struck a house. "The most extraordinary occurrence was what happened to a basket of eggs lying on the floor of the room. The eggs were shattered, so that they fell off when the eggs were put in boiling water, but the inner membrane was not broken." (R7, R8)

X6. July 1894. Harrogate. England. "A flash of lightning passed me, which was immediately followed by another flash and a very loud peal of thunder. The second flash went in the direction of the deceased, and I saw it strike him at the back of the head.... The deceased was lifted up about 4 to 5 feet in the air in a perpendicular direction, and was then carried out against the wind for 8 or 10 yards, away from the stack, and fell to the ground like a stone." (R9)

X7. June 24, 1935. Dodge, Nebraska. A phenomenal hole was excavated by lightning. "The dirt was thrown back from all sides of the 8-inch hole at the center for a distance of about 3 feet and heaped up about 6 inches above the level of the surrounding ground. The hole extended downward 8 feet almost vertically, with an average diameter of 8 inches, then became smaller for the next 7 feet and varied somewhat from the perpendicular. At a depth of 15 feet the diameter had been reduced to about 4 inches and branched out in three directions into 2-inch holes, which were followed for 3 or 4 feet into the bank where they disappeared.... The clay showed signs of fusion at a number of points, and the inside of the hole had a corrugated appearance, as though the moist clay had been forced back by high pressure." (R10)

X8. June 8, 1972. Edinburgh, Scotland. "The damage to the window pane occurred between 1230 and 1400 BST on 8 June 1972 during which time no one was in the room but a heavy thunderstorm passed overhead. The result was an almost circular hole of dimensions 4.9 cm by 4.6 cm, and an irregular crack which ran across the pane from the bottom of the hole; the pane broke along this crack while it was being removed. The 'missing' circle of glass was found intact inside the room; the edge of this piece and also that of the glass surrounding the hole had a fused appearance, and was smooth to the touch, on the inner side of the pane." (R11)

X9. No date given. Jefferson, Iowa. After lightning had struck a house. "After the occurrence, a pile of dinner plates, twelve in number, was found to have every other plate broken. It would seem as if the plates constituted a condenser under the intensely electrified condition of the atmosphere." (R12)

X10. July 1907. Ireland. A heavy thunderstorm leaves behind dead and withered vegetation. Sometimes only one side of a plant or tree would be affected. (R13)

References

R1. English Mechanic, 27:214, 1878. (X1)

R2. Hovey, Horace C.; "Death by Lightning," American Journal of Science, 3:6:157, 1873. (X2)

R3. "Singular Case of Lightning Stroke," Scientific American, 42:51, 1880. (X3)

R4. Marriott, W.; "Mortality and Accidents from Lightning," Knowledge, 12:197, 1889. (X3)

R5. Nature, 33:396, 1886. (X4)

R6. Popular Science Monthly, 29:719, 1886. (X4)

R7. Scott, Robert H.; "Note on a Lightning Stroke Presenting Some Features of Interest," Royal Meteorological Society, Quarterly Journal, 17:18, 1891. (X5)

R8. Scott, R.H.; "Note on a Lightning Stroke," American Meteorological Journal, 8:34, 1891. (X5)

R9. Marriott, William; "Thunder and Hail Storm over England....," Royal Meteorological Society Quarterly Journal, 20:31, 1894. (X6)

R10. Jensen, J.C.; "The Dodge, Nebraska, 'Fireball,'" Science, 83:574, 1936. (X7)

R11. McIntosh, D.H.; "Lightning Damage," Weather, 28:160, 1973. (X8)

R12. Nature, 66:158, 1902. (X9)

R13. Battersby, T. Preston; "Vegetation Killed by a Thunderstorm," English Mechanic, 86:195, 1907. (X10)

# GLL12 Hot-Air Blasts Following Lightning Strokes

Description. Blasts of hot air observed during thunderstorms. They are synchronized with the lightning strokes and follow the peals of thunder.

Data Evlauation. A single observation. Rating: 3.

Anomaly Evaluation. Lightning strokes possess ample energy to create blasts of hot air. At issue here is the precise accelerating mechanism. Whatever it may be, it would seem to be well within the capabilities of prevailing theories. Rating: 3.

Possible Explanation. Jets of hot air may be accelerated electrostatically and magnetically by the lightning stroke. (R4)

Similar and Related Phenomena. Powerful electrical discharges may also be involved in the production of hydrometeors (GFI) and the jets of matter seen emanating from stars, galaxies, and other astronomical objects. (R4)

Examples of Hot-Air Blasts

X1. July 1971. France. "I was standing on the summit of a mountain in central France, Puy Mary, 1786 metres, at about 3 pm in July 1971 and could see lightning below me in a band of black clouds to the north. I saw the lightning strike, heard the thunder about 5-6 seconds later and felt very few seconds afterwards a rather strong blast of hot air. This happened several times in quick succession. The blast was so forceful that I had to lean against it, and it came from the direction of the black clouds. It was a hot and sultry day, the temperature of the blast was quite a bit higher; it did not singe my hair or clothing. It was not raining on the top of the Pur Mary. The blast was associated with the lightning-thunder sequence, it travelled at a much lower speed than the thunder and made very much less noise, it was almost silent. It certainly was not just a drift of hot air up the mountain slope." (R1-R4) See also GLL10-X1.

References

R1. Worth, L.H.; "Lightning 'Blast,'" New Scientist, 57:330, 1973. (X1)
R2. Worth, L.H.; "Atmospheric Mystery," Weather, 28:86, 1973. (X1)
R3. Worth, L.H.; "Atmospheric Mystery," Nature, 236:413, 1972. (X1)
R4, Crew, E.W.; "Meteorological Flying Objects," Royal Astronomical Society, Quarterly Journal, 21:216, 1980. (X1)

# GLL13 Unusual Geographical Preferences of Lightning

Description. On a global basis, the much higher incidence of lightning over land than sea. On a local scale, lightning's preference for certain areas after discounting height and other obvious factors.

Background. It is well known that lightning prefers some tall buildings and certain rocky outcrops. Farmers and sheepherders can often point out certain spots, even in treeless regions, where men and animals are in much more danger from lightning. Here, the intent is to record examples where it is not readily apparent why lightning favors specific areas.

Data Evaluation. Global-scale observations from satellites are of high quality, but the local claims of "lightning districts" are not backed by credible statistics. Rating: 2.

Anomaly Evaluation. Both global and local anomalies in the distribution of lightning strikes are probably explainable in terms of meteorology and geological conditions. However, the land-over-sea preferences of lightning are so strong that even these factors may be inadequate. Rating: 3.

Possible Explanation. Satellite maps of lightning incidence show strong concentration over tropical areas, as one would expect meteorologically from known ground temperatures, humidity, etc. Meteorologists, however, are puzzled as to why land areas are favored by a factor of ten over the oceans. (R6) Uncharted local moisture and mineral concentrations can probably account for local lightning "hot spots" that persist after obvious lightning attractors, such as trees and prominences, are taken into account.

Similar and Related Phenomena. Lightning's preference for specific trees (GLL8), tornado alleys (GWT), horizontal lightning (GLL5), meandering lightning (GLL25).

Examples of Unusual Geographic Preferences of Lightning

X1. General observation. California. "The San Francisco Chronicle says there is a tract of country in Butte County, Cal., about fifteen miles long by half a mile in width, where lightning strikes trees nearly every time a storm passes over. Outside of this strip there is no such damage. The line can be plainly traced by dead timber." (R1)

X2. General observation. Van Wert, Ohio. One small area of the city is hit more frequently by lightning. (R2)

X3. General observation. A nine-year survey along a power line proved that certain sections were hit more often by lightning and that these sections seemed to be over underground springs. (R3)

X4. General observation. From satellite observations, lightning is roughly ten times more frequent over land areas than the oceans. (R4-R6)

References

R1. "Local Lightning," Scientific American, 45:146, 1881. (X1)
R2. "Electrical Districts," Monthly Weather Review, 25:249, 1897. (X2)
R3. Goodlet, B. L.; "Lightning," Institute of Electrical Engineers, Journal, 81:1, 1937. (X3)
R4. "Zeus Expends His Wrath More upon Land Than Sea," New Scientist, 47:512, 1970. (X4)
R5. "Thunderstorms Fall Mainly on the Land," Science News, 98:207, 1970. (X4)
R6. "Patterns of Thunderbolts," New Scientist, 92:102, 1981. (X4)

# GLL14 Black Lightning

Description. Lightning strokes that appear black or dark rather than bright either to the eye or photographic film.

Background. Early photographs of lightning frequently showed dark as well as bright strokes. (R1, R2) The question naturally arose as to whether black lightning had any objective existence in nature. After all, lightning could be red, blue, and other colors (GLL3). As the Nineteenth Century drew to a close, photographic experiments proved to everyone's satisfaction that the phenomenon of film reversal accounted for all photos of dark lightning. (R3)

Data Evaluation. Rating: 1.

Anomaly Evaluation. No anomaly appears to exist here. The phenomenon is included for the record and for comparison with the similar, and apparently genuine black auroras and black ball lightning. Rating: 4.

Possible Explanations. Photographs of dark or black lightning stem from film reversal. Visual observations of black lightning are due to retina fatigue.

Similar and Related Phenomena. Black auroras, which may be due to obscuring matter (GLA22), black ball lightning, which may be no more than smoky occurrences of normal ball lightning (GLB11).

References

R1. "Ribbon and Dark Lightning," Nature, 60:423, 1899.
R2. Wood, R.W.; "Dark Lightning," Nature, 60:460, 1899.
R3. Lockyer, William J.S.; "Dark Lightning Flashes," Nature, 60:570, 1899.

# GLL15 Slow or Prolonged Lightning

Description. Lightning flashes that persist for one or more seconds.

Data Evaluation. A limited number of observations exist, and some of these may be confused with sheet lightning. Rating: 3.

Anomaly Evaluation. Lightning flashes normally last for but a small fraction of a second. Those flashes that persist for a second or more pose serious problems in explanation. Rating: 2.

Possible Explanations. None, other than possible confusion with sheet lightning or some type of eye fatigue or afterimage.

Similar and Related Phenomena. Sheet lightning (GLL4), ball lightning which usually endures for several seconds (GLB1).

Examples of Slow or Prolonged Lightning

X1. 1916. Surrey, England. Observing a sudden squall. "But instead of a flash this was more like a deluge of light, hardly varying in intensity during the three seconds or more that it lasted, lightning up a house some thirty yards distant with a curiously even glow." (R1)

X2. 1916. Somerset, England. An extremely long flash. (R1)

X3. July 12, 1931. Herts, England. Lightning flashes persisted up to one second a and were markedly different from normal flashes. No thunder heard. (R2)

X4. October 17, 1956. North Atlantic Ocean. Slow, unusually wide stroke of lightning hits a cloud, from which proceeds a normal quick, narrow flash. (See GLL6-X4) (R3)

References

R1. Clark, J.E.; "Peculiar Lightning," Symons's Meteorological Magazine, 51:152, 1916. (X1, X2)

R2. Beckett, H.E., and Dufton, A.F.; "Unusual Lightning," Nature, 128:189, 1931. (X3)

R3. Smith, N.W.; "Lightning," Marine Observer, 27:208, 1957. (X4)

# GLL16 Correlation of Lightning and Cosmic Rays

Description. The correlation of increases in cosmic-ray intensity with increases in lightning (thunderstorm) frequency, suggesting a causal relationship.

Background. During the early 1930s, thunderclouds were found to produce a type of penetrating radiation. (R4). This relatively low energy radiation was shown to be produced by electron acceleration within thunderclouds and unrelated to extraterrestrial cosmic rays. However, this storm-produced radiation, while not anomalous in itself, may be related to crown flash (GLL7) and the aurora-like displays sometimes observed above thunderstorms (GLA9). In addition to the above radiation, thunderstorms by virtue of their strong electrostatic accelerating fields constitute a sort of natural atom smasher, which can generate neutrons and other nuclear radiation. The neutrons, in particular, may interact with carbon and upset carbon-dating measurements. Fortunately, lightning-produced neutrons have proven to be of minor importance. (R5) With these preliminaries out of the way, the rest of the discussion deals with the possible initiation of lightning (thunderstorms) by extraterrestrial cosmic rays.

Data Evaluation. Several modern studies suggest a causal relationship. Rating: 2.

Anomaly Evaluation: The possible initiation of thunderstorms by cosmic rays, an extraterrestrial influence, is only one element in the hotly debated subject of weather and astronomy. Generally speaking, extraterrestrial influences on weather are not yet assimilated into accepted theory. Rating: 2.

Possible Explanations. On a broad geographical scale, cosmic rays influence significantly the global electrical circuit and, by yet unknown mechanisms, may increase thunderstorm frequency. On a local scale, intense showers of cosmic rays might trigger specific storms or even specific lightning strokes, although this is sheer speculation. The ionizing ability of the cosmic rays is, of course, the probable triggering effect.

Similar and Related Phenomena. Correlation of auroras and thunderstorms (GLA9), the correlation of ball lightning with solar activity (GLB17).

Examples of Correlations between Cosmic Rays and Lightning (Thunderstorms)

X1. General observation. "An increase in the cosmic-ray component (E > 500 kev) was found to accompany thunderstorms. No correlation was found between nearby individual lightning strokes and count rate, but a general count increase with a time constant of some 10 min was found for all thunderstorms that passed nearby. Flux increases were typically 5%. The nature of the count increase in not known." (R1)

X2. General observation. "A significant maximum in cosmic rays occurs one day before solar sector boundary crossings. Therefore, previous work showing a maximum in thunderstorm frequency one day after solar sector boundary crossings may result from high cosmic ray counts. This is evidenced by a maximum in thunderstorm frequency three days after high cosmic rays and, also, by a maximum in thunderstorm frequency three days after minimum $K_p$." (R2)

X3. General observation. Electrification of the terrestrial atmosphere is strongly affected by cosmic rays. There also seems to be a thundercloud electrical mechanism that is sensitive to atmospheric electrification, so that thunderstorms (and lightning) are influenced by cosmic rays. (R3)

X4. General observation. Cosmic-ray showers may ionize enough air to trigger lightning discharges. (R6)

References

R1. Shaw, Glenn E.; "Background Cosmic Count Increase Associated with Thunderstorms," Journal of Geophysical Research, 72:4623, 1967. (X1)

R2. Lethbridge, M.D.; "Cosmic Rays and Thunderstorm Frequency," Geophysical Research Letters, 8:521, 1981. (X2)

R3. Markson, Ralph; "Modulation of the Earth's Electric Field by Cosmic Radiation," Nature, 291:304, 1981. (X3)

R4. Schonland, B.F.J., and Viljoen, J.P.T.; "On a Penetrating Radiation from Thunderclouds," Royal Society, Proceedings, A140:314, 1933.

R5. Fleischer, Robert L., et al; "Are Neutrons Generated by Lightning?" Journal of Geophysical Research, 79: 5013, 1974.

R6. "Do Cosmic Rays Trigger Lightning Discharges?" New Scientist, 77:88, 1978. (X4)

# GLL17 Lightning Superbolts

Description. Lightning bolts approximately 100 times more powerful than ordinary lightning. Most superbolts have been observed in winter storms over Japan and the northeast Pacific. They seem to originate in positively charged portions of the clouds.

Data Evaluation. Many satellite observations. Rating: 1.

Anomaly Evaluation. The much greater energy of the superbolts and their apparent origin in positively charged sections of thunderclouds may signify a different, hithertofore unrecognized, mechanism of lightning generation. Rating: 3.

Possible Explanation. None.

Examples of Lightning Superbolts

X1. General observation. "Unusually intense lightning strokes have been observed by optical sensors on the Vela satellites. These lightning flashes are over 100 times more intense than typical lightning. The lightning superbolts are characterized by optical power in the range of $10^{11}$ - $10^{13}$ W, have a duration on the order of 1 ms, and have a total radiant energy greater than $10^9$ J. In conjunction with sferics data the Vela trigger rates indicate that about two lightning flashes in $10^3$ exceed an optical power of $10^{11}$ W and five flashes in $10^7$ exceed an optical power of $3 \times 10^{12}$ W." Winter storms over Japan and the northeast Pacific have produced most of the superbolt observations. Lightning in these storms is very powerful, infrequently hits the ground, and seems to originate in positively charged regions of the clouds. (R1, R2)

X2. General observation. Superbolts detected in Oklahoma prairie storms. (R3)

References

R1. Turman, B.N.; "Detection of Lightning Superbolts," Journal of Geophysical Research, 82:2566, 1977. (X1)

R2. "Super Lightning Detected by Satellite," Science News, 112:15, 1977. (X1)

R3. "NOAA Finds Lightning Superbolts in Oklahoma Storms," American Meteorological Society, Bulletin, 62:1060, 1981. (X2)

# GLL18 Cyclic Flashing of Lightning

Description. The cyclic waxing and waning of the frequency of lightning flashing in a thunderstorm. During its lifetime, a thunderstorm may go through several cycles of lightning activity with periods of roughly 10-30 minutes.

Data Evaluation. Several reinforcing studies exist. Rating: 2.

Anomaly Evaluation. It is not clear why a single thunderstorm cell would wax and wane cyclically. Usually, they grow and decay---lasting one cycle only. Rating: 3.

Possible Explanation. Rather than single cells varying in their rates of activity, a composite cyclic effect may result when individual cells in a storm grow and decay sequentially.

Examples of Cyclic Flashing of Lightning

X1. General observation. Three thunderstorms were observed with a noise meter. The average periods of flash rate varied between 17 and 31 minutes. (R1)

References

R1. Satyam, M.; "Cyclic Variation of the Rate of Flashing in Thunderstorms," Journal of the Atmospheric Sciences, 19:346, 1962. (X1)

# GLL19 Dual Lightning Discharges

Description. Two or more lightning discharges that occur essentially simultaneously several miles apart. The discharge channels are not usually connected in any way. A substantial fraction of lightning discharges in a storm may be dual in nature.

Data Evaluation. Only a single study found to date. Rating: 2.

Anomaly Evaluation. No interconnecting mechanism for producing simultaneous lightning discharges has been found as yet. Electrostatic induction seems a likely prospect; and this process is well-understood. Rating: 3.

Possible Explanations. One discharge might simultaneous stimulate another, although the distances involved are admittedly large. More speculatively, cosmic ray showers might initiate two or more discharge channels.

Similar and Related Phenomena. The apparent induction of ball lightning inside enclosures (GLB10) is a good example of electrostatic induction over long distances.

## Examples of Dual Lightning Discharges

X1. General observation. "30% of the intra-cloud lightning discharges and 35% of the cloud-to-cloud lightning discharges that occurred in the late stages of a convective thunderstorm are shown to possess two major,unconnected channel portions several kilometers apart." (R1) This is an excerpt from the abstract. (WRC)

## References

R1. Teer, T. L.; "Dual Lightning Discharges," American Geophysical Union, Transactions, EOS, 55:1131, 1974. (X1)

# GLL20 Abnormally Long Lightning Strokes

Description. Lightning strokes on the order of 100 miles long.

Data Evaluation. A single report of radar observations. Rating: 3.

Anomaly Evaluation. Current theories of lightning discharges cannot easily account for lightning strokes 100 miles long. Rating: 2.

Possible Explanations. None.

Similar and Related Phenomena. Horizontal lightning (GLL5), superlightning (GLL17), meandering lightning (GLL25).

## Examples of Abnormally Long Lightning Flashes

X1. General observation. Reporting the results of radar observations of cloud-hidden flashes. "In Midwestern thunderstorms along the front separating masses of warm and cooler air, lightning strokes up to 100 miles long have been found. Theories of lightning discharge are expected to need revision because of the new observations. (R1)

## References

R1. "Spot Hidden Lightning," Science News Letter, 68:359, 1955. (X1)

# GLL21 Anomalous Electrical Fields and Currents during Lightning

Description. Measurements of near-zero electrical currents and fields during strong lightning strokes.

Data Evaluation. Repeated observations with one experimental setup. Rating: 3.

Anomaly Evaluation. Large electrical currents and strong fields would normally be expected during a lightning stroke. Rating: 2.

Possible Explanation. None.

Similar and Related Phenomena. None.

## Examples of Anomalous Electrical Fields and Currents during Lightning Strokes

X1. General observation. Toronto, Ontario, Results from current measuring instruments placed on the CN Tower, 1815 feet high. "In the past eight months, 31 flashes and their current measurements have been recorded to the tower, he said, but some puzzling observations---such as a zero electric field in the midst of a large flash---need to be explained." (R1)

## References

R1. "Flashy Research," Science News, 117: 376, 1980. (X1)

# GLL22 Lightning Shadowgraphs

Description. The production of shadowgraphs of objects on exposed surfaces by bright lightning flashes. The images often fade within a few minutes.

Background. Lightning shadowgraphs resemble the lightning figures said to appear on persons struck by lightning (GLL9). Whereas lightning figures are generally dismissed as misinterpretations of the dendritic patterns normally appearing on the bodies of lightning victims, the reality of lightning shadowgraphs might signify that some lightning figures, especially those involving metallic objects, have some basis in fact.

Data Evaluation. Several reports exist; some in good detail. Rating: 2.

Anomaly Evaluation. Although the precise photochemical reactions are not known, the reproduction of images on surfaces, particularly chemically treated surfaces, does not seem unreasonable. Rating: 3.

Similar and Related Phenomena. Lightning figures (GLL9).

## Examples of Lightning Shadowgraphs

X1. 1892. Errol, England. "A telephone wire had been broken or fused by lightning, and on an employe seeking to repair the break, he found on one of the white insulators a picture of the roof of a neighboring house. The picture had probably been flashed on the porcelain by the lightning." (R1)

X2. Circa 1904. North Atlantic Ocean. "While on a voyage recently from Hamburg to St. Thomas, the second officer of the Hamburg-America liner 'Galicia', being on the bridge during a terrific electrical display, observed the following phenomena, which he carefully noted, and which it is our privilege to present to our readers. In advance it may be remarked that all the

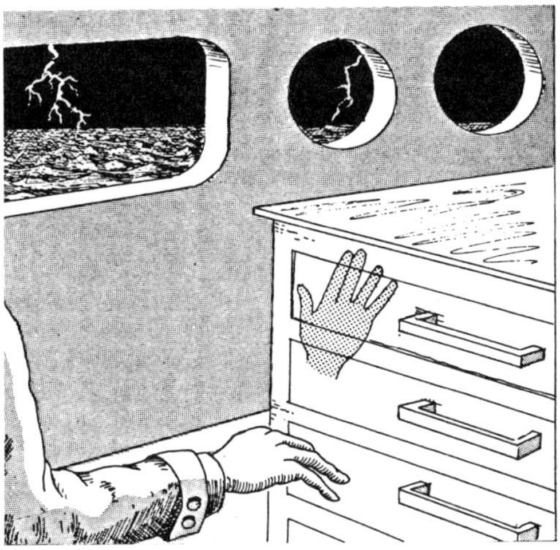

*Lightning shadowgraph on the bridge of the* Galicia, *1904*

wood and iron work about the bridge had been painted gray. In changing his position he casually removed his hand from a cabinet on the bridge immediately after a particularly brilliant flash of lightning, and what was his astonishment to notice was an exact counterpart of it in silhouette upon the cabinet, and to add to his amazement the picture remained imprinted fully five minutes. Such a spectacle was well calculated to inspire the officer to further observations, which he carried out with like results. Among others he placed an observation instrument upon the cabinet, and waiting his opportunity to remove it just after a vivid flash, to find the shadowgraph perfect in detail even to the cross-hairs over the objective plainly visible upon the surface." (R2)

X3. July 9, 1923. England. "During the storm of July 9th-10th a remarkable 'photograph' of a wastepaper basket was made by lightning on the floor of a Mincing Lane office, that of Messrs. Thompson Bros. & Co. The impression had been made on bare, unpolished, ordinary flooring boards, and consisted of a bleached image of the basket, which was apparently lying across the boards with its open end roughly towards south." A photograph of the shadowgraph accompanies the article. (R3)

References

R1. English Mechanic, 56:428, 1892. (X1)

R2. "Remarkable and Rare Effects of Lightning, Scientific American Supplement, 57:23679, 1904. (X2)

R3. "The Production by Lightning of a Shadow-Picture on Bare Boards," Meteorological Magazine, 58:166, 1923. (X3)

# GLL23 Wisps of Flame Left by Lightning Strokes

Description. Residual wisps of flame, resembling ignited material, left in lightning discharge channels. The wisps may seem to drift with the wind.

Background. This phenomenon possesses some characteristics of bead lightning (GLL2) and may in fact be a variety of this form.

Data Evaluation. A single but well-observed example. Rating: 3.

Anomaly Evaluation. The persistence of "ignited material" (probably incandescent gases) is curious as is their tendency to drift away from the discharge channel. Rating: 2.

Possible Explanation. Packets of incandescent gases detached from the discharge channel.

Similar and Related Phenomena. Bead lightning (GLL2), which sometimes seems to be composed of burning fragments.

Examples of Residual Wisps of Flame Left by Lightning Strokes

X1. July 27, 1936. Clarendon, Virginia. "The center of the storm kept somewhat west of the writer's position at Clarendon, Virginia, and at a point in the southwest, only a few miles away, there appeared a most spectacular frequency of discharges earthward. Some appeared as mere single sparks, but the majority were of the repeating or of the stream type of discharge of exceptional size, appearing like ribbons of flame searing the darkness. One of these, which seemed almost to hang in the sky only a few miles away, appeared to ignite something high in its course, leaving wisps of flame which persisted momentarily, detaching themselves from the stream and drifting eastward from the path of the discharge as if carried by the wind. This phenomenon was observed at several points along the discharge, and was also witnessed by my son, Howard F. Allard, who was observing the lightning beside me. A portion of the path of the discharge, at and below these points, was also indicated by a train of sparks persisting momentarily after the streak had gone." (R1)

References

R1. Allard, H.A.; "Remarkable Lightning Bolt," Science, 84:136, 1936. (X1)

# GLL24 Tubular Lightning

Description. Very broad lightning discharge channels possessing a tubular appearance. Channels up to 18 feet in diameter have been measured in photographs. These broad channels frequently meander and display a striated or broken structure.

Data Evaluation. Thick, stream-like strokes are often reported visually, but the basis of this entry is substantial photographic evidence. Rating: 2.

Anomaly Evaluation. Normal lightning channels are typically 5 inches in diameter, so that channels many feet in thickness sorely try prevailing models of lightning discharges. Rating: 2.

Possible Explanation. No physical explanation has been uncovered; however, camera motion during exposure is a possible source of error, although the examples reported below seem free of this problem. To a camera, ball lightning might appear as a tubular channel.

Similar and Related Phenomena. Bead lightning (GLL2) is sometimes thick and broken; meandering lightning (GLL25) may seem unusually broad. Rapidly moving ball lightning, which occasionally measures several feet in diameter, might be rendered photographically as tubular lightning (GLB1).

Examples of Tubular Lightning

X1. May 29, 1908. Chicago, Illinois. During lightning photography experiments. "The flash was a very bright one, but it was so sudden and vivid that I did not notice anything peculiar about it. The thunder accompanying it was sharp and sudden, like the report from a cannon. The interval between lightning and thunder cannot be given accurately; it was less than a second, and probably more than half a second. The picture of this flash is very remarkable; I have never seen any one resembling it, and would prefer to call it a tubular flash on account of its general shape and large diameter, measuring, as it does, over 3 mm at its widest part, and about 2 mm at its narrowest; this great width cannot be accounted for, to be caused by the movement of the camera; the uniformity of the width, both in

*Still camera photograph of meandering tubular lightning. (R2. Courtesy J. D. Barry)*

the vertical and horizontal portion of it, disproves that idea....If we assume that the nearest portion of this flash took place at a distance of 1,000 feet, which in my opinion would be a conservative estimate, we are confronted by the remarkable fact that the diameter of this flash would be over 18 feet. (The angle of the lens in 43°.) I am absolutely at a loss to account for this remarkable flash; it does not appear to be a ribbon flash, which can be accounted for by the movement of the lightning path, by air currents; so will have to defer my opinion until some future time, and leave it to others who may be able to give a plausible explanation." (R1)

X2. July 21, 1937. Deer Lodge, Manitoba, Canada. Photograph taken of a meandering, broad, tubular discharge channel. (R2)

References.

R1. Abbot, C.G.; "Remarkable Lightning Photographs," Smithsonian Miscellaneous Collections, 92:1, no. 12, 1934. (X1)
R2. Holzer, R.E., and Workman, E.J.; "Photographs of Unusual Discharges Occurring during Thunderstorms," Journal of Applied Physics, 10:659, 1939. (X2)

# GLL25 Meandering Lightning

Description. Intricate, looping, reversing, wandering lightning strokes. Such discharge channels often approach the ground closely but do not touch seemingly inviting objects. Meandering lightning is frequently broken and/or tubular.

Data Evaluation. Only a few of the many extant photos of meandering lightning are noted below. Rating: 1.

Anomaly Evaluation. It is difficult to conceive, in the framework of current thinking about the formation of lightning discharge channels, how such long, tortuous paths close to grounded objects could occur. It would seem that a multitude of paths of much lower resistance exists but are "ignored" by the lightning for some undetermined reason. Rating: 2.

Possible Explanations. Camera motion; but experienced investigators easily eliminate this. Rapidly moving ball lightning might appear as a meandering flash on film.

Similar and Related Phenomena. Ball lightning (GLB) habitually follows complicated courses. Meandering lightning is frequently striated or broken (GLL2) and tubular (GLL24). Horizontal lightning (GLL5) usually bypasses nearby, seemingly inviting targets.

Examples of Meandering Lightning

X1. July 17, 1908. Chicago, Illinois. Analyzing a lightning photograph. "It shows a meandering and very complicated flash, consisting of four distinct and separate rushes, following one another in the same path, opened up by the first discharge. It is almost incomprehensible how such a complicated flash can follow all those curves and bends which the photograph shows. The only reasonable explanation to my mind would be that the path of the flash was a partial vacuum with very low resistance, which the beaded or striated appearance of this flash would tend to confirm. How this partial vacuum can be accounted for is a difficult problem to solve. It is the first lightning photograph I have

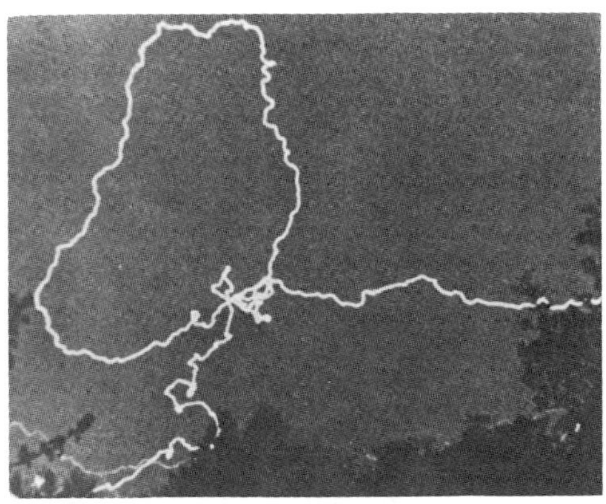

*Meandering lightning over Arnhem photographed in 1958. (Courtesy J. D. Barry)*

had the fortune of seeing that shows the path in broken lines, or striated." (R1)

X2. September 3, 1936. Santa Fe, New Mexico. Referring to a lightning photo. "The nost notable features of this discharge are: (1) its irregularity of path and rapid reversals in direction, (2) its proximity to ground objects with no apparent contact with the ground, (3) the beaded nature of the path, and (4) the progress of the discharge in two directions from a single point." (R2)

X3. July 21, 1937. Deer Lodge, Manitoba, Canada. A meandering, broad, tubular flash, close to but not touching ground objects. (R2)

## References

R1. Abbot, C.G.; "Remarkable Lightning Photographs," Smithsonian Miscellaneous Collections, 92:1, no. 12, 1934. (X1)

R2. Holzer, R.E., and Workman, E.J.; "Photographs of Unusual Discharges Occurring during Thunderstorms," Journal of Applied Physics, 10:659, 1939. (X2, X3)

# GLM LOW-LEVEL METEOR-LIKE LUMINOUS PHENOMENA

*Key to Phenomena*



## GLM0 Introduction

Most meteor anomalies occur at very high altitudes and belong, therefore, to the field of astronomy. As such, they are covered in another volume of this catalog (AYO). There remain, however, varieties of meteor-like luminous phenomena that apparently transpire at very low altitudes and can legitimately be included with other geophysical anomalies.

Predominant among these meteor-like phenomena are: (1) Luminous and nonluminous objects moving at very low altitudes that do not follow usual meteor protocol by disintegrating or burying themselves in the earth; and (2) Very fast, seemingly very near streaks of light that look a bit like meteors but may be auroral or electrical discharge in origin. Both phenomena are most perplexing because they do not obey the laws of meteoritics.

## GLM1 Low-Level Meteor-Like Objects

Description. Luminous objects, travelling below cloud level, often trailing fire and smoke and emitting considerable noise. Most such objects disintegrate or hit the earth, but others gain altitude and disappear from sight. Frequent accompanying phenomena are: a sensation of great heat, a smell of sulfur, the appearance of black clouds before the meteor proper, and a phenomenon time span of several minutes.

Data Evaluation. Most of the observations are rather old but presented with good detail. Rating: 2.

Anomaly Evaluation. All of the examples described below are unusual in some way, but the most anomalous features are: (1) The appearance of clouds minutes before the actual object is observed; and (2) the long lifetimes of the phenomena---lifetimes much longer than those of normal meteors. Other curious attributes, perhaps indicating that we actually have several causes acting, are: absence of sound, long horizontal paths close to the ground, and descent followed by regaining of altitude. Rating: 2.

Possible Explanations. Some of the examples recorded are undoubtedly true meteors, perhaps with the accounts garbled in some respects. Ball lightning and other electrical phenomena may be confused with meteoric objects in some cases. Some meteors may be accompanied by dust and gases that develop into clouds in the lower reaches of the atmosphere.

Similar and Related Phenomena. Some ball lightning manifestations (GLB) clearly resemble some of the phenomena cited here. The same is true for some of the electric discharge phenomena, especially GLD11. See AYO for long-duration high-altitude meteors. The similarity between low-level meteor-like objects and the classic thunderbolt (GFT) and UFO must be mentioned.

Examples of Low-Level Meteor-Like Objects

X1. March 4, 1822. Canajoharie, New York. About 10 p.m. "He first observed a sudden flash of light which appeared to extend from the heavens to the earth, and was followed by a momentary darkness, as if a cloud had passed over and intercepted the light. This darkness was soon dispelled, and the blazing meteor was in full view over his head, appearing to be twenty or thirty feet in diameter, and soon began to extend itself to the northeast and southwest, increasing in extension, and decreasing in its flaming appearance, until nothing was to be seen but two detached parts of it rapidly moving in different directions towards the northeast and southwest. Mr. Doty calculates that it was five or six minutes from the first appearance of the meteor until it finally vanished, and from six to ten minutes from its first appearance before the report of its explosion reached him, which resembled the noise of a distant cannon, and was followed by a strong sulphurous smell, that lasted fifteen or twenty minutes." (R1)

X2. January 2, 1825. Arezzo, Italy. "About a hundred paces off, at a height of ten fathoms, or less, from the ground, appeared, on a sudden, a luminous meteor, of the form of a truncated cone. This meteor appeared to be formed by a globe of fire situated in its fore part, which was the narrower, and which by its rapid motion, left behind a track of light, which gave it the appearance of a cone..... The whole length of the meteor appeared to be about two fathoms, and the diameter of the base half a fathom.... Its motion was very rapid, for in less than five seconds it traversed a space of about 350 paces. During this passage it shed a most brilliant light, so that a certain extent of land was illuminated, as in full day-light. The emanations of this luminous body were lost in the air, instead of being extinguished in the ground; it left behind no smell; produced no explosion or noise of any kind, not even that hissing made by artificial fireworks. The night in which this phenomenon occurred was calm, but very cold, and the sky clear. A great number of shooting stars were seen before and after the appearance of the meteor." (R2)

X3. 1846. Rangoon River, Burma. "It was about half-past seven in the evening, and then quite dark, when suddenly, without any warning, a tremendous sheet of light appeared to rush across the bows of the vessel in a horizontal direction. The light was not like lightning, but appeared to pass swiftly along, and had the aspect of a thick red flame, occupying the whole space left visible between the awning and the ship. The suddenness and terrific nature of the glare was such that she fell down upon the deck, believing, as she expresses it, that the world had come to an end, while the child screamed aloud with terror. As this frightful luminous appearance rushed by the ship a considerable accession of heat was felt by both the adult persons, and a strong sulphurous smell also accompanied it, though how long this feature lasted cannot now be stated with any certainty. But the phenomenon was unaccompanied by any sound, and sped noiselessly on; nor could they see where it went, or what finally became of it. The whole affair lasted but some seconds, the light having dashed past them with a speed only inferior to lightning." (R3)

X4. December 8, 1847. Forest Hill, Arkansas. "...the sky was perfectly clear, and had been so through the day; but at quarter past 3 P. M., it became suddenly very dark. The clouds, or what appeared to be such, whirled in the strangest contortions, and appeared like a solid black fleece lighted from above by a red glare as of many torches, but scarcely penetrating the masses of clouds below. In the presence of hundreds of spectators there was a deafening explosion, and the concussion shook the houses and caused the bell of the village church to toll several times. At the moment of the explosion an ignited body ---compared in size to a hogshead---'descended like lightning,' and struck the earth about twenty feet east of a cotton gin which stands near the village. Within twenty minutes, the sky was clear and the sun shone as brightly as ever." A hot aerolite was found in the hole. (R4) The cloud preceeding the fall of the meteor is curious here. (WRC)

X5. December 17, 1852. Dover, England. Squalls and red sheet lightning. "At about $4^h50^m$ A. M., however, these flashes constantly emerging from a dense, triangular and very remarkable cloud in the S. E., which perceptibly increased in size with great rapidity, he was induced to observe it with minute attention. At $4^h55^m$ A. M., Greenwich mean time, the cloud had assumed the form of a right-angled triangle, its hypotenuse, or longest side, trending east and west. At this instant he first heard a singular and extraordinary hissing sound in the air, not unlike that of a passing shot, which, although at first not very loud, was yet clearly distinguishable above the howling of the gale. At 5 A. M. the cloud had nearly doubled its original size, advancing steadily from the S. E. in a N. W. direction, or from nearly dead to leeward, towards the wind's eye; whilst the scintillations spoken of were emitted with increased rapidity. He also then first perceived in the centre of the cloud, a dull, red, obscure nucleus, or fireball, apparently about half the diameter of the moon, having a tail five or six times that length, from which the flashes mentioned were sent forth, of surpassing brilliancy, as the meteor clearly descended with great velocity through the air, accompanied by a detonating, hurtling hissing sound, impossible to describe, yet resembling that which precedes the shock of an earthquake. At three minutes past five o'clock A. M., the meteor having apparently spanned the Channel from S. E. to N. W., upon approaching the land---evidently throwing off portions of its substance as it passed through the atmosphere---the nucleus suddenly exploded with a report similar to a very heavy clap of thunder, giving out an intensely brilliant light, which rendered the minutest objects distinctly visible, although it rained violently and the sky was obscured by dark and threatening clouds. The dense body of the meteorite seemed to fall in the water about half a mile from the land, as indicated by a great volume of spray, which rose foaming in the distance." (R5) The clouds and duration of this phenomenon make the 'meteor' anomalous indeed. The observation resembles many thunderbolt tales. (WRC)

X6. November 13, 1872. Scilly Islands, England. "...at 2 A. M., a meteor burst against the 'Seven Stones' light-vessel belonging to this corporation, and moored about 9 1/2 miles E. by N. of the Scilly Islands; and that it has been reported that the watch were struck senseless for a short period, seeing nothing before the shock, but that, on recovery, balls of fire like large stars were falling in the water like splendid fireworks, and that the decks were covered with cinders, which crushed under the sailors' feet as they walked. It appeared, the men said, as if something was passing swiftly and met with the obstruction of the vessel and burst. The superintendent reports that the men say there was a decided smell of brimstone, but adds that they did not mention that until he asked them." (R6)

X7. February 28, 1904. North Atlantic Ocean. "The meteors appeared near the horizon and below the clouds, travelling in a group from northwest by north (true) directly toward the ship. At first their angular motion was rapid and color a rather bright red. As they approached the ship they appeared to soar, passing above the clouds at an elevation of about 45°. After rising above the clouds their angular motion became less and less until it ceased, when they appeared to be moving directly away from the earth at an elevation of about 75°and in direction west-northwest (true). It was noted that the color became less pronounced as the meteors gained in angular elevation. When sighted, the largest meteor was in the lead, followed by the second in size at a distance of less than twice the diameter of the larger, and then by the third in size at a similar distance from the second in size. They appeared to be travelling in echelon, and so continued as long as in sight. The largest meteor had an apparent area of six suns. It was egg-shaped, the sharper end forward. This end was jagged in outline. The after end was regular and full in outline.. The second and third meteors were round and showed no imperfections in shape. The second meteor was estimated to be about twice the size of the sun in appearance, and the third meteor about the size of the sun. When the meteors rose there was no change of relative position, nor was there at any time any evidence of rotation or tumbling of the larger meteor. I estimated the clouds to be not over one mile high. The near approach of these meteors

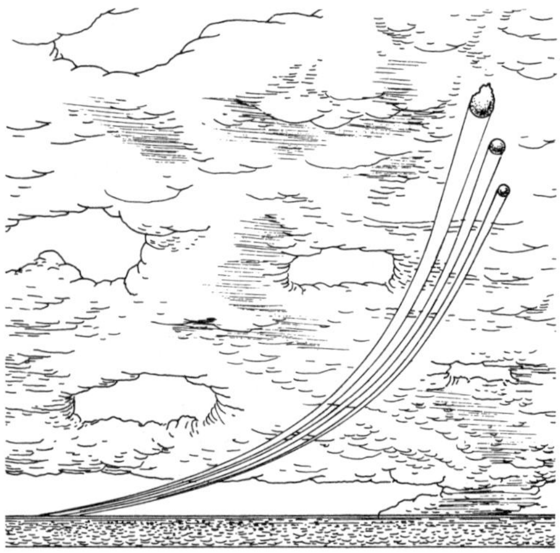

*Three meteors in formation dip below the clouds and swoop upwards out of sight.*

to the surface and the subsequent flight away from the surface appear to be most remarkable, especially so as their actual size could not have been great. That they did come below the clouds and soar instead of continuing their southeasterly course is also equally certain, as the angular motion ceased and the color faded as they rose. The clouds in passing between the meteors and the ship completely obscured the former. Blue sky could be seen in the intervals between the clouds. The meteors were in sight over two minutes and were carefully observed by three people, whose accounts agree as to details." (R7)

X8. October 29, 1936. Oshkosh, Wisconsin. "While duck hunting on Lake Butte des Mortes, a few miles northwest of Oshkosh, Wisconsin, my two companions, Arthur Nelson of Oshkosh, and F. E. Winans of Wauwatosa, and I were witnesses to a strange phenomenon.... We were facing the north watching a large raft of duck, when from the east, seemingly low over the water, came a rocket black formation throwing sparks of various colors, which shot across our front and seemed to disintegrate in the north-west." (R8)

X9. September 1977. Solomon Islands. "A mysterious flying object has been sighted in the Santa Cruz Island on three separate nights. The object was described as a 'very bright light' or 'ball of fire' which travelled at high speed across the sky just below cloud level lightning up the ground underneath. One of the objects was said to have landed but nothing was found when people rushed to the spot." (R9)

X10. March 19, 1887. North Atlantic Ocean. Two fireballs, one bright, the other dark, fell into the sea, making large waves. Heat and a suffocating atmosphere followed. (R10, R11)

References

R1. "The Meteor," American Journal of Science, 1:6:319, 1823. (X1)

R2. "Luminous Meteor," Edinburgh Philosophical Journal, 14:359, 1825. (X2)

R3. Collingwood, C.; "On a Remarkable Phenomenon Observed at Rangoon," Philosophical Magazine, 4:35:61, 1868. (X3)

R4. "Fall of an Aerolite at Forest Hill, Arkansaw," American Journal of Science, 2:5:293, 1848. (X4)

R5. Higginson, Francis; "An Explosive Meteorite," Royal Society, Proceedings, 6:276, 1853. (X5)

R6. Allen, Robin; "Account of a Meteor That Fell on the 'Seven Stones' Lightship," Royal Society, Proceedings, 21:122, 1873. (X6)

R7. Schofield, Frank H.; "Remarkable Meteors," Monthly Weather Review, 32:115, 1904. (X7)

R8. Keppler, F. W.; "A Strange Phenomenon, Evidently a Daylight Meteor," Popular Astronomy," 44:569, 1936. (X8)

R9. Burton, B. J.; "Bright Object over Solomons," Journal of Meteorology, U.K., 2:340, 1977. (X9)

R10. "Rare Electrical Phenomenon at Sea," Monthly Weather Review, 15: 84, 1887. (X10)

R11. "Rare Electrical Phenomenon at Sea," American Meteorological Journal, 4:98, 1887. (X10)

# GLM2 Darting Gleams of Light

Description. Very swiftly moving streaks or wisps of white or bluish-white light that appear suddenly and (apparently) very close to the observer. Called "sleeks" by some, most have been reported by expert meteor observers as a rare but persistent phenomenon.

Data Evaluation. Many observations by well-qualified observers. Rating: 1.

Anomaly Evaluation. The speed, apparent closeness, lack of sound and meteoric material, all suggest that sleeks are not true meteors. Rating: 2.

Possible Explanations. Some observers have proposed electrical or auroral origins, whatever these vague suggestions mean. Cosmic-ray stimulation of visual channels in the brain is a possibility, but one would then expect everyone to see the phenomenon indoors and out, rather than meteor watchers.

Similar and Related Phenomena. Anomalous and nebulous meteors (AYO).

Examples of Darting Gleams of Light

X1. January 9, 1870. Mediterranean Sea. "$14^h\ 50^m$; mag. 3; path from $169^\circ + 20^\circ$ to $157^\circ - 10^\circ$; length $31^\circ$; duration 0.1 second. An instantaneous flash; seemed to be in the air, quite near. Very curious." (R1)

X2. April 20, 1884. Bristol, England. "The meteor was about the third magnitude, but the singularity about it was its marvellous velocity and seeming nearness. It appeared to be in the air, a few yards distant, and I believe its path, extending (as it instantaneously impressed me) over some $16^\circ$ on the background of the sky, must have been traversed in less than the twentieth part of a second.....Now and then I have observed similar meteors before. They immediately strike one by their close proximity and enormous velocity. They are mere gleams of pale white light, which have little analogy to ordinary shooting stars, and suggest an electric origin, though I do not know whether the marvellous quickness with which they flash upon the eye is not to be held responsible for the sensation of nearness. They are somewhat rare, and one may watch through several whole nights without a single example, but, as far as my memory serves, I must have witnessed some scores of these meteoritic flashes." (R1)

X3. August 23, 1913. Bristol, England. "In late years I have not been able to pursue meteoric work to the same extent as formerly but I have occasionally caught one of these transient flashes. The last one seen here was on 1913 August 23 near the star Polaris and the meteor had a duration of less than the tenth of a second. It looked like a mere gleam of light; not a burning missile." (R2)

X4. September 13, 1913. England. (R2)
X5. December 1, 1913. England. (R2)
X6. February 1, 1914. England. (R2) Regarding these three examples. "This class of meteors is so different and distinct from ordinary ones that they are unmistakable when seen. They give one a little shock of almost fear, certainly surprise, at their seeming nearness and tremendous rapidity. They show no heads and I have never seen one with either a tail or an after streak."

X7. March 19, 1914. England. Two of them, "...appearing quite near, moving terribly rapid and suggesting little lines of light rushing close to one like a lighted wire." (R2)

X8. July 1929. In the Caribbean. Watching for shooting stars. "...and while I stood there looking for more a strange phenomenon took place. About fifty feet above the sea and say twenty feet from the ship's side a piece of incandescent matter of orange color, or perhaps a bit brighter, came down in a straight line and then was extinguished about ten feet before it fell into the sea. I heard no hiss as it struck the water owing to the noise of the water from the propellor (sic), nor could I see anything after its luminescence stopped." (R3)

X9. 1970. London, England. "I was lying in bed, in a darkened room, when I saw a bright streak of light to my extreme right. Had I seen this streak in the sky I would undoubtedly have called it a meteor and would have described it as: colour, white; magnitude, 11.5; speed, fast; length, $20^\circ$ arc; width, $1/2^\circ$ arc. However, it started from just below the ceiling and disappeared before it reached the floor. It did not move with my eye, but carried on falling as I looked up at it." (R4)

X10. General observation. "From time to time I have become aware of faint wisps of light that flashed into view, and startled me into thinking, for the instant, that I had seen a meteor. These wisps were extremely faint and nebulous, as if a piece of auroral streamer had swished across the sky. At first I took these appearances to be some psychological prank of the senses caused by the patient watch for meteors, or that they were faint meteors seen at such an extreme angle of vision that their light caused a distorted impression on the senses. When I especially watched for these appearances, they seemed quite real and to be just like the ones seen when I was not thinking of them while meteor observing. After observing these phenomena for some months in a somewhat desultory manner while making meteor observations, I came to the follow- conclusions. They appeared to assume many shapes, and their flight extended up to twenty degrees in .2 second, tho some may have travelled as long as .5 second. They were from one to ten degrees long, and one-half to three degrees broad, flashing into view quite in their entirity as to shape. They appeared from very few to twenty an hour, being especially numerous on three nights of slight auroral activity. One out of ten appeared directly in my line of vision. Lastly, they were more numerous in the early hours of the night, before midnight." (R5)

References

R1. Denning, W.F.; "A Peculiar Variety of Meteors," Royal Astronomical Society, Monthly Notices, 45:408, 1885. (X1, X2)

R2. Denning, W.F.; "A Curious Variety of Meteor," Popular Astronomy, 22: 404, 1914. (X3-X7)

R3. Speirs, Robert; "A Questionable and an Actual Meteorite," Popular Astronomy, 39:107, 1931. (X8)

R4. Galera, Delores J.; "Visible Cosmic Rays," New Scientist, 47:351, 1970. (X9)

R5. Anyzeski, Vincent; "Some Notes on a Possible Meteoric Phenomenon," Popular Astronomy, 54:203, 1946. (X10)

# GLN NOCTURNAL LIGHTS

*Key to Phenomena*



## GLN0 Introduction

Nocturnal lights are not associated in any obvious way with conventional meteors, electrical storms, or violent weather. The name "nocturnal light" had its origin in the UFO literature, but most nocturnal lights, such as the will-o'-the wisps, doubtless have very prosaic natural origins. The characteristics of the class are a flame-like or globular shape, a wide spectrum of colors, sizes that vary from an inch to many feet, and, most diagnostic of all, erratic motions, described as playful, elusive, inquisitive, and physically impossible.

Nocturnal lights are split here into two rather arbitrary categories: (1) low-level lights, which include will-o'-the-wisps, ghost lights, fireships, some curious natural flames, and a few UFOs; and (2) high-level lights, which seem to be predominantly UFOs.

Popular attention has long focussed on the high-level nocturnal lights because of the possibility they may have an extraterrestrial origin. This taint has prevented a concerted scientific attack on this fascinating problem. Low-level nocturnal lights, though ubiquitous today and in past centuries offer no alien overtones. Yet, they too are essentially ignored by science. The reason in this case is that it is commonly believed that the simple marsh gas theory explains all. This catalog section will demonstrate that the low-level lights pose many puzzles as to energy source, shape stability, and motion. It is a shame that such fascinating and common natural phenomena have received so little modern scientific attention.

## GLN1 Low-Level Nocturnal Lights

Description. Pale flames, softly glowing globes, and other luminous apparitions hovering and drifting erratically below tree-top level. These lights range from candle-size to radiating masses several feet in diameter. Colors vary widely: white, blue, and yellow being most common. Low-level nocturnal lights frequently vanish when approached closely only to reappear nearby. Their bobbing, swooping, elusive antics seem to beckon the observer to follow. Others, though, are essentially stationary and may be approached, especially the small flame-like species seen flickering over marshy areas. All low-level nocturnal lights, in fact, seem to prefer wet and swampy areas.

Background. Low-level nocturnal lights have been documented for centuries as will-o'-the-wisps, jack-o'-lanterns, ghost lights, fireships, corpse lights, and many other names. The standard explanation attributes them to spontaneously ignited gases escaping from swampy ground. The existence of this apparently reasonable explanation seems to have shut off all scientific research in this area. As the examples below demonstrate, the marsh-gas theory will simply not do in many instances. We actually have a complex, challenging phenomenon for science that may require several explanations.

Data Evaluation. Abundant testimony exists but modern scientific observation is essentially nonexistent. Rating: 1.

Anomaly Evaluation. No satisfactory mechanism has been demonstrated whereby gases escaping from marshy areas will spontaneously ignite. Furthermore, most low-level nocturnal lights are cold---not what one would expect from burning methane. Also, no one has explained how clouds of luminous gas can maintain size and shape while engaging in erratic maneuvers over many minutes. Rating: 2.

Possible Explanations. Some low-level nocturnal lights are undoubtedly spontaneously ignited gases escaping from the earth's surface. The cold flames, though, may be due to some unidentified kind of phosphorescence. The possibility that electromagnetic fields aid in generating light and in the long-term maintenance of size and shape should not be ruled out. There might, in fact, be a spectrum of electrochemical luminous phenomena beginning with highly energetic ball lightning, progressing through luminous aerial bubbles, and terminating with the weak and diffuse nocturnal lights. All of these phenomena seem to have similar life histories.

Similar and Related Phenomena. Ball lightning (GLB), St. Elmo's Fire (GLD), luminous aerial bubbles (GLD), and high-level nocturnal lights (GLN2).

Examples of Low-Level Nocturnal Lights

X1. Frequent occurrence. Lake Maracaibo, Venezuela. A luminous phenomenon seen frequently on the borders of the Cataumbo River near its junction with the Sulia. Can be seen 40 leagues away. Called the Faro or Taro of Maracaibo. Ascribed to ignited natural gas and electrical phenomena. (R1, R25, R30)

X2. May 26, 1821. Chapelle-aux-Planches, France. A pale red pyramid of light, 12 feet high, over marshy ground. (R2)

X3. Frequent occurrence. Along the Connecticut River. Lights often seen skipping along and sailing over the meadowlands. One observer saw at close range a solitary flame rising from the mud, flickering, waxing and waning. (R3, R4)

X4. Frequent occurrence. Newmark, Germany. "The water of the marsh is ferruginous, and covered with irridescent crust. During the day bubbles of air were seen rising from it, and in the night blue flames were observed shooting from and playing over its surface. As I suspected that there was some connexion between these flames and the bubbles of air, I marked during the day-time the place where the latter rose up most abundantly, and repaired thither during the night; to my great joy I actually observed bluish-purple flames, and did not hesitate to approach them. On reaching the spot they retired, and I pursued them in vain; all attempts to examine them closely were ineffectual.... On another day, in the twilight, I went again to the place, where I waited the approach of night; the flames became gradually visible, but redder than formerly, thus shewing that they burnt also during the day; I approached nearer, and they retired. Convinced that they would return again to the place of their origin, when the agitation of the air ceased, I remained stationary and motionless, and observed them again gradually approach. As I could easily reach them, it occurred to me to attempt to light paper by means of them, but for some time I did not succeed in this experiment, which I found was owing to my breathing. I therefore held my face from the flame, and also interposed a piece of cloth as a screen; on doing which I was able to singe paper, which became brown-coloured, and covered with a viscous moisture. I next used a narrow slip of paper, and enjoyed the pleasure of seeing it take fire. The gas was evidently flammable, and not a phosphorescent luminous one, as some have maintained. But how do these lights originate?" (R5, R6, R26, R67)

X5. Frequent occurrence. Upper Silesia, Germany. Stationary flames observed, but could not ignite paper. (R5, R6, R26)

X6. 1811. Konski Forests, Poland. Stationary flames, which did not ignite paper but covered it with a viscous moisture. (R5, R6, R26)

X7. 1812. Rubenzahl Garden, Germany. Very pale flames, which could not be approached without extinguishing them. (R5, R6)

X8. Frequent occurrence. Walkenried, Germany. Gas causing the lights was collected and found to cloud lime-water. (R5, R6)

X9. August 3, 1814. Minden, Germany. Rockets fired from the summit of a mountain caused small red flames to appear around the base of the summit. They quickly faded. (R5, R6, R26)

X10. September 19-20, 1836. Shetland Isles, Great Britain. Fishermen reported a light like a furnace on the surface of the water. It lasted two nights. Rays from it rose to great heights. (R7) This light may have been auroral. (WRC)

X11. Spring 1813. Near Lincoln, England. "... a Jack-o'-lantern caught my attention, proceeding in the same direction as I was travelling. Its motion was irregular, sometimes near the surface of the ground, and then suddenly rising to the height of five or six feet. I followed very cautiously for some distance, being determined, if possible, to obtain a near view of my luminous guide. As the night was rather dark, I had everything favourable for observation. At length it rested just at an angle in the road. I dismounted, and proceeded very cautiously, in the hope of capturing it; but in this I was disappointed; for, on my near approach, whether from the noise I made, or some other cause, it suddenly rose from its resting-place, about two feet from the ground, cleared a high bank, and pursued its course in a direct line over the adjoining fields. The broad and deep dikes rendered pursuit fruitless; but my eyes followed its almost butterfly motion till the glimmering taper was lost in the distance." (R8)

X12. November 1818. Schnepfenthal, Germany. Observing a boggy meadow. "I perceived a number of reddish-yellow flames in different parts of the expanse of almost level ground. I descried, perhaps, no more than six at a time; but dying away, and appearing in other places so rapidly, that it was impossible to count them; but I should say, on a rough estimation, that there were about 20 or 25 within a second. Some were small, or burned dimly; others flashed with a bright flame, in a direction almost parallel with the ground, and coinciding with that of the wind, which was rather brisk. After having for some time looked with amazement at this brilliant scene as a whole, I tried to study its details; and soon found that the flames which were nearest originated in a quagmire, whose position I knew exactly, by a solitary cluster of willows; and I could trace a succession of flashes from that spot to a certain point of the margin of the wood, across the rivulet and meadow. The distance of the two points from each other was more than half a mile, and the flames travelled over it perhaps in less than a second. The first flash was not always observed in the immediate neighborhood of the quagmire; but the succession of flames lay always in the same straight line, and in the direction of the wind; whilst other sets were, though not with the same distinctness, observed in the more distant parts of the meadow-ground." (R8)

X13. Frequent occurrence. Bologna, Italy. Near a pool. "About 11 o'clock I saw the light appear which I was desirous to observe; and I instantly seized the stick which I always kept ready for the purpose, and which had some flax attached to its extremity, and speedily repaired to the spot indicated. When I was not more than about twenty feet from the light, I stopped a moment to observe it. It had the form and colour of an ordinary flame, with a slight discharge of smoke. Its thickness was about a decimeter; and it was moving slowly in a direction from south to north. When I approached nearer it changed its direction, retired from me, and began to rise upwards. I hurried forward with my stick, and thrust it into the flame, which kindled the flax. Soon after, the Jack-o'-lantern became extinct at a height of about two or three feet above the stature of a man. It soon reappeared of smaller size (for I was led to believe it was the same), on another pool placed at a little distance. I ran immediately towards it, but in vain, as it vanished in a few second." (R9)

X14. Circa 1840. Rome, New York. A lambent flame was seen playing on the surface of a sand bank. Digging revealed a irregular tube of fused sand nearly 50 feet long. The tube resembled a fulgerite except that it was much larger. (R10)

X15. December 2, 1845. Ryook Phyoo. A large flame appeared out on the sea, flickering 15 or 20 minutes. (R11)

X16. No date given. Liege, Belgium. A blue flame rose from a well and burned for some time. The ground in this region is calcined and hot in places. (R12)

X17. 1860. Cleveland, Ohio. A luminous phenomenon appearing like a vessel on fire out on Lake Erie. Such apparitions called frequent. (R13)

X18. June 15, 1872. Nandidroog, India. "...the plain to the east, north, and north-east of Nandidroog was covered with 'many thousands' of lights, which have been observed occasionally in former years. The correspondent compares the appearance to that of a large city brilliantly illuminated, and in one direction the scene, through a binocular glass, 'looked like a view of part of the starry heavens, each flame being like a star.' As many of the lights were from ten to fifteen miles distant from the reporter's point of view, he conjectures that each flame must have been five to six feet in length." (R14, R31)

X19. October 5, 1872. The Lizard, England. Lights seen bouncing over a marshy field reached lengths of 30 feet. (R15)

X20. February 1875. Wales. "Some few days ago we witnessed here what we have never seen before---certain lights, eight in number, extending over, I should say, a distance of 8 miles; all seemed to keep their own ground, although moving in horizontal, perpendicular, and zigzag directions. Sometimes they were of a light blue colour, then like the bright light of a carriage lamp, then almost like an electric light, and going out altogether, in a few minutes would reappear again dimly, and come up as before." (R16) See GLN1-X39 for similar lights seen during the great Welsh religious revival. (WRC)

X21. 1832. Fifeshire, England. Lights seen frequently over a bog moved in straight and zigzag paths for 20-30 yards. (17)

X22. August 11, 1880. Whitby, England. Tongues of flame 15-30 feet high seen along a distant seashore. (R18)

X23. General observation. A long review of the will-o'-the-wisp in folklore and tradition indicates a persistent, world-wide phenomenon. (R19)

X24. Circa 1880. Off Montauk Point, New York. A light moved along the surface of the sea like a phantom ship. (R20) This is the general locale of the famed Palatine Light, a fireship. (WRC)

X25. No date given. On the River Parana, South America. While investigating a mysterious light reported by villagers. "Only twice was it seen by me, but then very distinctly; the first time some little distance off, but the second quite close. On the first occasion the light started up from the ground with the brightness and speed of a rocket, and then again descended to the earth with equal velocity but less splendour; on the second we caught sight of it as it directly, but gently approached along the road, upon which, running to intercept it, and stumbling at every step over rough and swampy ground, we managed to arrive within three yards of the glowing vision as it slowly glided on a level of about 5 feet from the earth. It presented a globular form of bluish light, so intense that we could scarcely look at it, but emitted no rays and cast no shadows; and when about to actually grasp the incandescent nothingness, suddenly elongating into a pear-shape tapering to the ground, it instantly vanished; but on looking round up it rose again within 50 yards, but this time we could not overtake it, as it bounded over a hedge, then over trees, and finally disappeared in an inpenetrable swamp. According to the testimony of the soldiers, on another occasion, they beheld it rise from the swamp, and perch for some minutes on the top of the roof of a neighboring rancho without walls, after which it pierced the roof and subsided in the ground underneath...." (R21)

*Pursuing a mysterious light in South America*

X26. No date given. England. A large will-o'-the-wisp moved up and down, back and forth in a high wind. (R21)

X27. No date given. Italy. While walking along a road surrounded by water-covered rice fields, flickering lights appeared at eye level in all directions. They always kept a fixed distance away. The observer thought he was in a faintly luminous mist, which appeared flame-like where it was most dense. (R22)

X28. No date or place given. A waving flame-like light 5 feet high, 18 inches wide, moved slowly across a graveyard against the wind. (R23)

X29. July 11, 1881. At sea off Sydney, Australia. A strange red light on the sea was seen by 13 crewmen and two nearby ships. Some thought it a phantom ship. (R24)

X30. No date given. Sussex, England. Often seen in the evening hours, this light sometimes makes a graceful sweep to quite a height. Bad weather frequently follows. (R27)

X31. February 24, 1893. North Pacific Ocean. "... some unusual lights were reported by the officer of the watch between the ship and Mount Auckland, a mountain 6,000 feet high. It was a windy, cold, moonlit night. My first impression was that they were either some fires on shore, apparently higher from the horizon than a ship's masthead, or some junk's 'flare up' lights raised by mirage. To the naked eye they appeared sometimes as a mass; at others, spread in an irregular line, and, being globular in form, they resembled Chinese lanterns festooned between the masts of a lofty vessel. They bore north (magnetic), and remained on that bearing until lost sight of about midnight. As the ship was passing the land to the eastward at the rate of seven knots an hour, it soon became obvious that the lights were not on the land, though observed with the mountain behind them. On the following night, February 25, about the same time, 10 p. m., the ship having cleared Port Hamilton, was steering east, on the parallel of 34°, when these curious lights were again observed on the same bearing, at an altitude of 3° or 4° above the horizon. It was a clear, still, moonlit night, and cold. On this occasion there was no land in sight on a north bearing when the lights were first observed, but soon afterwards a small islet was passed, which for a time eclipsed the lights. As the ship steamed on at a rate of seven knots an hour, the lights maintained a constant bearing (magnetic) of N. 2°W., as if carried by some vessel travelling in the same direction and at the same speed. The globes of fire altered in their formation as on the previous night, now in a massed group, with an outlying light away to the right, then the isolated one would disappear, and the others would take the form of a crescent or diamond, or hang festoon-fashion in a curved line. A clear reflection or glare could be seen on the horizon beneath the lights. Through a telescope the globes appeared to be of a reddish colour, and to emit a thin smoke. I watched them for several hours, and could distinguish no perceptible alteration in their bearing or altitude, the changes occurring only in their relative formation, but each light maintained its oval, globular form." (R28)

X32. No date given. Bay of Islands, New Zealand. Balls of pale light floated a yard off the ground or attached themselves to pointed objects. (R29)

X33. April 1897. Lee, Michigan. A ball of

*Group of low-level lights over the North Pacific*

light the size of a hen's egg emerges from a river, floats through the air 10 feet above the ground with a zigzag motion and a whizzing sound. (R32) The sound hints at ball lightning. (WRC)

X34. August 1897. Croisic, France. Numerous bright bubbles of light rose from the waters of the port. (R33, R34)

X35. August 1900. Scotland. Rapidly moving lights climbed a hillside and went down to a loch shore. (R35)

X36. June 5, 1902. Blundellsands, England. "The evening was dull and grey, a strong north-westerly wind was blowing in from the sea and the tide was flowing in. In the distance we first saw smoke with frequent jets of fire bursting forth from the mud of a shallow canal. Drawing near, we perceived a strong sulphurous odour, and saw little flames of fire and heard a hissing sound as though a large quantity of phosphorus was being ignited. It was impossible to detect anything which caused the fire, only the water where the flames appeared had particles of a bluish hue floating on the surface. The area over which the tiny flames kept bursting forth was

*Flames erupt from English mud flat in 1902.*

about 40 yards. A gentleman present stirred up the mud with his walking stick, and immediately large yellow flames nearly 2 feet in length and breadth burst forth. The phenomenon lasted some time, until the tide covered the part and quenched the fire.. As we returned from our walk the atmosphere was impregnated with a strong odour of sulphur." (R36)

X37. Circa 1900. Towyn, Wales. Four or five white lights the size of a railway lamp move across sandhills. (R37)

X38. Winter 1694. Harlech, Wales. "A pestilential vapour resembling a weak blue flame arose during a fortnight or three weeks out of a sandy, marshy tract called Morfa Byden, and crossed over a channel of 8 miles to Harlech. It set fire on that side to 16 ricks of hay and 2 barns, one filled with hay, the other with corn. It infected the grass in such a manner that cattle, etc., died, yet men eat of it with impunity. It was easily dispelled; any great noise, sounding of horns, discharging of guns, at once repelled it. Moved only by night, and appeared at times, but less frequently; after this it disappeared." (R37)

X39. 1905. Wales. Many who were caught up in the excitement of the Welsh religious revival of 1905 claimed they saw unusual lights. (R37)

X40. 1877. Wales. Lights of various colors moved across the estuary of the Dysynni. (R37)

X41. September 1, 1905. Kittery Point, Maine. "On the evening of Friday, September 1, the guests at the Hotel Parkfield were startled by the appearance of flames rising from the beach and from the surface of the water, an event of so remarkable and unusual a character as to excite great curiosity and some alarm. The conflagration occurred between seven and eight o'clock in the evening, and lasted for upwards of forty-five minutes. The flames were about one foot in height. They were accompanied by a loud and continuous crackling noise which could be distinctly heard one hundred yards away, while at the same time there was a very strong liberation of sulphurous acid fumes which penetrated the hotel, drove the proprietor and his staff from the office and filled the other rooms to such an extent as to cause great inconvenience to the guests. One guest of an investigating turn of mind secured some of the sand in his hand, but was obliged to drop it on account of the heat. When some of the sand was taken into the hotel and stirred in water, bubbles of gas were liberated and produced flame as they broke at the surface in contact with the air." (R38, R40)

X42. General observations. Bay Chaleur, Canada. The summary of Ganong's study of the reputed Bay Chaleur Fireship. "After an examination of all the evidence it appears to the author plain (1) that a physical light is frequently seen over the waters of Bay Chaleur and its vicinity; (2) that it occurs at all seasons, or at least in winter and summer; (3) that it usually precedes a storm; (4) 'that its usual form is roughly hemispherical with the flat side to the water, and that at times it simply glows without much change of form, but that at other times it rises into slender moving columns, giving rise to an appearance capable of interpretation as the flaming rigging of a ship, its vibrating and dancing movements increasing the illusion. '" (R39, R41) There may be a connection between the supposed fireships and the weather lights (GLA16).

X43. General observations. Norfolk, England. The marshmen and wherrymen fire guns at the will-o'-the-wisps. The detonations usually extinguish them. (R42)

X44. No date given. Southern Nevada. The writer was tramping across a desert area that was usually dry but had just been flooded by one to several inches of water. "On this night the writer was tramping from Goodsprings, Nevada, to Ivanpah, California, and at 2 A.M. encountered a water-covered area. The flitting lights did not begin to appear until after some progress had been made through the water. The night was very dark and suddenly one of these lights appeared to float in the air about five feet above the ground. However, owing to no knowledge regarding its size, it could readily be imagined to be the light from a cabin window at some distance but for the fact that there was no habitation within twenty miles of the spot. Suddenly this light ceased to float complacently and sailed swiftly away to the left for some distance, then stopped. It was easy to imagine that it was waiting and coaxing---living true to its name in attempting to 'lure benighted travelers to destruction.' Soon others appeared, some floating apparently stationary, others shooting here and there. Little wonder that the peasantry of ancient times viewed this phenomenon with superstitious awe! When the display was at its height, hundreds of individual lights were visible simultaneously. The display was continous during the hour or more that the water-covered area was being traversed. It possibly continued for days. At first it appeared to be very chaotic, but after observing it for a time during which the writer continued walking, the individual lights seemed to progress rather jerkily or spasmodically in the same general direction." (R43)

X45. General observation. The curious motions of will-o'-the-wisps may be due to an autokinetic illusion, based on a personal experience. (R44)

X46. Frequent occurrence. Shipton, England. A peculiar light is often seen moving along a road 3-4 feet from the ground. The light may swing back and forth, emulating a hand-held lantern. (R45)

X47. No date given. Keswick, England. White lights seen near a road. One approached, slowed, stopped, quivered, and slowly went out. It was globular, white, about 6 feet in diameter and high enough to just clear a man's head. (R46)

X48. December 2, 1807. No place given. Numerous little bluish flames observed over a peat marsh. Some were stationary, but others moved horizontally in coordinated groups. (R47)

X49. No date given. Germany. In a wet area: greenish-white lights the size of a hen's egg, standing quietly amidst the grass. Waving a finger at them would usually make them disappear with a faint pop. No heat could be felt. (R47)

X50. No date given. Germany. Little flames 1 inch high in a railway cut. They were 3 inches above the ground, emitted neither smoke nor odor. (R47)

X51. No date given. Germany. Among the marsh grass, a cylindrical light 5 inches high, 1 1/2 inches in diameter. No odor or smoke. Not easily disturbed by air motion. (R47)

X52. 1921. England. Lights seen moving around inside a building attributed to will-o'-the-wisps created by the accidental release of producer gas. (R48)

X53. April 7, 1921. Cayutaville, New York. Five globules of bluish light the size of a half dollar were suspended, waveringly, 1 inch over marshy ground. (R49)

X54. Frequence occurrence. Brown Mountain, North Carolina. "Since about 1850 there have been reports of 'lights' seen at night over the mountain. These lights have been described as varying in size, but to somewhat resemble 'toy balloons' and to be pink, orange, or reddish in color. They are said to rise into view over the mountain and to hover for periods of one to fifteen minutes duration before fading out. Sometimes, it is claimed, as many as three lights can be seen simultaneously at widely separated points. They are alleged to be bright enough to be visible from Blowing Rock, which is twenty miles away. Sightings are said to be frequent in fine weather, but also occur when the sky is overcast or the mountain hidden in mist.... On the night of 17 September I maintained a watch from the chosen position on the neighboring hill from 7.30 p.m. until 1 a.m. It had rained heavily all day. The rain stopped at about 8 p.m.,, but the sky was heavily overcast and the air exceedingly damp and misty.... from 8.30 p.m. until 11.00 p.m. I witnessed the lights which showed remarkable regularities. There were three distinct types.... Each light appeared to have its own characteristic color, frequency of occurrence, and elevation.... Light 1 was yellowish white and appeared at intervals of 8-11 seconds, having a duration of about half a second. Light 2 was reddish orange in color, and appeared at intervals of 12-15 seconds, having a duration of about half a second. Light 3 was violet in color. When it flashed the duration was only about one eighth of a second. Its occurrence was much less regular that that of the others. Sometimes there would be a series of several flashes mutually separated by time intervals of approximately 35 seconds. These bursts would be followed by lengthy pauses with no activity until the next burst. The duration of the pauses varied between 5 minutes and 30 minutes. According to my visual observation the lights seemed to be electrical in nature, and had a distinct resemblence to lightning. However, the somewhat rhythmic cycles as described above were quite unlike what is normally expected of lightning." (R50, R51, R64)

X55. November 1927. Bohemia, Germany. In a marshy area near an iron mine. Four small bluish flames 1 inch in diameter were constantly hopping up and down. Nearby a yellowish-green flame measured 3 feet high, 1 foot in diameter. The flames produced no heat and persisted about 15 minutes. (R52)

X56. September 1914. France. Thin, pale, greenish flames, 12-18 inches high seen shooting up and down over a marshy area. (R52)

X57. August 4, 1924. Mandeville, Louisiana. In hot, very dry weather, many marsh fires ignited spontaneously. (R53)

X58. Summer 1840. Argylshire, England. Several candle-size flames were seen flitting across the surface of a pond. They seemed to leap from place to place and disappear. The flames rose 1-2 feet above the surface of the water. (R54)

X59. September 1858. South Wales. Many bright lights swung about over a bog. (R54)

X60. No date given. Grupont, Belgium. In some marshy woods, white clouds rose from the ground and formed into luminous globes, while attaining an altitude of a dozen yards. They remained visible several minutes. (R55)

X61. Circa 1812. New London, Connecticut. On December 2, as Commodore Decatur was planning to run the British Blockade, blue lights appeared on the shores of the harbor. They were never explained. (R55)

X62. Frequent occurrence. Esperanza Creek, Texas. "This light on the Esperanza Creek, this Will O' the Wisp, appears to be about the size of the headlight on an automobile. It travels at various speeds, and at a distance of from one to three feet above the ground. It has been seen in all kinds of weather conditions, and at all seasons of the year, and at all hours from dusk till dawn.... Only travelers on horseback or in horse-drawn vehicles see the light, which may approach the traveler from either side. It comes within twenty steps before it stops. When the traveler's horse stops, blinded by the light, the light retreats and meanders along, crossing the road back and forth in front of the traveler. Sometimes the light slowly recedes until it can scarcely be seen, but it returns at great speed. After crossing the creek, the light continues to follow for a few miles, but always returns to the river at the mouth of the creek and goes straight up and then down and disappears. (R56) There are scores of such 'ghost lights' described in the folklore and Fortean literature. (R65, R66)

X63. Frequent occurrence. Northeastern Oklahoma. The famous Tri-State or Hornet 'spooklight.' "In the northeastern corner of Oklahoma, some 14 miles from Joplin, Mo., is a lonesome stretch of country road called the 'Devil's Promenade.' Some mighty strange people have lived along this road, and some very strange things have happened there. The best of the 'haunted road' stories cannot be told at this time, but there is no longer any secret about the phenomena of the 'Indian lights,' which have been seen by thousands of tourists and discussed in newspapers as far off as St. Louis and Kansas City.... One has only to drive slowly along the road any night after dark to see the 'jack-o'-lantern' come bobbing along, always traveling in an easterly direction. Sometimes it swings from one side of the road to another, sometimes it seems to roll on the ground, sometimes it rises to the tops of the scrubby oak trees at the roadside, but it never gets more than a few feet from the road on either side. I have seen this light myself, on three occasions. It first appeared about the size of an egg, but varied until sometimes it looked as big as a washtub. It was about as powerful as an ordinary automobile headlight with the dimmer on. It is hard to judge the distance, but it seemed about a quarter of a mile off when I first saw it, and disappeared when it approached to a distance of perhaps 75 yards. I saw only a single glow, but other witnesses have seen it split into two, three and four smaller lights. It looked yellowish to me, but some observers describe it as red, green, blue or even purple in color. One man swore that it passed so close to him that he could 'plainly feel the heat,' and a woman saw it 'burst like a bubble, scattering sparks in all directions,' A fellow who drove his car straight at the dancing phantom lost sight of it, but others standing a little way off said that they saw the light hovering impishly above the pursuer's car, out of sight but plainly visible to everyone else in the neighborhood." (R57) Several who have carefully studied the Hornet Light are convinced that the refraction of distant automobile headlights is the cause. (R60, R61, R63)

X64. No date given. England. Channels cut through a peat bog released huge quantities of marsh gas, but there was no ignition and no will-o'-the-wisps. (R58)

X65. August 7, 1957. North Atlantic Ocean. "At 2150 G.M.T. what appeared to be the loom of three lights appeared on the eastern horizon. The central loom was the most distinct and bore 098$^{\circ}$, the others were 2$^{\circ}$ on each side of it. It was thought at first that the phenomenon was due to the reflection of moonlight from the clouds, but the source of light appeared to be below the horizon and was not steady. The moon bore 168$^{\circ}$. No echoes showed on the radar screen on this bearing though the echo of a ship at a distance of 15 miles showed clearly, and also those of rain squalls over 30 miles away, so it would appear unlikely that the source of the lights was associated with a ship or ships. As the nearest land, Cape St. Vincent, was 200 miles distant, it was difficult to relate the lights to anything on the land. The central light persisted for 40 min, until 2230; its bearing became more N'ly and was 080$^{\circ}$ when it disappeared. The lesser lights lasted 30 min, until 2220." (R59)

X66. Frequent occurrence. Silver Cliff, Colorado. Dim, round spots of blue-white light glowed among the gravestones. They vanished when approached. (R62)

X67. General observations. A chemical study of the gases evolved from decaying organic matter provides no support for the popular belief that will-o'-the-wisps are spontaneously ignited marsh gas. (R67)

X68. No date given. Caspian Sea. Subterranean disturbances release immense quantities of naptha from underground deposits. The naptha ignites and thousands of square miles of the Caspian are set afire. (R68)

References


R1. "Meteoric Phenomenon Called the Lantern of Maracaybo," Edinburgh Philosophical Journal, 1:424, 1819. (X1)

R2. "Meteoric Fire in the Marsh of the Chapelle-aux-Planches," Edinburgh Philosophical Journal, 6:381, 1822. (X2)

R3. Mitchell, John; "Observations on Ignis Fatuus," American Journal of Science, 1:16:246, 1829. (X3)

R4. Mitchell, John; "Observations on Ignis Fatuus," Franklin Institute, Journal, 8:73, 1829. (X3)

R5. Blesson, L.; "Observations on the Ignis Fatuus,..." Edinburgh New Philosophical Journal, 14:90, 1832. (X4-X9)

R6. Blesson, L.; "Observations on the Ignis Fatuus,..." Franklin Institute, Journal, 15:408, 1833. (X4-X9)

R7. "Luminous Appearance at Sea off the Shetland Isles," Edinburgh New Philosophical Journal, 22:191, 1836. (X10)

R8. White, W.H., and Weissenborn, W.; "Ignis Fatuus," Magazine of Natural History, 1:551, 1837. (X11, X12)

R9. "Ignis Fatuus....," Edinburgh New Philosophical Journal, 34:383, 1843. (X13)

R10. West, Charles E.; "Notice of Certain Siliceous Tubes (Fulgerites) Formed in the Earth," American Journal of Science, 1:45:220, 1843. (X14)

R11. Mahon, Viscount; Royal Society, Proceedings, 5:627, 1846. (X15)

R12. "Curious Phenomenon," Scientific American, 13:318, 1858. (X16)

R13. "Optical Illusion on Lake Erie," Scientific American, 3:123, 1860. (X17)

R14. Nature, 6:270, 1872. (X18)

R15. Fox, Howard; "Will-o'-the-Wisps," Nature, 7:222, 1873. (X19)

R16. "Strange Lights in Wales," Notes and Queries, 5:3:306, 1875. (X20)

R17. Elder, David; "Will-wi'-the-Wisp," Symons's Monthly Meteorological Magazine, 15:156, 1880. (X21)
R18. "Atmospheric Phenomenon," Nature, 22:362, 1880. (X22)
R19. Dyer, T. F. Thiselton; "The Will-o'-the-Wisp and Its Folklore," Popular Science Monthly, 19:67, 1881. (X23)
R20. "Phantom Lights at Sea," Scientific American, 47:56, 1882. (X24)
R21. "Will-o'-the-Wisp: A Confession," Journal of Science, 19:447, 1882. (X25, X26)
R22. Williams, W. Mattieu; "Will-o'-the-Wisp Again," Journal of Science, 19:533, 1882. (X27)
R23. Journal of Science, 20:247, 1883. (X28)
R24. "The Flying Dutchman," Scientific American, 54:279, 1886. (X29)
R25. "Asphalt and Petroleum in Venezuela," Popular Science Monthly, 35:142, 1889. (X1)
R26. Tomlinson, Charles; "Some Remarks Suggested by a Popular Article on the Ignis Fatuus," Philosphical Magazine, 5:32:464, 1891. (X4, X9)
R27. "The 'Ignis Fatuus,' or, Will o'the Wisp," Royal Meteorological Society, Quarterly Journal, 17:260, 1891. (X30)
R28. Norcock, Chas. J.; "An Atmospheric Phenomenon in the North China Sea," Nature, 48:76, 1893. (X31)
R29. Tomlinson, Charles; "The Ignis Fatuus," Knowledge, 16:89, 1893. (X32)
R30. "Infernito," Nature, 40:429, 1889. (X1)
R31. Walhouse, M. J.; "Ghostly Lights," Folk-Lore, 5:293, 1894. (X18)
R32. "Ignis Fatuus or Jack-o'-Lantern," Monthly Weather Review, 25:211, 1897. (X33)
R33. Bleunard, M. A.; "Extraordinary Will-o'-the-Wisps," English Mechanic, 64:586, 1897. (X34)
R34. Bleunard, M. A.; "Will-o'-the-Wisps," English Mechanic, 68:433, 1898. (X34)
R35. Lang, A.; "Spectral Lights," Folk-Lore, 18:343, 1901. (X35)
R36. Dixon, H. T.; "Flames from Mud on a Sea-Shore," Nature, 66:151, 1902. (X36)
R37. Fryer, A. T.; "Psychological Aspects of the Welsh Revival," Society for Psychical Research, Proceedings, 19:80, 1905. (X20, X37-X40)
R38. Penhallow, D. P.; "A Blazing Beach," Science, 22:794, 1905. (X41)
R39. "The Fire-Ship of Bay Chaleur," Science, 24:501, 1906. (X42)
R40. Penhallow, D. P.; "A Blazing Beach," Popular Science Monthly, 70:557, 1907. (X41)
R41. Ganong, W. F.; "Mysterious Lights," Society for Psychical Research, Journal, 13:8, 1907. (X42)
R42. "Jack-o'-Lantern," English Mechanic, 96:454, 1912. (X43)
R43. Luckiesh, M.; "An Unusual Experience with Ignes Fatui," Scientific American Supplement, 82:384, 1916. (X44)
R44. Peterson, Joseph; "Some Striking Illusions of Movement of a Single Light on Mountains," American Journal of Psychology, 28:476, 1917. (X45)
R45. "A Mystery," English Mechanic, 110:128, 1919. (X46)
R46. Sington, T.; "A Mystery," English Mechanic, 110:152, 1919. (X47)
R47. Sanford, Fernando; "Ignis Fatuus," Scientific Monthly, 9:358, 1919. (X48-X51)
R48. Smith, David J.; "Will-o'-the-Wisp?" English Mechanic, 113:178, 1921. (X52)
R49. Hausman, Leon Augustus; "The Mystery of the Will-o'-the-Wisp," The Mentor, 13:58, May 1925. (X53)
R50. "The Queer Lights on Brown Mountain," Literary Digest, 87:48, 1925. (X54)
R51. Chater, Melville; "Motor Coaching through North Carolina," National Geographic Magazine, 49:51, 1926. (X54)
R52. "Will-o'-the-Wisp," Science, 68:sup xii, October 5, 1928. (X55, X56)
R53. Viosca, Percy, Jr.; "Spontaneous Combustion in the Marshes of Southern Louisiana," Science, 75:461, 1932. (X57)
R54. Talman, Charles Fitzhugh; "What Is Will-o'-the-Wisp?" Nature Magazine, 20:7, 1932. (X58-X60)
R55. Fortean Society Magazine, 1:16, October 1937. (X61)
R56. Blackwell, John W.; "Will-o-the-Wisp of the Esperanza," Texas Folklore Society, Publications, 17:118, 1941. (X62)
R57. Randolph, Vance; Ozark Magic and Folklore, New York, 1947, p. 233. (X63)
R58. Thomson, G.; "The Will-o'-the-Wisp," Weather, 6:63, 1951. (X64)
R59. Crane, T. C.; "Unidentified Phenomenon," Marine Observer, 28:81, 1958. (X65)
R60. "Spook Light," Science News Letter, 77:189, 1960. (X63)
R61. Gannon, Robert; "Balls O'Fire," Popular Mechanics, 124:116, September 1965. (X63)
R62. Linehan, Edward J.; "The Rockies' Pot of Gold," National Geographic Magazine, 136:157, 1969. (X66)

R63. Hynek, J. A., and Powers, R. T.; "The Tri-State Spook Light," New Horizons, 1:6, Summer 1972. (X63)
R64. Bessent, Malcolm; "The Lights of Brown Mountain," New Horizons, 1:3, Summer 1972. (X54)
R65. Arnold, Larry E.; "Ahoy, Mate! Which Flamin' Phantom Ship Sails Thar?" Pursuit, 11:144, 1978. (X62)
R66. Hall, Mark A.; "A Ghost Light Survey," INFO Journal, 3:19, 1972. (X62)
R67. Mills, A. A.; "Will-o'-the-Wisp," Chemistry in Britain, 16:69, February 1980. (X4, X67)
R68. "The Caspian Sea on Fire," Eclectic Magazine, 10:380, 1869. (X68)

# GLN2 High-Level Nocturnal Lights

Description. Stationary and moving lights, usually well above tree-top level, that cannot be correlated with known objects or phenomena. These lights may move erratically, often at great speeds and with accelerations seemingly incompatible with the characteristics of human-built vehicles. The color of the lights may vary during the observation. Radar and visual contacts are sometimes made together.

Background. The high-level nocturnal light is a classic type of UFO. Few observations of this phenomenon are to be found in the mainstream scientific literature, which is the focus of the Sourcebook Project. However, many thousands of cases exist in newspapers, UFO journals, and in the files of UFO investigators. Some of this material will be incorporated in future editions of this catalog.

Data Evaluation. High-level nocturnal lights are reported almost every day in some part of the world, as any UFO clipping service clearly demonstrates. The real question is not the existence of such lights but whether they can be correlated with known vehicles and/or physical processes. Rating: 1.

Anomaly Evaluation. If it is actually impossible to identify all high-level nocturnal lights, a major scientific anomaly must exist, for these lights seem to possess properties (accelerations, color changes, velocities, materialization and disappearance) that lie far outside the explanatory capabilities of present-day science. Rating: 1. (Note: if high-level nocturnal lights are ultimately identified as extraterrestrial vehicles, the anomalousness of the situation will decrease, because current thinking leans towards a universe teeming with other intelligent civilizations.)

Possible Explanations. Terrestrially built vehicles, including aircraft, spacecraft, missiles, and balloons; misidentified astronomical objects, such as meteors and planets; anomalous electrical discharges; ball lightning (GLB); antimatter meteorites; physiological and psychological phenomena associated with the observer; extraterrestrial vehicles; holograms.

Similar and Related Phenomena. Low-level nocturnal lights (GLN1), ball lightning (GLB), erratic meteors (AYO), enigmatic astronomical objects (AEO), unrecognized life forms (BXU), unrecognized artifacts (XSU).

Examples of High-Level Nocturnal Lights

X1. Summer 1896. Iowa. Looking out the door of the mail car on a moving train. "The darkness was intense; not a ray of light was visible from any point, except from the train. When a few miles out from Princeton, and while traveling almost due north, I observed a peculiar light low down on the western horizon. It appeared to be perfectly round and about a foot in diameter, of a dull rose color, or, possibly, like a piece of live coal. When first observed it seemed to be floating within a hundred feet of the earth, but soon rose to a height about midway between the horizon and the zenith. For a time it floated very steadily, but soon began to oscillate up and down, at times even dropping out of sight behind hills. The wind was quite strong from the east, but the light traveled in an almost due north course. Its speed varied, sometimes seeming to outrun the train considerably, and at others it would fall behind, but never far enough to be lost to sight. Most of the time it appeared to be nearly abreast of the train and apparently from half a mile to a mile distant. Soon after it was first observed by me, my companion arose, and we both watched it closely until the town of Lineville, Iowa, was reached. There it passed out of sight behind the depot, and we saw it no more." (R1)

X2. July 1952. Florida. "On one night several airmen independently observed a light approach at a very low speed, come to a halt nearly overhead, then reverse direction with no apparent turn. On two other nights, three other lights appeared in other sections of the sky, of similar appearance, but maneuvering more rapidly. They were observed for some 10 minutes by 9 airmen, including a control tower operator, an aircraft dispatcher, and two pilots from Wright Field. In the words of one of the men, 'For the next fifteen minutes, we watched this light and speculated on what it might be. It was not a sharp light like a bare bulb but more like a light shining through frosted glass. No shape of any kind was discernible. It appeared to blink, but with no regularity whatever." (R2)

X3. July 1952. New Mexico. "Our station was notified that an unconventional aircraft had been picked up with both electronic and visual contact. Our station made electronic contact with the object and two of our men and I had gone outside the building and saw it hovering under a cloud layer to the east of us. It appeared as a large light, at an uncertain distance, and was hovering at the time. A minute or so later, it moved rapidly toward the north for a short distance and stopped as suddenly as it had begun to move." (R2)

X4. 1952. No place given. "Early last month shortly before dawn colored lights were observed in the sky southeast of the radar station. At the same time and the same azimuth, unidentified targets appeared on the scope. Only a very slight temperature inversion was present, $1^{o}$ at 25 000 feet. No more than two lights appeared at one time. They were observed to be moving in a rather erratic pattern and changing colors occasionally. The last thirty minutes of observation revealed the lights remaining yellow---prior to that they were red, green, and blue. They moved in no apparent formation but mostly appeared in one area and disappeared in another, when either the light went out or the objects dived behind clouds. They were starlike objects and appeared to develop long, white vapor trails, when they dived. They were motionless at times and moved rapidly at other times. This corresponded to similar movements observed on the radar scope. One light went out as it changed direction and continued as a black silhouette against the dawn sky. Observation was for a period of about an hour and was made by two airmen and a radar operator---all three observers were experienced aircraft control and warning operators. Objects were observed at 20 to $40^{o}$ above horizon. Radar gave distances of 50 to 80 miles. This implies a height of about 40 miles. There was no air traffic on radar within 100 miles." (R2) Except for the altitude, this observation closely resembles GLN1-X31.

X5. No time or place given. The first observation was a series of localized light flashes in a clear, blue sky, estimated to be less than 10 miles away. Following this 3-4 second display, an off-white light 'switched on' and proceeded slowly across the sky. Apparently it traveled a straight path at an altitude of less than 10 000 feet. When viewed through binoculars, the latter light appeared as a single, spherical shape unattached to a material object. Assuming that the angular resolution of the eye is one second of arc, a solid object greater than about two feet in its largest dimensions would be resolved in the binocular field, assuming a distance of 10 miles between the observer and the object." (R3)

X6. No time or place given. "The second observation was of an off-white light that 'switched on' above a long, narrow cloud. The intensity of the light varied at a rate of about one 'blink' per second. Within a few seconds, this was accompanied by the transient 'random flash-bulb effect,' all visible in the seven-degree binocular field. Upon depressing the microphone switch of a 37.1 MHz radio, the blinking light extinguished." (R3)

X7. 1978. Ocala National Forest, Florida. "Unexplained lights in the night sky over a remote sector of Florida forest were sighted and tracked on radar at a Navy electronic center, naval personnel have confirmed. They say one light hovered in midair, then sped away after eluding computerized tracking gear." (R4) There are thousands of similar reports in newspapers and the UFO literature. (WRC)

X8. April 11, 1978. North Atlantic Ocean.

"At approximately 2200 GMT three cone-shaped lights as shown in the sketch emerged from a large pale-yellow glow in the sky. The lights appeared to be equally spaced and were travelling at great speed; they were brightest at the apex. Each light was observed for about one minute---the time taken to travel from the yellow glow to a point just above the layer of alto-stratus cloud---before fading rapidly and disappearing at an altitude just above the moon." (R5)

References

R1. "Ball Lightning," Monthly Weather Review, 27:358, 1899. (X1)
R2. Hynek, J. A.; "Unusual Aerial Phenomena," Optical Society of America, Journal, 43:311, 1953. (X2-X4)
R3. Rutledge, Harley D.; "Light Flashes in the Sky," Physics Today, 27:11, September 1974. (X5, X6)
R4. "Odd Lights Elude Radar," Baltimore Sun, May 17, 1978, p. A-8. (X7)
R5. Butler, M. J.; "Unidentified Flying Objects," Marine Observer, 49:71, 1979. (X8)

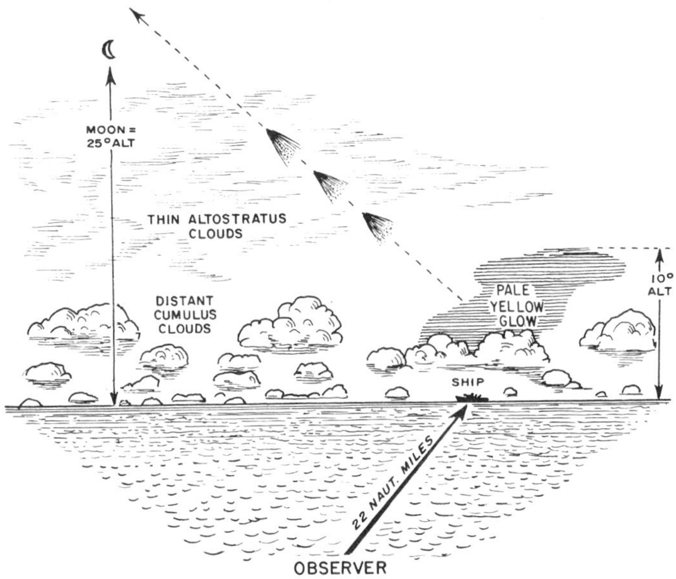


*Three cone-shaped lights travel at great speeds over the North Atlantic.*

# GLW MARINE PHOSPHORESCENT DISPLAYS

### *Key to Phenomena*



## GLW0 Introduction

Ships that ply the Indian Ocean, particularly the waters leading to the oil-sodden lands around the Persian Gulf, frequently encounter dazzling phosphorescent seas. As Kipling described it, the ship's wake is "a welt of light that holds the hot sky tame." Huge globes of light rise from the depths and burst on the surface. Wavetops sparkle, porpoises resemble luminous torpedoes, and broad geometrically precise corridors of light stretch from horizon to horizon. Buckets lowered into these glowing seas prove that marine organisms seem to cause most of the phosphorescent displays.

Phosphorescent ship wakes are mundane and unimpressive compared to the vast rotating wheels of light and the other fantastic luminescent displays encountered from the Persian Gulf, across the Indian Ocean, and into the South China Sea. Ridiculed as wild sailors' tales for centuries, modern ships have reported scores of bona fide geometrical displays. Mariners tell of great spoke-like bands of light seemingly spinning about some distant hub. Occasionally several wheels will overlap, while simultaneously turning in clockwise or counterclockwise senses, creating a vast tableau of moving spokes miles wide. Expanding rings of light and bright whirling crescents (the latter radar-stimulated) may also decorate the ocean surface. Crews that see these fantastic apparitions do not soon forget them. Scientists, alas, have generally ignored these awe-inspiring apparitions.

One's first reaction is to explain the wheels of light and related geometrical displays in terms of marine bioluminescence stimulated by natural forces that, like the wake of a ship, leave behind glowing evidence of their passage. Sound waves emanating from submarine disturbances have been the most popular type of disturbance in this explanation. But what combination of seismic waves could stimulate overlapping, counterrotating wheels or hundreds of spinning phosphorescent crescents? Furthermore, there are several well-attested cases where the luminous displays were seen in the air well above the sea's surface. This fact plus the persistence of the phenomena (about half an hour) and the complex nature of the displays suggest that we look for other stimuli and nonbiological sources of light.

The physical forces that create the auroras and the Andes Glow may be at work near the ocean's surface, unlikely as it may seem. To illustrate this possibility, the luminous mist seen during some low-level auroras closely resembles the aerial phosphorescence seen in some marine displays. Some ship captains have, in fact, noted the similarities between auroral and marine phosphorescent displays. The curious interaction of radar with marine phosphorescence is also suggestive. Another potential explanation would use the collective behavior of marine bioluminescent organisms. Travelers in the tropics, for example, tell amazing accounts of the synchronized flashing of immense assemblages of fireflies. Could marine bioluminescent organisms indulge in similar cooperative action? If so, how do they communicate pattern geometries and why?

Many other questions can be asked about marine phosphorescent displays. Why are most concentrated in the Indian Ocean and South China Sea when other seas also teem with bioluminescent organisms? Where does the mysterious underwater lightning called te lape by the Polynesians fit in? Unfortunately only a few scientists have deigned to notice this fertile field of research.

# GLW1 Long, Parallel, Stationary Phosphorescent Bands

Description. Horizon-to-horizon bands of bright phosphorescence in single or parallel array. The bands may vary from a few feet to a half mile in width. The phosphorescence may be steady or broken up into flashing patches. Blue and green are common colors.

Background. The wide bands of constant brightness might be a different phenomenon than the narrow strips of pulsing patches.

Data Evaluation. Several excellent descriptions exist. Rating: 1.

Anomaly Evaluation. Assuming the light is biological in origin, the geometry and large-scale organization of the display must be explained. In this instance, oceanic internal waves provide a reasonable mechanism for stimulating bioluminescence in long, straight lines where they intersect the surface. Rating: 3.

Possible Explanations. Stimulation of bioluminescence by internal waves---especially in the case of the broad lanes of constant brightness. The periodic switching on and off of the bright patches may be influenced by the ship's engines.

Similar and Related Phenomena. Moving bands of phosphorescence (GLW2), the milky sea (GLW9),

## Examples of Long, Parallel, Stationary, Phosphorescent Bands

X1. November 24, 1908. Gulf of Mexico. "A remarkable marine phenomenon was observed by the steamship Dover, Capt. Yon A. Carlson, as that vessel steamed to Tampa from Mobile. When at a point 35 miles from Mobile light, at 7 o'clock in the evening of the 24th, the ship ran suddenly in a streak of light coming from the water which alternated blue and green, the colors being so brilliant that the vessel was lighted up as if she were covered with arc lights with colored globes. A half mile

*One of two parallel corridors of luminescence encountered in the Gulf of Mexico in 1908*

streak of dark water, and a blackness that settled like a pall over the ship followed, and a second streak of the same brilliant-hued waters was encountered. The second streak was about as wide as the first one, and when the ship ran out of it the same black waters and a night of exceptional blackness were also encountered.... Each of the streaks and the intermediate streak of black water was about half a mile wide." (R1)

X2. May 30, 1926. Arabian Sea. "On May 30th, about 1 a.m. when standing to N'd between Quoin Is. and Larak I. in about Latitude 26° 40' N., Longitude 56° 33' E., passed through a phenomenal, scintillating, phosphorescent belt of water. It was first sighted as a line of phosphorescent water stretching across the horizon ahead from east to west. As the ship approached the area, it presented a curious scintillating effect. On passing through it, it was found to be a belt about half a mile in width extending to the horizon in an east and west direction. The effect at close quarters was as though thousands of powerful beams of light directed upwards from under water, each illuminating a patch of some twenty to thirty square yards of sea surface, were being switched on and off alternately, independently of each other. If any one of these patches were watched, the intervals of light and darkness were found to be of surprising regularity about 1 to 1 1/2 seconds. The belt gradually receded astern and the display continued until it was lost over the horizon to the southward." (R2)

X3. February 19, 1968. Bismarck Sea. "At 1756 GMT, when 7 1/2 miles off Cape Gordon, a line of bright pulsating patches was observed 2 miles away on the starboard bow. Each patch was about 35 feet long and pear-shaped, flashing in unison every 1/2 sec. The engine revs. were 112 per min and the radar was not switched on. The patches became brighter as the ship approached them and the radar was switched on without producing an effect. The ship did not cut through any patches and another line of them was seen off to port in about the same line as the starboard one, possibly marking the edge of a local current." (R3)

X4. July 5, 1926. South Atlantic Ocean. Five bands stretching as far as the eye could see. The bright bands seemed to make the sea thick and sluggish. (R4)

X5. April 24, 1928. China Sea. Six parallel streaks. The phosphorescent regions presented an oily appearance. (R4)

X6. November 9, 1955. South African Waters. Four long streaks of bright blue phosphorescence. (R5)

X7. December 21, 1958. New Zealand waters. Strong lines of phosphorescence 4 feet wide, running east and west. A nauseating stench accompanied the display. (R6)

X8. August 3, 1929. Indian Ocean. Band of phosphorescence 1/4 mile wide. No other phosphorescence noted. (R7)

X9. January 10, 1953. Gulf of Aden. "The ship passed through several bands of brilliant phosphorescence, each band approximately 1/2 to 3/4 mile wide (edges not well-defined) and about 2 1/2 miles between each band. A glow in the sky could be seen before the bands were visible, which were continually flashing with pinpoints of light, and stretched from horizon to horizon in a 120°-300° direction. At 2307 the vessel passed through a band about 20 ft wide of exceptional brilliance (a steady not a flashing light), which temporarily illuminated the whole ship like a greenish-blue searchlight. The band had sharply defined edges and curved slightly from horizon to horizon resembling a tide line, although no floating debris was visible. The glow from this band could be seen for about 6 miles." (R8)

X10. 1880. Indian Ocean. A river of silvery white light in the sea, 1/4 mile wide. The water was full of tiny phosphorescent fish. (R9)

X11. July 8, 1977. North Atlantic Ocean. "At 2230 GMT the vessel passed through parallel bands of bioluminescence which were approximately 65 centimetres wide with a distance of about 8 metres between each band; owing to the darkness it was not possible to determine the length of the bands. Each band lay along the direction of the wind which, at that time, was SW'ly. The phenomenon was observed for 12 minutes and it was interesting to note that throughout this time the calls of seabirds could be heard and once or twice they were observed in the glow from the navigation lights. At no other time during the night were these birds heard or observed." (R10)

R12. August 23, 1977. South Atlantic Ocean. "At 0330 GMT in a position 35 n. mile southwest of the Fernando de Noronha Archipelago the vessel entered a large area of bioluminescence; it took the form mainly of parallel bands, but there were also some patches of white water and some rapid flashes on the sea surface. The bands were about 200 metres apart and they appeared to be about 5 n. mile in length and about 4 metres in width. They seemed to be moving with the wind in a north-westerly direction. The size of the individual luminous flashes varied in diameter from 15-60 centimetres. A sample of the water was taken and its temperature was found to be 26°C. The glow from the bioluminescence was considerable, so much so that it was not necessary to switch on a torch to examine the sample of the sea-water taken. The phenomenon was chiefly white in colour but there were emerald-green patches on the parallel bands and in the wake of the vessel. The vessel steamed about 40 n. mile before clearing the area of bioluminescence." (R10)

X13. December 29, 1929. Off Cape Verde, North Atlantic Ocean. (R11)

## References

R1. "Brilliant Gulf Waters," Monthly Weather Review, 36:371, 1908. (X1)

R2. Mordue, J. A.; "Phosphorescence," Marine Observer, 3:70, 1926. (X2)

R3. Hilder, Brett; "Luminescence," Marine Observer, 39:13, 1969. (X3)

R4. Clarke, R. E. D.; "Phosphorescence of the Sea," Nature, 139:592, 1937. (X4-X5)

R5. Moffatt, B. Forbes; "Phosphorescence," Marine Observer, 26:197, 1956. (X6)

R6. Bailey, F. W.; "Phosphorescence," Marine Observer, 29:176, 1959. (X7)

R7. Dawson, E. E. N.; "Phosphorescence," Marine Observer, 7:169, 1930. (X8)

R8. Jenkins, C. N.; "Phosphorescence," Marine Observer, 24:8, 1954. (X9)

R9. "White Water," English Mechanic, 77:149, 1903. (X10)

R10. Morton, K., and White, A. H.; "Bioluminescence," Marine Observer, 48:118, 1978. (X11, X12)

R11. Evans, J. O.; "Phosphorescence of the Sea," Marine Observer, 8:230, 1931. (X13)

# GLW2 Moving, Parallel Bands of Phosphorescence

Description. Long, parallel bars or ripples of light moving swiftly across the sea, usually at speeds of 60-100 miles per hour. The luminous bands are generally 10-30 feet in breadth, separated by about the same distance, and anywhere from a few feet in length to several miles. Large variations exist in all dimensions. In one type of display, the bands come first from one direction for about 15 minutes, then cease for several minutes, and conclude with another 15 minutes of bands from a different direction. In other occurrences, the direction changes frequently. The parallel bands often precede and follow the wheel type of display (GLW4).

Background. As in the case of stationary bands (GLW1), two distinct phenomena may exist here: (1) the common, very rapid bands; and (2) the rare, slowly moving variety, exemplified by GLW2-X2. It is also possible that the fast bands are merely manifestations of a distant wheel-type display, one so far away that the spokes seem parallel.

Data Evaluation. This phenomenon is rather variable, but many excellent observations are on file. Rating: 1.

Anomaly Evaluation. The stimulus and mechanism for generating the bands, maintaining their geometry, and effecting apparent motion are all unknown. Rating: 2.

Possible Explanations. It is generally assumed that the luminous bands owe their existence to bioluminescent organisms, although there is even some doubt here (GLW3). The illusion of high-speed bands is probably created by sequentially pulsing groups of luminous organisms; something akin to the moving patterns on a theater marquee. Just what initiates this cooperative biological action is a mystery, but some suspect the ship's radar or engine noise. It is interesting to note in this context that tropical fireflies put on spectacular rhymthmic displays (BIB). Electromagnetic stimuli should not be ruled out. Slowly moving internal waves.

Similar and Related Phenomena. Phosphorescent wheels and other geometries (GLW); synchronously flashing fireflies (BIB); ground-level, electrified, luminous patches (GLD4). luminous fogs (GLA21) which occasionally exhibits a wave-like structure.

Examples of Moving, Parallel Bands

X1. April 11, 1870. Gulf of Siam. "On both sides obliquely in front of us, long white waves of light were seen floating toward the ship, increasing in brightness and rapidity till at last they almost disappeared, and nothing was observed but a white lustreless, whirling (schwirrendes) light upon the water. After gazing for some time it was impossible to distinguish between water, sky, and atmosphere, all which were but just now clearly distinguishable, and a thick fog in long streaks appeared to be driving upon the ship with furious swiftness. The phenomenon of light was somewhat similar to that which would be produced by the whirling round of a ball striped black and white so rapidly that the white stripes seem to be lost and blended with the dark ones. The light was just as if we were enveloped in a thick white fog. The direction of the waves of light upon the ship was always on both sides obliquely from the front. The phenomenon lasted about five minutes and repeated itself once more afterwards for about two minutes." (R1)

X2. 1883. Greenland waters. "We saw suddenly behind the vessel on the surface of the sea a broad but clearly defined band of light. It shone with a steady, yellowish light, somewhat like that of phosphorescent elements, while, in spite of the speed maintained, viz, four to six knots, the band came nearer and nearer. When it reached the ship it seemed as if we were steaming through a sea of fire or molten metal. After a while the light traveled beyond the vessel, and we saw it at last disappear on the horizon.... It was beyond doubt of a different nature to the bluish-white phosphorescent light, which throughout its appearance was seen distinctly in our wake; and as the light was perfectly steady it cannot have been caused by the phosphoresence from a passing shoal of fish." (R2)

X3. Circa 1906. Gulf of Oman. "When we got to within twenty yards of the whitish water I saw it break into life and light---shafts of brilliant light came sweeping across the ship's bow at a prodigious speed, which might be put down as anything between sixty and two hundred miles an hour. Then we steamed into the light, and the effect was weird to a degree. It was just as if a large gun with a rectangular muzzle were shooting bars of light at us from infinity. These bars of light were about twenty feet apart and most regular; their brilliancy was dazzling. They first struck us on our broadside, and I noticed on our leeside that an intervening ship had no effect on the light beams; they started away from the leeside of the ship just as if they had traveled right through it. The direction from which the light bars came gradually veered round the whole compass. After a quarter of an hour of this harmless bombardment we ran into the night's ordinary darkness. But in another five minutes we were again attacked by the light bars, now traveling in a direction the exact opposite of the ship's course. This second attack lasted for about five minutes, and then the entertainment was over for the night. It was impossible to say whether the light bars travelled through the water or on its surface. I collected a bucketful of the water and examined it under the microscope, but could not detect anything abnormal." (R3)

X4. August 8, 1950. Arabian Sea. "The sea was wholly covered with bright, flashing, closely spaced patches of phosphorescence. They took the form of circles about 30 ft. in diameter and oblongs 30 ft. by 10 ft. The patches 'flashed' alternately and gave the appearance of the whole sea shimmering with lines of light running criss-cross in all directions. (R4)

X5. October 23, 1954. Java Sea. "Numerous closely-spaced patches of phosphorescence, about 3 ft in diameter, were observed extending all around the ship to a distance of at leat 1/4 mile. Each patch appeared to flash regularly at 1/2-sec intervals. After 30 min the patches slowly increased in brilliance and bands of light were seen continuously during the subsequent 15 min, appearing as straight lines except for one short period when the seemed to radiate from a position approximately two points abaft the port beam, although no definite centre could be seen. For most of the time the bands appeared to come from no definite direction, but were continually changing, staying in one direction for only short periods, and never coming from more than one point at a time. In the observer's opinion the bands consisted of closely packed illuminated areas; they had the same frequency as the flashing of the separate patches, which remained visible throughout. The observer considers that the apparent motion was caused by the illumination of alternate bands." (R5)

X6. September 27, 1959. East Indian Archipelago. "The first indication of anything unusual was the appearance of white caps on the sea here and there, which made me think that the wind had freshened, but I could feel that this was not so. Then flashing beams appeared over the water, which made the Officer on watch think that the fishing boats were using powerful flashlights. These beams of light became more intense and appeared absolutely parallel, about 8 ft wide, and could be seen coming from right ahead at about 1/2 sec intervals. At this time, I thought I could hear a swish as they passed, but decided that this was imagination. They did not appear like rings or arcs of a circle, unless it was a circle so big as to make them appear as straight lines. It was like the pedestrian's angle of a huge zebra crossing passing under him whilst he is standing still. While this part of the phenomenon was at its height it looked as if huge seas were dashing towards the vessel, and the sea surface appeared to be boiling, but it was more or less normal around a fishing vessel which we passed fairly close. The lights of various fishing vessels were visible through the beams of light, though dimmed by the brightness of the latter. The character of the flashes changed and took on the appearance of beams from a lighthouse situated about two miles on the starboard bow, or as if the centre of a giant wheel was somewhere on the starboard bow with the beams as its spokes. As the beams from the wheel on the starboard

*High-speed bars of light observed in 1959 in the East Indian Archipelago. Phosphorescent wheel in the distance*

bow weakened, the same pattern appeared on the port bow at the same distance and regularity. The wheel on the starboard bow revolved anticlockwise and the one on the port bow revolved clockwise, i.e. both wheels were revolving towards the ship. The wheel on the starboard bow diminished as the one on the port bow increased; when the latter was at its peak the one on the starboard bow had disappeared. The next change was that the beams appeared to be travelling in the exact course of the ship, like a following sea, i.e. the beams now seen were a reversal of those seen at first." (R6)

X7. April 1875. Gulf of Mexico. A line of light appeared on the horizon and swept past the ship in a series of luminous pulsations. (R7)
X8. April 4, 1901. Persian Gulf. Two luminous bands per second. Speed about 60 miles per second. First from one direction, then reversed. (R8, R9)
X9. April 9, 1901. Persian Gulf. Like GLW2-X8. Bands 50 feet wide. (R8)
X10. 1905. Indian Ocean. Flashes of light passed over the sea. (R10)
X11. October 27, 1924. Sunda Straits. Bands passed at the rate of two per second. Kept changing direction. (R11)
X12. March 21, 1925. Off the West Coast of India. Bands about one second apart. Anticlockwise light wheel followed. (R12)
X13. November 1, 1926. Gulf of Siam. Bands 5 feet wide, 12-16 feet apart. (R13)
X14. November 19, 1928. China Sea. Three waves of light per second for 15 minutes, then changed direction. (R14)
X15. November 19, 1930. Gulf of Oman. (R15)
X16. May 21, 1936. Gulf of Siam. Two bands per second. Duration: 30 minutes. (R16)
X17. November 9, 1953. China Sea. Long, straight, greyish bands passed at the rate of 3 per second for 15 minutes, then 25 minutes of darkness, then 15 minutes more of display. (R17)
X18. April 2, 1954. Red Sea. (R18)
X19. May 31, 1955. Gulf of Siam. (R19)
X20. April 2, 1956. Persian Gulf. Waves of light. Echometer detected shoals of fish. (R20)
X21. October 17, 1961. China Sea. (R21)
X22. April 22, 1971. Malacca Strait. Pale-green bands, 2 miles long, moving at 24 knots. (R22)
X23. December 8, 1974. Arabian Sea. Green bands, 50 x 20 meters. Many bioluminescent organisms in water. (R23)
X24. No date or place given. Sixty bands per minute, 2 miles wide. Two light wheels seen at the same time rotating in opposite directions. (R24)
X25. April 23, 1955. Persian Gulf. Medley of moving bands, wheels, and expanding circles. See description in GLW5-X3. (R25)
X26. May 13, 1955. Gulf of Aden. Waves of light 12 feet above the sea. See description in GLW3-X4. (R26)
X27. January 5, 1880. Malabar Coast, Indian Ocean. Moving bars of luminous mist above water surface. See description in GLW3-X1. (R27)
X28. May 23, 1906. Bender Abbas, Indian Ocean. A display of moving, parallel bands changes to wheel. (R28)
X29. April 5, 1953. Arabian Sea. Parallel bands change into wheel display. See description in GLW4-X23.
X30. September 6, 1955. Gulf of Aden. Displays of moving, parallel bands alternate with wheels. (R30)
X31. April 3, 1956. Persian Gulf. Moving bands change into wheel display. See description in GLW4-X35. (R31)
X32. May 7, 1963. Persian Gulf. Display of moving bands changes into two rotating wheels. (R32)
X33. May 30, 1956. Persian Gulf. (R33)
X34. March 27, 1976. Gulf of Siam. Moving band system changes into wheel display. (R34)
X35. June 12, 1980. South China Sea. (R35)
X36. April 11, 1870. Gulf of Siam. (R36)
X37. July 23, 1954. Arabian Sea. (R37)

## References

R1. "Illumination of the Sea," Nature, 2: 165, 1870. (X1)
R2. Nordenskjold, A. E.; "Nordenskjold's Greenland Expedition of 1883," Scientific American Supplement, 17:6740, 1884. (X2)
R3. Carnegie, A. A.; "Remarkable Display of Phosphorescence," Royal Meteorological Society, Quarterly Journal, 32: 280, 1906. (X3)
R4. Stormont, E.; "Phosphorescence," Marine Observer, 21:155, 1951. (X4)
R5. Evans, J. W.; "Phosphorescence," Marine Observer, 25:210, 1955. (X5)
R6. Rutherford, W.; "Phosphorescent Wheel," Marine Observer, 30:128, 1960. (X6)
R7. Moss, Edward L.; "Report of an Unusual Phenomenon Observed at Sea," Nature, 20:428, 1879. (X7)

R8. Hoseason, W. S.; "Remarkable Phosphorescent Phenomenon Observed in the Persian Gulf, April 4 and 9, 1901," Royal Meteorological Society, Quarterly Journal, 28:29, 1902. (X8, X9)
R9. "Phosphorescent Sea," Nature, 125: 513, 1930. (X8)
R10. Nature, 72:106, 1905. (X10)
R11. Findlay, J.; "Phosphorescence," Marine Observer, 2:156, 1925. (X11)
R12. Brisley, P. L.; "Phosphorescence," Marine Observer, 3:40, 1926. (X12)
R13. Vaughan, G. E.; "Phosphorescent Beams," Marine Observer, 4:209, 1927. (X13)
R14. Randell, G. G.; "Phosphorescence," Marine Observer, 6:238, 1929. (X14)
R15. Putt, R. O.; "Phosphorescence," Marine Observer, 8:227, 1931. (X15)
R16. "Luminous Phenomena on the Sea During a Thunderstorm," Nature, 138:278, 1936. (X16)
R17. Wilks, R. E.; "Phosphorescence," Marine Observer, 24:205, 1954. (X17)
R18. Stewart, R. L.; "Phosphorescence," Marine Observer, 25:94, 1955. (X18)
R19. Clarke, N.; "Phosphorescence," Marine Observer, 26:78, 1956. (X19)
R20. Nettleship, J. G.; "Phosphorescence," Marine Observer, 27:91, 1957. (X20)
R21. Lewis, M. W.; "Phosphorescence," Marine Observer, 32:181, 1962. (X21)
R22. Wright, D.; "Luminescence," Marine Observer, 42:62, 1972. (X22)
R23. Winn, J. K.; "Bioluminescence," Marine Observer, 45:159, 1975. (X23)
R24. Palmer, A. R.; "The White Water," Nature, 104:563, 1920. (X24)
R25. Baxter, R. R.; "Phosphorescent Wheels," Marine Observer, 26:78, 1956. (X25)
R26. Austen, G. A.; "Phosphorescence Seen in the Air," Marine Observer, 26: 78, 1956. (X26)
R27. "A Strange Phenomenon," Nature, 21:409, 1880. (X27)
R28. Kalle, Kurt; "Wheels of Light," Sea Frontiers, 15:116, 1969. (X28)
R29. Henney, A.; "Phosphorescent Wheels," Marine Observer, 24:73, 1954. (X29)
R30. Henney, A.; "Phosphorescent Wheel," Marine Observer, 26:138, 1956. (X30)
R31. Evans, H.; "Phosphorescent Wheels," Marine Observer, 27:90, 1957. (X31)
R32. Harrison, P. P. O.; "Phosphorescent Wheels," Marine Observer, 34:69, 1964. (X32)
R33. Arthur, D. E. O.; "Phosphorescence," Marine Observer, 27:92, 1957. (X33)
R34. Cowie, R. E.; "Bioluminescence," Marine Observer, 47:17, 1977. (X34)
R35. Rundle, A. A.; "Bioluminescence," Marine Observer, 51:66, 1981. (X35)
R36. "Remarkable Illumination of the Sea," Scientific American, 23:102, 1870. (X36)
R37. Dennis, T.; "Phosphorescence Seen in the Air," Marine Observer, 25:119, 1955. (X37)

# GLW3 Aerial Phosphorescent Displays

Description. Marine phosphorescent displays observed in the air well above the surface of the water.

Background. Marine phosphorescent displays are believed to be caused by bioluminescent organisms in the sea; and indeed the bulk of the displays do seem confined to the ocean surface and just below it. In some instances, however, the observer cannot decide whether the display occurs on the surface or just above it. The situation is complicated by the possible illumination of surface mist and suspended particles by light sources in the water proper. Such borderline cases are not included here, where all accounts describe displays that seem at least several feet above the sea's surface.

Data Evaluation. Only a handful of good cases. Rating: 2.

Anomaly Evaluation. Since bioluminescent organisms are not known to enter and remain suspended in the atmosphere for long periods, the aerial phosphorescent displays would seem to have no recognized source of light. Rating: 2.

Possible Explanations. The long-term suspension of bioluminescent organisms in the air is always possible, but seems unlikely. The aerial displays may result from the simple illumination of mist and fog by bright light sources in the water, although some observers specifically deny this. The strong resemblence between aerial phosphorescent displays and low-level auroras and luminous surface fogs is intriguing. Some luminous fogs, for example, seem to have a wave structure. (GLA21-X4)

Similar and Related Phenomena. Low-level auroras (GLA4), luminous fogs and mists (GLA21).

## Examples of Aerial Phosphorescent Displays

X1. January 5, 1880. Malabar Coast, Indian Ocean. "At 10 P.M. we were steaming along very comfortably; there was a perfect calm, the water was without a ripple upon it, the sky was cloudless, and, there being no moon, the stars shone brightly. The atmosphere was beautifully clear, and the night was one of great quietude. At the above-named hour I went on deck, and at once observed a streak of white matter on the horizon bearing south-south-west. I then went on the bridge and drew the third officer's attention to it. It a few minutes it had assumed the shape of a segment of a circle measuring about 45° in length and several degrees in altitude about its centre. At this time it shone with a peculiar but beautiful milky whiteness, and resembled (only in a huge mass, and greater luminous intensity) the nebulae sometimes seen in the heavens. We were steaming to the southward, and as the bank of light extended, one of its arms crossed our path. The whole thing appeared so foreign to anything I had ever seen, and so wonderful, that I stopped the ship just on its outskirts, so that I might try to form a true and just conception of what it really was. By this time all the officers and engineers had assembled on deck to witness the scene, and were all equally astonished and interested. Some little time before the first body of light reached the ship I was enabled, with my night glasses, to resolve in a measure what appeared, to the unassisted eye, a huge mass of nebulous matter. I distinctly saw spaces between what again appeared to be waves of light of great lustre. These came rolling on with ever-increasing rapidity till they reached the ship, and in a short time the ship was completely surrounded with one great body of undulating light, which soon extended to the horizon on all sides. On looking into the water it was seen to be studded with patches of faint, luminous, inanimate matter, measuring about two feet in diameter. Although these emitted a certain amount of light, it was most insignificant when compared with the great waves of light that were floating on the surface of the water, and which were at this time converging upon the ship. The waves stood many degrees above the water, like a highly luminous mist, and obscured by their intensity the distant horizon; and as wave succeeded wave in rapid succession, one of the most grand and brilliant, yet solemn, spectacles that one could ever think of was here witnessed. In speaking of waves of light I do not wish to convey the idea that they were mere ripplings, which are sometimes caused by fish passing through a phosphorescent sea, but waves of great length and breadth, or in other words, great bodies of light. If the sea could be converted into a huge mirror and thousands of powerful electric lights were made to throw their rays across it, it would convey no adequate idea of this strange yet grand phenomenon. As the waves of light converged upon the ship from all sides they appeared higher than her hull, and looked as if they were about to envelope her, and as they impinged upon her, her sides seemed to collapse and expand. Whilst this was going on the ship was perfectly at rest, and the water was like a millpond. After about half an hour had elapsed the brilliancy of the light some-

*Off the Malabar Coast of India, a vessel is engulfed in great waves of light floating on the sea's surface.*

what abated, and there was a great paucity of the faint lustrous patches which I have referred to, but still the body of light was great, and, if emanating from these patches, was out of all proportion to their number." (R1)

X2. July 24, 1908. Gulf of Siam. "...the steamer past (sic) thru a small field of remarkable phosphorescent patches in the form of a kind of vapor lying above the surface of the water in lengths of 500 to 1,000 feet and breadths of 100 feet approximately, and about 15 to 20 feet in depth to the surface of the water. At distances of 1 to 2 miles these 'streaks' appeared like shining silver (no moon shining), and at first were taken to be shoals of fish, but on passing directly thru one it had all the effect of a slight luminous fog. No disturbance or presence of any fish appeared in the water, which is only about 25 to 30 fathoms in depth, and no unusual color appeared in the contents of a draw-bucket taken at the time." (R2)

X3. July 23, 1954. Arabian Sea. Pale white bands of light moved just above the surface of the water. (R3)

X4. May 13, 1955. Gulf of Aden. "Shortly after leaving Aden for Suez bright pulsating waves of light having a period of 1.5 sec were observed emanating from the sea in close proximity to the ship. They were travelling in a direction of 070° for some distance and extended to a height of about 12 ft above the sea. This phenomenon was definitely not caused by the proximity of shore lights; there was no sign of phosphorescence in the sea at the time of observation." (R4)

X5. March 31, 1958. North Atlantic Ocean. "Between 2250 and 2315 G.M.T., what appeared to be phosphorescence in the air was observed. Small phosphorescent particles passed upwards from the sea to a height of about 3 1/2 ft all round the ship. At no time before 2250 or after 2315 was any phosphorescence visible either in the sea or in the air above it." (R5) Since this event occurred far from the regions where most of the marine phosphorescent displays take place, it may be unrelated to others in this category. (WRC)

X6. July 10, 1979. Arabian Sea. Phosphorescence seemed at eye level to those aboard the ship. (R6)

X7. No date or place given. Spokes of a phosphorescent wheel seemed about 8 feet high. (R7)

X8. November 5, 1953. Gulf of Oman. The spokes of a wheel display appeared 1 meter above the surface. (R8)

X9. November 14, 1920. China Sea. "...a phosphorescent wheel was observed in the air above the sea, in a stratum more or less parallel to the sea surface.... Broadly speaking the phenomenon was identical with the phosphorescent wheel observed on board the Valentijn, but the spokes appeared to rotate at a level of about 30 ft. above the sea surface. It was observed, when the wheel appeared, that the water was covered with a haze, and as soon as the wheel disappeared the haze was no longer there. It was also observed that the wheel did not only turn anti-clockwise, but occasionally changed its movement into a clockwise direction. During the whole phenomenon the water was but slightly phosphorescent." (R9, R10)

References

R1. Harris, R. E.; "A Strange Phenomenon," Nature, 21:409, 1880. (X1)

R2. Turner, George A.; "Luminous Fog," Monthly Weather Review, 36:371, 1908. (X2)

R3. Dennis, T.; "Phosphorescence Seen in the Air," Marine Observer, 25:119, 1955. (X3)

R4. Austen, G. A.; "Phosphorescence Seen in the Air," Marine Observer, 26:78, 1956. (X4)

R5. Robertson, C.; "Phosphorescence,: Marine Observer, 29:14, 1959. (X5)

R6. Penman, B.; "Bioluminescence," Marine Observer, 50:114, 1980. (X6)

R7. "Luminous Wheels Puzzle Seamen," New Scientist, 33:447, 1967. (X7)

R8. Rodewald, M.; "Phosphorescent Wheel," Marine Observer, 24:233, 1954. (X8)

R9. "Phosphorescent Wheel," Marine Observer, 20:229, 1950. (X9)

R10. Tydeman, G. F.; "Phosphorescent Wheels," Marine Observer, 9:32, 1932. (X9)

# GLW4 Marine Phosphorescent Wheels

Description. Spoke-like bands of light rotating around a central hub. The spokes may be straight, curved, or S-shaped. Rotation is in either direction and may change during the display. In some cases, the outer part of the wheel seems to spin in a different sense from the central part. Illustrating the illusory character of the phenomenon, different observers sometimes see the same wheel rotating in opposite directions. Wheel sizes range from tens of feet to several miles, with spoke widths of 5-50 feet being common. Generally, the sources of light seem to be on or just beneath the surface, but several examples exist of wheels composed of luminous mist spinning well above the sea's surface. Spoke colors are whitish and greenish in most cases. Several wheels may appear simultaneously, rotating in various senses with overlapping patterns. Phosphorescent wheel displays are frequently preceded and followed by displays of moving, parallel bands (GLW2). Like the moving band displays, the phosphorescent wheels are most common in the Indian Ocean, especially the Persian Gulf, and the China Sea. The few wheel-like structures seen in other waters are usually poorly formed and stationary. The duration of a wheel-type display lasts from a few minutes to more than an hour.

Data Evaluation. Scores of well-observed examples. Rating: 1.

Anomaly Evaluation. As with most of the organized phosphorescent displays, the major problem seems to be explaining the origin and long-term stability of rather complex geometrical patterns. If the light source is not bioluminescence, as generally supposed, the anomaly is even stronger. Rating: 2.

Possible Explanations. Certainly bioluminescence is the most likely source of light, although observers frequently remark that the ship's wake is not luminous during wheel-type displays. The aerial wheels of luminous mist, if not illusory, would require air-borne organisms in cases where no wheel is visible in the water proper. Earthquake tremors may stimulate bioluminescence, with the interference patterns created by multiple sources accounting for the complex display geometries. The persistence of intricate geometries over many minutes seems to militate against this theory. Again, as in GLW3, the strong similarity of some marine phosphorescent displays to the so-called low-level auroras (GLA4) is striking. Some wheel observers have noted this, and electromagnetic forces should not be dismissed off hand.

Similar and Related Phenomena. The other marine phosphorescent displays (GLW), low-level auroras(GLA4).

Examples of Phosphorescent Wheels

X1. May 15, 1879. Persian Gulf. "I noticed luminous waves or pulsations in the water, moving at great speed and passing under the ship from the south-south-west. On looking towards the east, the appearance was that of a revolving wheel with centre on that bearing, and whose spokes were illuminated, and looking towards the west a similar wheel appeared to be revolving, but in the opposite direction. I then went to the mizen top (fifty feet above the water) with the first lieutenant, and saw that the luminous waves or pulsations were really travelling parallel to each other, and that their apparently rotatory motion, as seen from the deck, was caused by their high speed and the greater angular motion of the nearer than the more remote part of the waves. The light of these waves looked homogeneous and lighter, but not so sparkling, as phosphorescent appearances at sea usually are, and extended from the surface well under water; they lit up the white bottoms of the quarter-boats in passing. I judged them to be twenty-five feet broad, with dark intervals of about seventy-five between each, or 100 from crest to crest, and their period was seventy-four to seventy-five per minute, giving a speed roughly of eighty-four English miles an hour. From this height of fifty feet, looking with or against their direction, I could only distinguish six or seven waves; but, looking along them as they passed under the ship, the luminosity showed much further. The phenomenon was beautiful and striking, commencing at about 6 h. 3 m. Greenwich mean time, and lasting some thirty-five minutes. The direction from which the luminous waves travelled

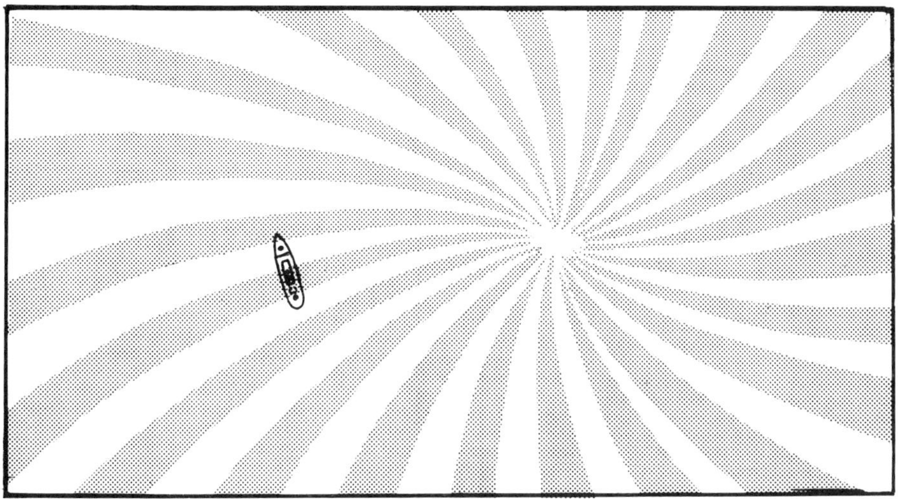

*General configuration of a typical phosphorescent wheel*

changed from south-south-west by degrees to south-east and to east. During the last five minutes concentric waves appeared to emanate from a spot about 200 yards east, and these meeting the parallel waves from south-east did not cross, but appeared to obliterate each other at the moving point of contact, and approached the ship, inclosing an angle about 90°." (R1)

X2. May 1880. Persian Gulf. "In May, 1880, on a dark, calm night, about 11.30 p.m., there suddenly appeared on each side of the ship an enormous luminous wheel whirling round, the spokes of which seemed to brush the ship along. The spokes would be 200 or 300 yards long, and resembled the birch rods of the dames' schools. Each wheel contained about 16 spokes, and made the revolution in about twelve seconds. One could almost fancy one heard the swish as the spokes whizzed past the ship, and, although the wheels must have been some 500 or 600 yards in diameter, the spokes could be seen distinctly all the way round. The phosphorescent gleam seemed to glide along flat on the surface of the sea, no light being visible in the air above the water. The appearance of the spokes could be almost exactly represented by standing in a boat and flashing a bull's-eye lantern horizontally along the surface of the water round and round." (R2)

X3. May 23, 1906. Indian Ocean. A parallel band display changes into a wheel. The author theorizes that the wheels are seismically stimulated bioluminescent interference patterns. (R3)

X4. March 14, 1907. Malacca Strait. A wheel with spokes 300 yards long. Duration: 30 minutes. (R4)

X5. March 22, 1908. Persian Gulf. Phosphorescent wheel seemed to be revolving just above the water's surface. Duration: 20 minutes. (R5)

X6. August 1908. South China Sea. An anticlockwise wheel. (R6)

X7. June 19, 1909. Malacca Strait. A clockwise wheel with curved spokes concave in the direction of rotation. Duration: 15 minutes. (R6, R7)

X8. August 12, 1910. China Sea. Wheel observed, but no normal phosphorescence in the water. (R7, R8)

X9. July 3, 1919. Indian Ocean. Clockwise wheel with curved spokes. (R9)

X10. November 14, 1920. China Sea. Phosphorescent wheel visible on the underside of a thin layer of mist 30-35 feet above the surface of the water. No display seen in the water at all. The wheel changed its direction of rotation during the observation. (R8, R10)

X11. April 18, 1923. Gulf of Oman. Pale blue spokes. Duration: almost 60 minutes. (R11)

X12. March 21, 1925. Off west coast of India. Two separate wheels. (R12)

X13. September 30, 1926. Arabian Sea. Eddies of light formed discontinuous spokes of a rotating wheel. (R13)

X14. December 3, 1927. Malacca Strait. Two clockwise wheels on opposite sides of ship. Spokes concave in direction of rotation. Duration: 20 minutes. (R14)

X15. December 19, 1927. Indian Ocean. Anticlockwise wheel observed for 5 minutes; it disappears; a clockwise wheel appears for 5 minutes; it disappears; then, another anticlockwise wheel is observed for 5 minutes. (R15, R16)

X16. September 9, 1929. Malacca Strait. Bands of light wheel over the sea. Duration: 1 hour. Display combined with vivid lightning. (R17)

X17. December 28, 1929. North Indian Ocean. S-shaped spokes pass ship; hub about 5 miles away. During the display, the ship's engines were straining. (R16, R18)

X18. October 5, 1932. Andaman Sea. (R19)

X19. December 29, 1932. Andaman Sea. Duration: 26 minutes. (R19)

X20. February 25, 1949. Bay of Bengal. (R20)

X21. July 23, 1949. Malacca Strait. Duration: 20 minutes. (R21)

X22. April 16, 1950. Gulf of Siam. A number of wheels seen simultaneously. Duration of display: 30 minutes. (R22)

X23. April 5, 1953. Arabian Sea. "Commencing from about NNW, shafts of pale white diffused light appeared, apparently travelling on the surface of the water at a great speed. Each shaft was several feet wide and they stretched as far as the eye could see. At first they appeared in perfectly parallel lines, equally spaced, passing the ship at about one every second, but after five minutes they wheeled round in perfect formation and approached the ship from all points of the compass. They came from only one compass point at a time and each change of direction was swift and definite, though not abrupt. The most frequent directions were from NNW and SSE. After about 15 minutes the shafts occasionally formed into a rotating radial movement in which they retained their equal geometrical precision and the frequency of about one per second. At this time the pattern was continually changing about every 20-30 seconds from the parallel lines to the wheel. The periods of transition were hardly noticeable, but they were not abrupt. Each time the wheel appeared it was in a different place. On one occasion there were two distinct wheels visible at the same time. Throughout the period the wheels appeared they varied in direction of rotation, some clockwise and some anticlockwise. Five minutes later the pattern became still more complicated but remained perfectly regular and at 2150 the light faded out over a period of 30 seconds. Although the light appeared to be on the surface of the water it was completely unaffected by the wind and no disturbance of the water was produced. The most notable feature of the phenomenon was the effortless speed and mathematical precision of movement." (R23)

X24. April 24, 1953. Gulf of Thailand. "Faint flashes of light with oscillating movements were observed on the sea. The flashes gradually increased in strength until at 0230 they suddenly changed into rather intensive rays of light moving around centres lying near the horizon. Three groups of rays were present, as shown in the sketch. (a) One on the port bow having a bearing of about 300° with the rays rotating anticlockwise. (b) One on the port bow having a bearing of about 230°, rotating clockwise. (c) One on the starboard bow having a bearing of about 95°, rotating anticlockwise. The

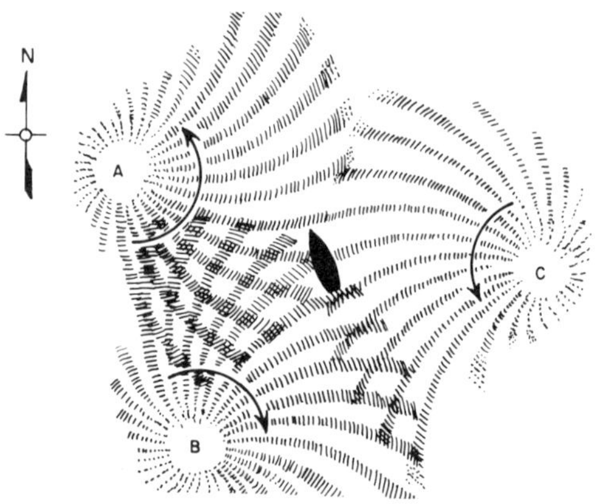


*Three phosphorescent wheels turning simultaneously in the Gulf of Thailand*

beams were curved with the concave side in the direction of the movement, and were passing the ship continuously with a frequency of about three a second; they looked more like glowing shafts than beams of light. Reflections on the ship were clearly visible.... The phenomenon lasted till 0250, and it had been clear by the increasing strength of the group ahead and decreasing strength of the groups astern, that the ship was advancing through the area of phosphorescence. Soon only the oscillating flashes could be seen and they also disappeared shortly afterwards. At 0300 the situation was normal again." (R23)

X25. September 13, 1953. Off Sumatra, Indian Ocean. Duration: 20 minutes. (R24)

X26. October 30, 1953. Gulf of Siam. Display with curved spokes, radius 2 miles. Spokes 100 yards apart one mile from hub. Wheel rotating at 20 rpm. Duration: 10 minutes. (R25)

X27. November 5, 1953. Gulf of Oman. Clockwise fog-like spokes, 1 meter above the sea's surface. Spoke radius: 2 nautical miles; duration: 2 minutes. (R26)

X28. November 12, 1953. China Sea. Close study of wheel revealed that its rotation was due to a marquee effect. (R25)

X29. May 3, 1954. Atlantic equatorial waters. A stationary wheel. (R27) It appears that wheel displays outside the usual habitat (Indian Ocean and China Sea) are rather degenerate forms. (WRC)

X30. February 18, 1955. Indian Ocean. Four pale-green bands, 3 feet wide, 5 feet apart, converged just astern and fanned out ahead. No rotation. (R28)

X31. April 23, 1955. Persian Gulf. "When approaching Jazirat Tunb Island a bright flashing light was observed on the port bow, distant about one mile. Almost simultaneously another was observed on the starboard bow. On approaching it was seen that these were two revolving phosphorescent wheels. The ship passed between them, the centres being about 1/4 mile distant on either side. The wheel on the port side appeared to revolve anticlockwise, and that on the starboard side clockwise. The spokes of radius 1/4 mile were from 6 to 12 ft broad at the tips with about 15 ft between them. They passed with a frequency of 1 1/2 sec with a colour similar to that of a dull electric light. Immediately we had passed these two wheels, further phenomena were observed. On the port side concentric circles were seen to radiate from a centre, with an effect similar to that of dropping something in still water. On the starboard side there appeared lines, apparently moving away from the ship in a manner similar to the wake. As the ship passed they gradually faded from sight astern and had been in sight 10 or 15 min. At 1700 a similar phenomenon was observed to the SE., distant about 7 miles. The Master comments: The sky was overcast and the atmosphere appeared to have more than the usual amount of particles in suspension, but the all-around visibility was very good. Jazirat Tunb Light was about 20 miles away and its rotating beams were visible in the air throughout their complete revolutions. Above the sea surface there was apparently a layer of mist a yard deep. The bands of phosphorescent light appeared to float on top of this layer, but on closer examination of the beams as they passed vertically under the observer it could be seen that the sea was affected to a considerable depth. Each band was similar in colour and appearance to the Milky Way, myriads of particles of light dust with brighter and larger specks here and there. My impression, especially during the concentric ring phenomenon, was of shock waves causing millions of organisms to light up as the wave passed through them, then going dark until the next wave struck them. I do not believe the organisms themselves were on the move. The effect on the onlookers seems to have been a feeling of weirdness, bordering on fear, similar to that experienced by people ashore during earthquake tremors." (R29)

X32. June 25, 1955. Gulf of Oman. Moving bands changed into a display of five wheels; three single wheels rotating anticlockwise, and two superimposed wheels revolving in opposite directions. (R29)

X33. September 6, 1955. Gulf of Aden. White wheels alternated with formations of white, parallel bands. The bands were 30-40 feet wide and moved unusually slowly---passing at the rate of one every 15 seconds. (R30)

X34. October 11, 1955. South Pacific Ocean. The ship was the center of about 16 converging, stationary bands. (R31)

X35. April 3, 1956. Persian Gulf. Following a disordered display. "Ten minutes later the display was resumed in the form of parallel bars of light travelling towards the ship from about four points on the port bow, leaving her on the starboard quarter. The bars passed the ship at regular intervals of about 3 sec. Then a brighter patch was seen ahead, and as this was approached it was seen to be an area of jumbled blotches of light, the parallel bar formation having ceased. As this area was watched the profusion settled itself out and began to resemble a slowly revolving wheel. Each 'spoke' of the wheel was disjointed, consisting of several separate bars of light, the component bars decreasing in intensity of light with distance from the centre of rotation. The spokes were not straight, but curved slightly in a clockwise direction. The centre round which all revolved was a brighter spot which showed no movement. This was passed by the ship at a distance of about a mile to port. The spokes could also been seen extending far off on the starboard side, not coming to an abrupt end but fading away in the distance. The wheel showed two types of motion, that of rotation and an outward rippling movement. In addition, each spoke flashed regularly in sections so that not all of it would be visible at once. The flashing occurred at a constant rate of one per second. A number of other wheels were subsequently seen, for the most part separately but a few were intermingled. At times wheels could be seen all around the ship. Each wheel appeared to keep a fixed position relative to the others. As each wheel passed astern it became distorted and indistinct, finally appearing as just a glow in the distance. All the wheels that were seen near the ship had the same character as the one described above, with visible centres and curved spokes. A difference of opinion arose as to which way the first wheel was rotating. It appeared to Mr. Youngs to be turning anticlockwise, with some distant bars moving clockwise, while the Chief Officer and the two other apprentices said the direction was clockwise. Independent opinions were therefore sought from three members of the watch, each of whom gave the movement as anticlockwise." (R32)

X36. July 2, 1956. Persian Gulf. Two clockwise wheels. (R33)

X37. February 24, 1957. Gulf of Siam. An anticlockwise wheel with curved spokes. A second wheel appeared 3 minutes later. After 3 more minutes two oppositely rotating wheels with close centers were seen. Bright flashes appeared where the spokes crossed. (R34)

X38. April 10, 1958. Persian Gulf. Two clockwise wheels that seemed like waves of light 3 feet above the sea. (R35)

X39. April 14, 1958. Persian Gulf. Many confused wheel displays. The apparent direction of rotation of some wheels could be changed by looking away and then back again. (R35)

X40. September 27, 1959. East Indian Archipelago. Moving parallel bands preceded a wheel display. (R36) See GLW2-X6 for description.

X41. September 10, 1960. South Pacific off Tocopilla, Chile. Lines spinning in a circular movement. (R37)

X42. September 9, 1961. Indian Ocean. Three phosphorescent wheels, each 80 feet in diameter, all moving northeast. (R38)

X43. May 11, 1962. South China Sea. An anticlockwise wheel changed to clockwise at its peak of brightness. (R39)

X44. May 7, 1963. Persian Gulf. Straight, moving band display changed into two phosphorescent wheels. (R40)

X45. June 17, 1963. Gulf of Siam. (R40)

X46. November 22, 1967. Malacca Strait. (R41)

X47. March 23, 1968. Arabian Sea. Anticlockwise wheel changes to clockwise. (R42)

X48. April 23, 1968. Persian Gulf. Phosphorescent wheels with S-shaped spokes seen rotating in both directions. (R43)

X49. April 13, 1970. South China Sea. Six or seven phosphorescent wheels seen, all rotating clockwise at the same speed simultaneously. (R44)

X50. April 22, 1971. Gulf of Siam. Two wheels observed rotating anticlockwise. Three minutes later, they changed to clockwise. Later, they spun in opposite directions. (R45)

X51. February 10, 1973. Manila Bay. A phosphorescent wheel seemed to rotate anticlockwise on the near side, clockwise on the far side. (R46)

X52. January 3, 1976. Southern North Pacific. Rotating circles (?) with radii of 3 nautical miles. (R47)

X53. March 27, 1976. Gulf of Siam. "At 1917 GMT pulsating bands of parallel light were observed in the sea moving towards the vessel from 045°T. After two to three minutes the bands took on a definite spoke formation, the centre of which was not seen but lay in the direction of 315°T. The spokes passed the vessel at an ever-increasing rate, two spokes per second at the fastest. At this time they were about 22 metres in width and there was 22 metres between each spoke. The light given off from the spokes was white to light green in colour, it increased in intensity with the speed of rotation. The direction of rotation was clockwise. By 1925 the centre of the spokes had shifted from 315° to 360°T and gradually reverted back to advancing bands of parallel light. Shortly after this the parallel bands gave way to a counter-clockwise spoke rotation. This was observed in a direction centred along 315°T from the vessel, the spokes moved across the bow to 045°T, at which point they became parallel bands which diminished in intensity. By 1934 they had completely disappeared." (R47)

X54. February 14, 1977. Malacca Strait. Clockwise phosphorescent wheel display with curved spokes convex in the direction of rotation. Changed to a V-type display. (R48) See description in GLW12.

X55. April 8, 1978. Arabian Sea. Greenish anticlockwise wheel. Duration: 10 minutes. A second wheel was seen 2 hours later. (R49)

X56. November 6, 1978. South China Sea. "At 1750 GMT bioluminescence in the form

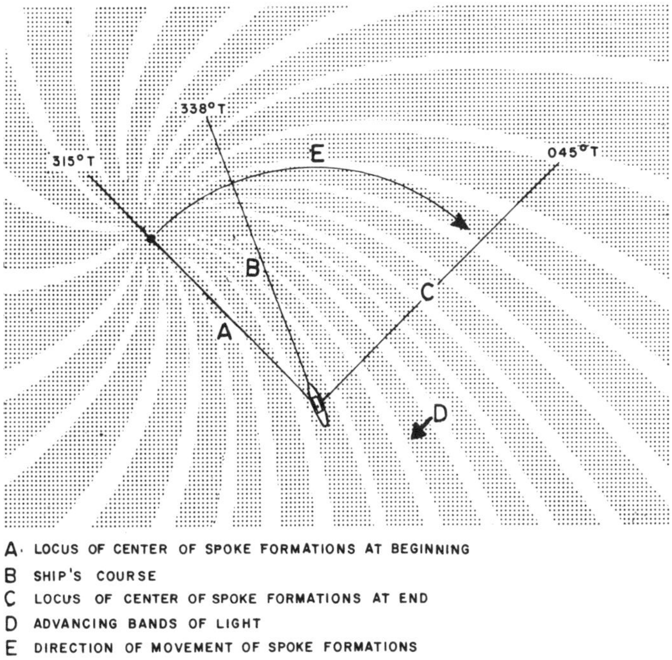


*Wheel phase of a complex sequence of moving bars and wheels observed in the Gulf of Siam*

of a system of rotating bands with no central hub visible was observed. The phenomenon, which was observed on both sides of the vessel, had a milky-white appearance and pulsated at the rate of two to three flashes every second. On the port bow the bands were observed about 20° off the bow, moving towards the vessel and rotating in a clockwise direction. They disappeared from view they reappeared abaft the beam coming into the vessel amidships. On the starboard bow the bands rotated anti-clockwise and moved in towards the vessel. After moving into the vessel's side the bands sheered off astern then disappeared altogether. The phenomenon was observed for about 10 minutes." (R50)

X57. March 6, 1980. Arabian Sea. "At 1552 GMT bioluminescence in the form of diffused white light in 'whirlpool' and 'cartwheel' formations was observed; within 3 minutes it completely encircled the vessel and extended to the horizon. The 'cartwheel' formations were brightest at the centre with a halo effect surrounding the outer edges. As the vessel passed over 2 such formations the 'spokes' were estimated to be 2-2 1/2 metres in width and the entire concentration, which was more than the width of the vessel (approximately 27 metres), was observed on both sides of the bridge-wing simultaneously. The 'whirlpool' formations, with a distinct central hub, varied from 1 1/4 to 2 metres in width and from 1 to 15 metres in length. The phenomenon was observed for 40 minutes." (R51)

X58. No date given. Persian Gulf. (R52)

X59. No date or place given. A wheel of milky white mist 30 feet wide and 30 feet apart and 8 feet deep passed at the rate of two per second. (R53)

X60. General observation. Lights wheels are mostly confined to the continental shelves of Asia and Middle East. They may be caused by earth tremors. (R54)

X61. General observations similar to X60. (R55)

X62. General observation. Persian Gulf. Fire circles seen flitting over the surface of the water at 100 miles per hour. (R56, R57) Classification is ambiquous here. (WRC)

References

R1. Pringle, J. Eliot; "Report of an Unusual Phenomenon Observed at Sea," Nature, 20:291, 1879. (X1)

R2. Brace, Lee Fore; "Strange Phenomenon," Knowledge, 4:396, 1883. (X2)

Obviously a pseudonym. (WRC)
R3. Kalle, Kurt; "Wheels of Light," Sea Frontiers, 15:116, 1969. (X3)
R4. Patterson, S. C.; "Display of Phosphorescence," Royal Meteorological Society, Quarterly Journal, 33:294, 1907. (X4)
R5. "A Remarkable Atmospheric Phenomenon Observed at Sea," Geographical Journal, 61:66, 1923. (X5)
R6. Nature, 86:90, 1911. (X6, X7)
R7. "Curious Light Phenomenon of the Indian Seas," Scientific American, 106:51, 1912. (X7, X8)
R8. "Phosphorescent Wheels," Marine Observer, 20:229, 1950. (X8, X10)
R9. "Mysterious Luminous Phenomenon at Sea," Scientific American, 121:551, 1919. (X9)
R10. Tydeman, G. F.; "Phosphorescent Wheels," Marine Observer, 9:32, 1932. (X10)
R11. Cartmer, G. E.; "Phosphorescence," Marine Observer, 1:54, 1924. (X11)
R12. Brisley, P. L.; "Phosphorescence," Marine Observer, 3:40, 1926. (X12)
R13. Brooks, E. G.; "Phosphorescent Wheel," Marine Observer, 4:169, 1927. (X13)
R14. Wallace, W. K.; "Phosphorescent Wheels," Marine Observer, 4:231, 1927. (X14)
R15. Duncan, S. Sinclair; "Phosphorescence," Marine Observer, 5:245, 1928. (X15)
R16. Smith, H. T.; "Phosphorescence of the Sea," Marine Observer, 8:230, 1931. (X15, X17)
R17. Whitehead, J. S.; "Phosphorescence," Marine Observer, 7:187, 1930. (X16)
R18. Hocking, R. W.; "Phosphorescent Wheel," Marine Observer, 7:240, 1930. (X17)
R19. Jackman, C.; "Phosphorescent Beams, Marine Observer, 10:122, 1933. (X18, X19)
R20. Glover, F. D.; "Phosphorescent Wheel," Marine Observer, 20:6, 1950. (X20)
R21. Maclaren, J. B.; "Phosphorescent Wheel," Marine Observer, 20:139, 1950. (X21)
R22. Paul, W. B. B.; "Phosphorescence," Marine Observer, 21:75, 1951. (X22)
R23. Henney, A., et al; "Phosphorescent Wheels," Marine Observer, 24:73, 1954. (X23, X24)
R24. Naismith, A.; "Phosphorescence," Marine Observer, 24:137, 1954. (X25)
R25. Last, J. C. W., and Naysmith, G. A.; "Phosphorescent Wheel," Marine Observer, 24:206, 1954. (X26, X28)
R26. Rodewald, M.; "Phosphorescent Wheel," Marine Observer, 24:233, 1954. (X27)
R27. Mitchinson, E. L.; "Phosphorescence," Marine Observer, 25:93, 1955. (X29)
R28. Weston, D. L.; "Phosphorescence," Marine Observer, 26:7, 1956. (X30)
R29. Baxter, R. R., and Gilchrist, J. A.; "Phosphorescent Wheels," Marine Observer, 26:78, 1956. (X31, X32)
R30. Henney, A.; "Phosphorescent Wheel," Marine Observer, 26:138, 1956. (X33)
R31. Nimmo, W.; "Phosphorescence," Marine Observer, 26:199, 1956. (X34)
R32. Evans, H.; "Phosphorescent Wheels," Marine Observer, 27:90, 1957. (X35)
R33. Bruce, S.; "Phosphorescence," Marine Observer, 27:142, 1957. (X36)
R34. Thomas, J. H.; "Phosphorescent Wheels," Marine Observer, 28:11, 1958. (X37)
R35. Bradley, E. E., and Armstrong, G. R.; "Phosphorescent Wheels," Marine Observer, 29:54, 1959. (X38, X39)
R36. Rutherford, W.; "Phosphorescent Wheel," Marine Observer, 30:128, 1960. (X40)
R37. Bryant, R. B.; "Phosphorescence," Marine Observer, 31:123, 1961. (X41)
R38. Clarke, H. S.; "Phosphorescent Wheels," Marine Observer, 32:120, 1962. (X42)
R39. Champion, G. S. H.; "Phosphorescent Wheel," Marine Observer, 33:68, 1963. (X43)
R40. Harrison, P. P. O., and Francis, G. G.; "Phosphorescent Wheels," Marine Observer, 34:69, 1964. (X44, X45)
R41. McIntosh, D. D.; "Luminescence," Marine Observer, 38:180, 1968. (X46)
R42. Wray, J. E.; "Phosphorescent Wheel," Marine Observer, 39:13, 1969. (X47)
R43. Brown, A. C.; "Phosphorescent Wheels," Marine Observer, 39:66, 1969. (X48)
R44. Brown, E. M.; "Phosphorescent Wheels," Marine Observer, 41:59, 1971. (X49)
R45. Fyfe, T.; "Phosphorescent Wheels," Marine Observer, 42:61, 1972. (X50)
R46. Thomas, M. G.; "Phosphorescent Wheel," Marine Observer, 44:20, 1974. (X51)
R47. Rowntree, C., et al; "Bioluminescence," Marine Observer, 47:17, 1977. (X52, X53)
R48. Banna, J.; "Bioluminescence," Marine Observer, 48:16, 1978. (X54)

R49. Sudarshan, R. D.; "Bioluminescence," Marine Observer, 49:69, 1979. (X55)
R50. Brown, D. G.; "Bioluminescence," Marine Observer, 49:189, 1979. (X56)
R51. Messinger, P. A.; "Bioluminescence," Marine Observer, 51:13, 1981. (X57)
R52. Palmer, A. R.; "The White Water," Nature, 104:563, 1920. (X58)
R53. "Luminous Wheels Puzzle Seamen," New Scientist, 33:447, 1967. (X59)
R54. "Do Earthquakes Cause Glowing 'Wheels' on the Sea?" New Scientist, 10:528, 1961. (X60)
R55. Turner, R. J.; "Marine Bioluminescence," Marine Observer, 36:20, 1966. (X61)
R56. "Mysteries of the Persian Gulf," Scientific American, 76:331, 1897. (X62)
R57. "Mysteries of the Persian Gulf," English Mechanic, 64:516, 1897. (X62)

# GLW5 Expanding Phosphorescent Rings

Description. Whitish or greenish ripples of phosphorescence expanding from a central point at high speeds. Several centers may be active at the same time with different velocities of expansion. Ripples are usually circular, but one elliptical observation exists. This phenomenon appears in the same regions as the wheel-type display.

Data Evaluation. A rare phenomenon with only a handful of observations. Rating: 2.

Anomaly Evaluation. The simple geometry would suggest a simple disturbance that triggers bioluminescence; but the expanding rings travel too fast for water waves and too slowly for sound waves in water. The unknowns include: light source (usually assumed to be bioluminescence), and the origins of the geometrically precise, repetitive figures. Rating: 2.

Possible Explanations. Interference patterns from seismic sources are the stimuli most frequently suggested, with bioluminescent marine organisms as the basic light producers. This explanation is strained by the observation of multiple sources synchronously emitting ripples. Electromagnetic action remains a possibility as does cooperative biological activity similar to that in tropical firefly displays (BIB).

Similar and Related Phenomena. Other marine phosphorescent displays (GLW), synchronized firefly displays (BIB), flashing volcanic arcs (GLD9-X2).

## Examples of Expanding Phosphorescent Rings

X1. April 17, 1957. Persian Gulf. "...a strange effect was noticed on the sea surface about 2 miles away on the port bow. On approaching the area large bands or ripples of faint light appeared to be travelling in a confused way across the surface. Further inspection showed that these bands were spreading out from several central points distributed fairly evenly over an area about 2 miles across, in the same way as the circular ripples produced by raindrops in a puddle of water. Each band was about 70 yd wide and gave a faint white light. They spread, in unison, from the various central points, at intervals of little more than 1 sec. They travelled at a considerable speed, so that by the time one circular band was formed the previous one was vanishing, at a radius of about 1/2 mile. When first seen it was supposed that the effect was being produced by moon shadows, as the moon was at an altitude of about 35° in the west, and there was a large number of small misty clouds. The clouds were, however, all moving in the same direction and could not have produced the effect of localised circular ripples. There was some general phosphorescence in the water but the presence of moonlight made observation of this difficult." (R1)

X2. October 14, 1960. Gulf of Oman. "At 2130 GMT three very fast outward-moving rings of light were seen emanating suddenly from three separate vortices which were spaced about three cables apart in a straight line parallel to the ship's fore and aft line, and about five cables distant on the starboard side: the positions of the vortices in the water appeared to remain unchanged. The first two appeared almost simultaneous-

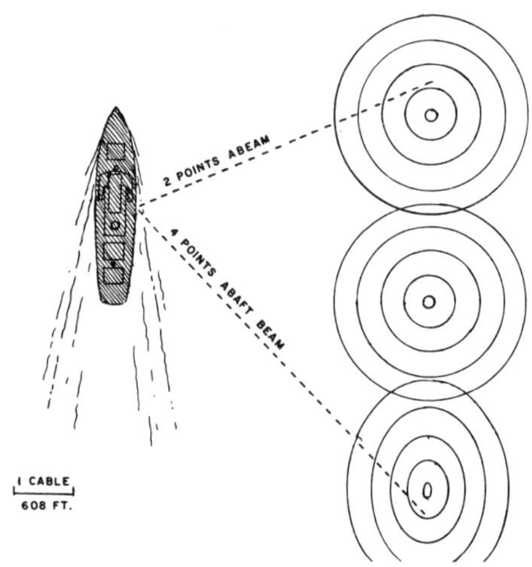


*Three sets of expanding rings—two circular, one elliptical*

ly and produced circular rings, at a frequency of about one second or slightly less. The third appeared a minute or so later and produced elliptical rings which were moving much faster than the others. They appeared about 2 points forward of the starboard beam and disappeared about 4 points abaft the beam. Their disappearance was more or less simultaneous. Although the individual patterns tended to cut across and overlap each other at places, they did not become confused, nor were they stopped by the ship's hull, as the rings could be clearly seen to pass from starboard to port when in line with the bridge. There was bright moonlight, and it was therefore fairly certain that there was no disturbance on the surface at any time, such as might be caused by jumping fish, etc." (R2)

X3. April 23, 1955. Persian Gulf. Concentric circles seen radiating from a center. Full description in GLW4-X31. (R3)

X4. March 12, 1970. South China Sea. Curved bands of luminescence seem to radiate from a center beyond the horizon. (R4)

References

R1. Whyborn, J. A.; "Phosphorescence," Marine Observer, 27:92, 1957. (X1)

R2. Bruce, S.; "Phosphorescence," Marine Observer, 31:184, 1961. (X2)

R3. Baxter, R. R.; "Phosphorescent Wheels," Marine Observer, 26:78, 1956. (X3)

R4. Curphey, F. N.; "Luminescence," Marine Observer, 41:18, 1971. (X4)

# GLW6 Phosphorescent Patches Moving in Circles

Description. Circular patches of light moving in large circular patterns. Several such displays may be active simultaneously.

Background. This phenomenon may be a degenerate form of the phosphorescent wheel (GLW4).

Data Evaluation. One observation of good quality. Rating: 2.

Anomaly Evaluation. See the discussion under marine phosphorescent wheels (GLW4), which is applicable here. It is, however, more difficult to conceive of interfering seismic waves creating patterns of several groups of circling circles of phosphorescence. Rating: 2.

Possible Explanation. This type of display makes seismic interference patterns look less likely, leaving some bizarre form of cooperative action of bioluminescent organisms as a better bet.

Similar and Related Phenomena. Marine phosphorescent wheels (GLW4).

## Examples of Phosphorescent Patches Moving in Circles

X1. May 30, 1956. Persian Gulf. After a display of moving parallel bands. "The parallel line formation disappeared and the patches were seen to be moving fairly slowly in an anticlockwise direction round circles of from 100 ft to 300 ft in diameter. When the vessel passed through a circle the patches of light disappeared on reach-the ship's port side and reappeared in the same formation on the starboard side. After about 20 min the whole effect ceased near the ship but was still seen in the distance on the starboard quarter.... After a further 15 or 20 min more circles appeared, of varying sizes. The patches on the starboard side were small, about 1-2 ft in diameter, and were rotating in circles of about 20 ft diameter. After 10 min these dropped astern and no further phosphorescence was seen." (R1)

References

R1. Arthur, D. E. O.; "Phosphorescence," Marine Observer, 27:92, 1957. (X1)

*Groups of phosphorescent patches moving in circles in the Persian Gulf*

# GLW7 Phosphorescent Spinning Crescents

Description. A bizarre phosphorescent display consisting of rotating crescents of light centered on the ship and sweeping toward the ship or rotating around it. Observed on both sides of the observing ship, the crescents may reverse their directions of rotation at some point in their trajectory. The display characteristics seem to be bilaterally symmetric; i.e., mirror images bisected by the ship's axis. The ship's radar may influence this phenomenon.

Data Evaluation. Several observations, all slightly different, possess enough similar features to be collected in this category. The phenomenon is variable and not clear cut. Rating: 2.

Anomaly Evaluation. Most marine phosphorescent displays are not ship-centered. Here, some influence emanating from the ship seems to stimulate the phenomenon. Whatever the mechanism involved, it is not at all obvious how it creates this strange moving pattern of spinning crescents. Rating: 2.

Possible Explanations. One or more of three ship-created 'forces' may be involved: (1) the radar; (2) the engine noise transmitted through the water; and (3) the physical disturbance of the water (eddies, etc.) caused by the ship's motion. The last influence can probably be eliminated because the phenomenon commences far from the ship's wake. The engine noise is an unlikely candidate because the effect is rare in a region plied by thousands of ships with engines. This leaves the ship's radar as an attractive possibility, perhaps inferring that other phosphorescent displays may involve electromagnetic stimulation, perhaps of specific frequencies and/or repetition rates, natural or artificial. Certainly seismic tremors cannot account for all facets of these complex, ship-centered displays.

Similar and Related Phenomena. All other marine phosphorescent displays, especially GLW10.

Examples of Spinning Crescents

X1. November 30, 1951. Gulf of Oman. "The ship's radar apparatus had been switched on with a view to checking her position, when, in the same instant that this gear became operative, most brilliant boomerang-shaped arcs of phosphorescent light appeared in the sea, gyrating in a clockwise direction to starboard and anticlockwise to port, but all sweeping inwards towards the ship from points situated from five to six points on either bow and some two miles distant, and conveying the impression that they ricocheted from each other on meeting at the ship's bows and then turned and travelled away astern to similar points which were equidistant on either side and about four points on each quarter." (R1)

X3. July 11, 1980. Malacca Strait. "At approximately 2200 GMT a phenomenon similar to phosphorescent wheels was observed. The phenomenon did not seem to build up slowly, but just appeared. It took the form of curved lines of uniform crescent shape which were horizontal to the sea surface, moving in a circular path around the ship starting from just forward of the bow. The speed of the lines was hard to estimate but was in the order of three lines passing a point every second. This high speed gave the impression of pulsating light. The lines were about 100 metres long, 0.5 metres wide and appeared to some to be light green in colour although to the others they appeared silvery-white. There was some disagreement about the position of the phenomenon as to whether the lines were on the

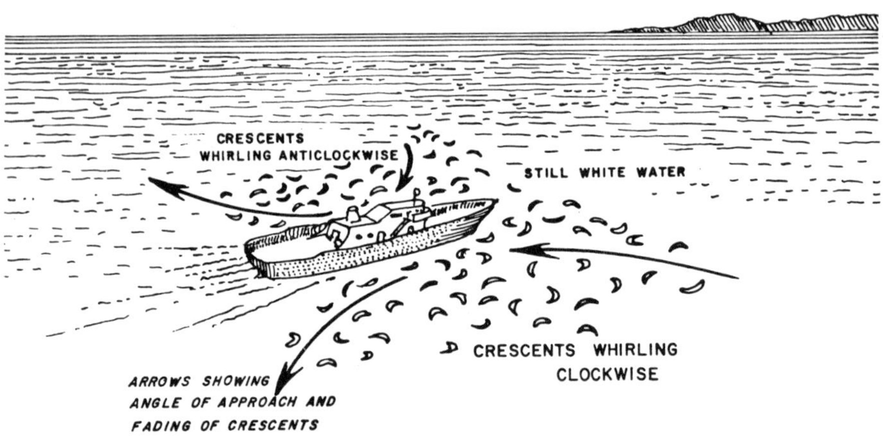


*Radar-initiated display of rotating crescents*

X2. July 10, 1979. Arabian Sea. "At 1200 GMT large patches of milky-grey bioluminescence were observed; the patches appeared to form circular patterns resembling cartwheels, some of the configurations, however, did not have the central hub, see sketch. The patches pulsated at regular intervals (3 or 4 times per second). They moved in an anticlockwise direction until about 3 points abaft the beam where the direction of movement was reversed. On the beam they appeared to be at eye level, at all other times they were just above the surface of the water. The average size of the 'wheels' was 35 metres. The phenomenon was observed for about 10 minutes, the pattern of the 'wheels' in the most dense areas was confused." (R2)

*Groups of phosphorescent crescents were observed rotating about several different centers in the Arabian Sea. The groups averaged 35 meters in diameter.*

surface of the sea or above it. The phenomenon seemed to be centred about the ship and could not be seen for more than 100 metres on either side and astern, also the whole system appeared to be moving along with the ship. After about 5 minutes the lines faded away over a period of 1/2 - 1 minute." (R3)

References

R1. Baker, F.G.; "Phosphorescence," Marine Observer, 22:190, 1952. (X1)
R2. Penman, B.; "Bioluminescence," Marine Observer, 50:114, 1980. (X2)
R3. Lardler, D.A.; " Bioluminescence," Marine Observer, 51:116, 1981. (X3)

# GLW8 Zigzag Phosphorescent Flashes

Description. Zigzag flashes of phosphorescence moving lightning-like across the surface of the sea.

Data Evaluation. So far, only a single observation. Rating: 3.

Anomaly Evaluation. The propagation of simple patterns of light by bioluminescent organisms, either through cooperative action or external stimulus, is not an unreasonable phenomenon. A simple circle of expanding light, for example, would be understandable. But the lightning-like propagation of well-defined zigzag flashes is hard-to-explain. Rating: 2.

Possible Explanations. Cooperative bioluminescent activity is about the only possibility, unless some form of low-altitude electromagnetic activity stimulus exists.

Similar and Related Phenomena. All other marine phosphorescent displays, particularly the te lapa (GLW11), heat lightning (GLL).

Examples of Zigzag Phosphorescent Flashes

X1. August 30, 1949. Gulf of Cutch. "In vicinity 22° 30'N., 68° 10'E., at mouth of Gulf of Cutch, brilliant phosphorescence in lightning-like zig-zag form was observed in patches of about a square mile each. These patches seemed to have a slight directional movement WNW, but the zig-zag phosphorescence flashed from left to right in a WNW-ESE direction." (R1)

References

R1. Spencely, T.A.; "Phosphorescence," Marine Observer, 20:139, 1950. (X1)

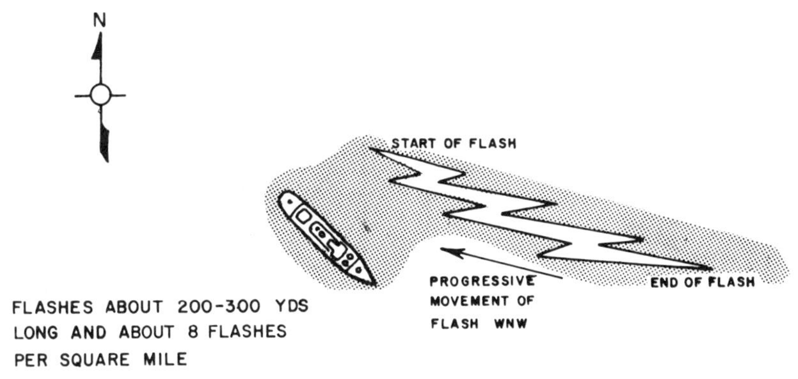


*Lightning-like flashes of phosphorescence recorded in the Gulf of Cutch, 1949*

# GLW9 White Water or Milky Sea

Description. A white phosphorescent display consisting of a soft but often brilliant luminous sea stretching for many miles. The milky sea has often been compared to a field of snow, during which the horizon may seem to disappear much as it does during arctic white-outs. The sea's surface often seems subdued during this eerie spectacle, although the motion of the ship indicates that this is not so. In some instances, the light seems to come from great depths as if from huge underwater searchlights. At other times, there seems to be a luminous fog or mist above the water surface. Milky seas can appear suddenly over a wide region. They may wax and wane over periods of several hours. The geographical distribution of milky seas is much wider than for the phosphorescent wheels, being found in all oceans at any season. Nevertheless, they seem to concentrate in the Arabian Sea during summer months.

Background. Milky seas should not be confused with ordinary bioluminescence of the sea which occurs when water is agitated by a vessel, fish, waves, etc.

Data Evaluation. Many excellent observations over two centuries. Rating: 1.

Anomaly Evaluation. The prevailing assumption is that the milky sea is produced by bioluminescent organisms, possibly bacteria, but no specific organism has ever been implicated. No one has suggested the names of any species known to be present in immense numbers and capable of producing bright, sustained light for hours. Further, water drawn from a milky sea frequently shows no bioluminescence whatever. To compound the problem, some milky seas seem lit from the depths while others derive from a luminous mist above the surface. Manifestly, there is much to learn about this phenomenon. Rating: 2.

Possible Explanation. Despite the deficiencies noted above, bioluminescence is certainly the most likely explanation.

Similar and Related Phenomena. All other marine phosphorescent displays (GLW), luminous fogs and mists (GLA21).

Examples of White Water or Milky Sea

X1. July 31, 1785. North Atlantic Ocean. "About a quarter past 7 P. M. the sea was observed to be remarkably white. The sky was every where clear, except around the horizon, for about 15$^{o}$, where it was covered with a dark haze, as is usual in such latitudes. The whiteness gradually increased till past eight. The sea then was as high coloured as milk, very much resembling the milky-way in the heavens; the luminous appearance of the sea resembling the brighter stars in that constellation. It continued in this situation till past midnight, and only disappeared as day-light advanced. The whiteness prevented us from being able to see either the break or the swell of the sea, although both were considerable, as we knew from the motion of the ship, and the noise. There was much light upon deck, as we could discern all the ropes much more distinctly than usual. We drew several buckets of water, in which, even when at rest, there appeared a great number of luminous bodies." (R1)

X2. September 7, 1826. Gulf of St. Lawrence. "The night was star lit, but suddenly the sky became overcast in the direction of the high land of Cornwallis county, and a rapid, instantaneous and immensely brilliant light, resembling the Aurora Borealis, shot out of the hitherto gloomy and dark sea on the lee bow, and was so vivid that it lighted every thing distinctly, even to the mast head. The mate, having alarmed the master, put the helm down, took in sail and called all hands up. The light now spread over the whole sea between the two shores; and the waves, which before had been tranquil, now began to be agitated. Capt. B. describes the scene, as that of a blazing sheet of awful and most brilliant light. A long and vivid line of light, superior in brightness to the parts of the sea not immediately near the vessel, showed us the base of the high, frowning and dark land abreast of us; the sky became lowering and intensely obscure. The oldest sailors on board had never seen anything of the kind to compare with it, except the captain, who said that he had observed something of the kind in the Trades. Long tortuous lines of light in a contrary direction to the sea, shewed us immense numbers of very large fish darting about as if in consternation at the scene. The spirit-sail yard and mizen-boom were lighted by the reflection as though gas lights had been burning immediately under them; and until just before day break, at four o'clock, the most minute objects in a watch were distinctly visible. Day broke very slowly, and the sun rose of a fiery and threatening aspect. Rain followed." (R2)

X3. 1871. South Atlantic Ocean. (R3)
X4. January 26, 1874. Arabian Sea. (R4)
X5. Fenruary 9-13, 1880. Arabian Sea. The sea resembled a field of snow on a clear night. (R5-R7)
X6. February 1, 1881. Indian Ocean. (R8)
X7. August 22, 1898. Indian Ocean. (R9, R10)
X8. May 1899. Indian Ocean. (R11)

X9. March 18, 1924. Indian Ocean. "A remarkable combination of submarine earthquake and phosphorescent seas has been reported to the Hydrographic Office. The stirring up of the ocean brought so many light-emitting organisms to the surface that the British steamer Trefusis on her way from Aden to Columbo seemed to be steaming across a snow-covered plain. It was during the middle watch of March 18, a dark night, quiet sea, and as usual in the northern Indian Ocean, the wake of the ship rather brilliantly phosphorescent while the rest of the sea was dark. Just before four o'clock a distant tremor was felt, followed quickly by others, as if a series of mines were being set off at great depths. Immediately, great patches of light rose to the surface and spread in all directions until for miles around the ship the whole ocean was a brilliant foaming glitter of phosphorecence, the pale blue glow lightning the deck of the vessel with a wierd light." (R12)

X10. August 27, 1925. In the East Indies. The whole sea was suddenly illuminated by a soft, lambent light. (R13)
X11. February 9, 1926. China Sea. The sea appeared to be illuminated from below. (R14)
X12. October 4, 1926. South Atlantic Ocean. A milky sea disappeared during two stormy periods, reappearing afterwards. (R15)
X13. August 12, 1928. Arabian Sea. (R16, R17)
X14. August 14, 1928. Banda Sea, South Pacific Ocean. (R16)
X15. July 31, 1929. Arabian Sea. (R17)
X16. August 2, 1929. Gulf of Aden. (R18)
X17. August 20, 1930. Gulf of Aden. (R19)
X18. August 28, 1930. Arabian Sea. (R19)
X19. June 10, 1931. Persian Gulf. (R20)
X20. August 28, 1933. South Pacific Ocean. The sea was lit from below---not surface phosphorescence. (R21)
X21. August 28, 1946. Arabian Sea. The sea appeared calm, but this was contradicted by the motion of the vessel. (R22)
X22. February 1, 1952. Arabian Sea. (R23)

X23. August 15, 1953. Arabian Sea. "At 1823 milky phosphorescence was observed of a slight greenish colour, which had the same light value as the sky so that the horizon was difficult to discern. At 1830 the phenomenon brightened suddenly to a full milk-white. Spray looked darker than the surrounding sea. At 2004 the luminous appearance of the sea ceased abruptly, although a glow could be seen in the sky ahead. The appearance of the sea at 1823 and 1830 was repeated at 2023 and 2045 respectively. The full milk-white appearance reached maximum brightness at 2103 when the horizon was very sharply defined. At 2133 the brightness commenced to fade again until the light values of sea and sky were the same. More variations of brightening and fading occurred again from 2200 to 2203, but at 2219 the phosphorescence brightened so that the horizon was easily discernible, and these conditions remained much the same until 2320 when it all ceased abruptly.... It was noted that the phosphorescence was not carried aboard by spray, this appeared dark against the background of the sea. During the brightest periods the phosphorescence completely masked the wave crests and gave the sea an appearance of unnatural calm at once belied by wind, motion of the ship and spray coming aboard." (R24)

X24. August 19, 1958. Arabian Sea. (R25)
X25. February 8, 1962. Indian Ocean. One could read a book on the bridge by the light of the sea. (R26)
X26. July 30, 1971. Arabian Sea. (R27)
X27. January 6, 1976. Indian Ocean. (R28)
X28. January 23, 1976. Indian Ocean. (R28)
X29. January 25, 1976. Indian Ocean. (R28)
X30. July 22, 1976. Arabian Sea. (R29)
X31. July 30, 1976. Indian Ocean. (R29)

X32. August 4, 1977. Indian Ocean. "At 1735 the vessel encountered a very large area of milky sea, the area stretched as far as the horizon. The intensity was so great that the deck appeared to be just a black shadow. There was an apparent increase in humidity and a small number of fish together with a small amount of seaweed were observed. During the phenomenon the Radio Officer reported a decrease in signal strength on HF and static on MF frequencies. The intensity of the phenomenon decreased for five minutes at about 1800, thereafter, it increased again and was observed until 1911." (R30) Radio reception may be suppressed during low-level auroras (GLA20-X2).

X33. August 12, 1977. Timor Sea, South Pacific Ocean. (R30)

X34. September 6, 1977. Indian Ocean. "At 1830 GMT an area of bioluminescence, which had the appearance of white sea fog, was observed to the west of the vessel. About 20 minutes later, when the vessel entered the affected area, a diffuse milky light effect was observed just above the sea surface. The sea was clearly visible, but the white horses were noticably reduced in brightness. No effect was observed on the bioluminescence when the Aldis lamp was switched on. At its brightest the phenomenon was sufficient to illuminate the clouds. There was no moon that night, the luminous effect was, therefore, due entirely to the phenomenon. The bioluminescence began to reduce in intensity at about 2100 and was no longer observed about half an hour later. A sample of sea-water was taken and was observed to be clear and normal. It was also noted that the water showed no response when exposed to a fluorescent ultra-violet light." (R30)

X35. September 16, 1977. Arabian Sea. (R30)

X36. General observation. White water phenomenon sometimes seems like a luminous fog in which sea and sky seem to join, and all sense of distance is lost. (R31)

X37. General observation. Observations of white water seem to be concentrated in the Arabian sea during August. (R32)

X38. General observations. (R33)

References


R1. Buchanan, Francis; "Account of an Extraordinary Appearance of the Sea...," Edinburgh Philosophical Journal, 5:303, 1821. (X1)

R2. "Phosphorescence of the Sea in the Gulf of St. Lawrence," American Journal of Science, 1:18:187, 1830. (X2)

R3. Ressorp; "Luminous Sea," Knowledge, 2:438, 1882. (X3)

R4. Vacher, H. P.; "A Luminous Sea," Knowledge, 2:282, 1882. (X4)

R5. "The Milky Sea," Popular Science Monthly, 17:573, 1880. (X5)

R6. "A Milky Sea," English Mechanic, 31:153, 1880. (X5)

R7. "A Luminous Sea," Scientific American, 42:359, 1880. (X5)

R8. Pidgeon, Dan.; "A White, or Milky Sea," Nature, 58:520, 1898. (X6)

R9. Barrett, James W.; "A White Sea," Nature, 58:496, 1898. (X7)

R10. "White Water," English Mechanic, 77:149, 1903. (X7)

R11. Gardiner, J. Stanley; "The White Water," Nature, 104:563, 1920. (X8)

R12. Science, 59:sup xiv, June 6, 1924. (X9)

R13. Edgell, J. A.; "Illumination of the Sea," Marine Observer, 3:132, 1926. (X10)

R14. Byers, Captain; "Sea Illuminated As If from Below," Marine Observer, 4:25, 1927. (X11)

R15. Taylor, F. C.; "Phosphorescence and Electrical Storm," Marine Observer, 4:190, 1927. (X12)

R16. Stout, G. L., et al; "Phosphorescence," Marine Observer, 6:170, 1929. (X13, X14)

R17. Smith, H. T.; "Phosphorescence of the Sea," Marine Observer, 8:230, 1931. (X13, X15)

R18. Brown, A. J.; "Phosphorescence," Marine Observer, 7:169, 1930. (X16)

R19. Willcon, J. H., and Griffin, G. A.; "Phosphorescence," Marine Observer, 8:165, 1931. (X17, X18)

R20. Dewar, M. D.; "Phosphorescence," Marine Observer, 9:114, 1932. (X19)

R21. Almond, J. G.; "White Water," Marine Observer, 11:90, 1934. (X20)

R22. Nettleship, J. C.; "Phosphorescence," Marine Observer, 17:11, 1947. (X21)

R23. Aplin, J. P.; "Phosphorescence," Marine Observer, 23:7, 1953. (X22)

R24. Readshaw, H. E.; "Phosphorescence," Marine Observer, 24:136, 1954. (X23)

R25. Bolton, F. G.; "Phosphorescence," Marine Observer, 29:111, 1959. (X24)

R26. Johnston, R.; "Phosphorescence," Marine Observer, 33:16, 1963. (X25)

R27. Morrison, J.; "Luminescence," Marine Observer, 42:111, 1972. (X26)

R28. Saunders, D., et al; "Bioluminescence," Marine Observer, 47:17, 1977. (X27-X29)

R29. Howe, D., and Crump, R.; "Bioluminescence," Marine Observer, 47:109, 1977. (X30, X31)

R30. Innes, D., et al; "Bioluminescence," Marine Observer, 48:118, 1978. (X32-X35)

R31. Nature, 67:423, 1903. (X36)
R32. Turner, R. J.; "Marine Bioluminescence," Marine Observer, 36:20, 1966. (X37)
R33. Collingwood, Cuthbert; "On the Luminosity of the Sea," Journal of Science, 4:500, 1867. (X38)

# GLW10 Radar-Stimulated Phosphorescent Displays

Description. The initiation or enhancement of marine phosphorescent displays by radar.

Data Evaluation. The few examples that exist involve somewhat different circumstances, but a tentative causal connection seems justified. Rating: 2.

Anomaly Evaluation. Since there is no recognized connection between radar and marine phosphorescence, this phenomenon has more than usual interest. Rating: 2.

Possible Explanation. The bioluminescent organisms which supposedly cause marine phosphorescent displays may, in fact, be stimulated by electromagnetic fields. If this is the case, this phenomenon supports the suggestion that some displays may be controlled by natural electromagnetic forces.

Similar and Related Phenomena. The detection of marine phosphorescence by radar (GLW14), the resemblance of marine phosphorescent displays to some auroral phenomena. (GLA4, GLA21)

## Examples of Radar-Stimulated Phosphorescent Displays

X1. February 9, 1953. Gulf of Oman. "Between 0130 and 0200 white patches of light were observed on the sea surface. Milky-white patches were first noticed on the starboard beam about 2 cables away and appeared to 'flash' about once every second. Later they moved closer to the ship, being as bright as a phosphorous patch, although there was no indication of phosphorescence in the water even when the ship's wake broke into the patches. The patches had many different movements, each one continuing for a minute or so---rotary, clockwise and anticlockwise---towards the ship in waves and away from it in waves parallel to the ship's course. During the entire time of observation the period of reaching maximum brilliance and fading was about 1 sec, giving a regular flashing appearance. At 0152 the waves reached their maximum brilliance, appearing to travel from the starboard quarter to the port bow. On switching off the radar the phenomenon ceased abruptly close to the ship, but it was still faintly discernible on the port beam about 2 cables away. At 0157 the radar was switched on again, the phenomenon did reappear close to the ship but only faintly, and soon disappeared altogether. Nothing was observed on the radar screen during this time that was out of the ordinary." (R1)

X2. March 4, 1955. Solomon Islands. "At 1707 GMT the vessel was bound north and in the vicinity of Savo Island when it passed through an area of oval patches of phosphorescence. An amber light appeared to come from beneath the surface which was found to be pulsating 94 times per minute in direct time to the ship's engine revolutions, but not simultaneously with them. The progressive form of the phenomenon was as follows: 1700 GMT Radar in use, no phosphorescence observed; 1705 First patch sighted on starboard bow, about 1 mile distant, with appearance of breaking reef; 1707 Radar in use, an increasing number of flashes seen, covering a field up to 2 miles from the ship; 1710 Vessel surrounded by flashes of intense light and visibility reduced; 1715 Radar in use, fewer lights, less intense and dying out; 1725 Radar off, no phosphorescence visible; 1726 Radar in use, phosphorescence observed close to, and encircling, ship; 1730 Radar in use, phosphorescence diminishing; 1800 Radar in use, no phosphorescence observed. At no time was any effect visible on the radar screen. The patches were about 50 ft. in length, and the strongest light came from the end furthest away." (R2, R3)

X3. November 30, 1951. Gulf of Oman. The instant the ship's radar was switched on, rotating, crescents of phosphorescence appeared. See GLW7-X1. (R4)

References

R1. Paice, J. M.; "Phosphorescence," Marine Observer, 24:8, 1954. (X1)

R2. Hilder, B.; "Phosphorescence," Marine Observer, 31:184, 1961. (X2)

R3. Hilder, Brett; "Radar and Phosphorescence at Sea," Nature, 176:174, 1955. (X2)

R4. Baker, F. G.; "Phosphorescence," Marine Observer, 22:190, 1952. (X3)

# GLW11 Te Lapa: Underwater Lightning

Description. Streaks of light, flashes, and glowing plaques appearing well below the surface of the ocean, apparently emanating from distant land masses. Supposedly used by Polynesian navigators.

Data Evaluation. One account in a popular publication. Rating: 3.

Anomaly Evaluation. So little is known about this phenomenon that it is difficult to estimate its challenge to current theories. If te lapa is simply bioluminescence due to wave action, it is primarily a curiosity. But some unrecognized process may be involved. Rating: 2.

Possible Explanations. Oceanic internal waves reflecting off land masses might possibly produce deep patterns of bioluminescence.

Similar and Related Phenomena. Some marine phosphorescent displays seem to originate in the depths rather than near the surface. Similar in appearance are the zigzag flashes of surface phosphorescence (GLW8).

Examples of Te Lapa

X1. General observation. Pacific Ocean. A conversation with a Polynesian at sea. "'Of course,' he said rather hesitantly, looking back at me over his shoulder, 'you must know all about te lapa.' I truthfully denied knowing anything about it at all. 'Then look.' Tevake pointed over the side. 'No, not on top, deep down. You see him all same underwater lightning.' The phrase was apt. Streaks, flashes, and momentarily glowing plaques of light kept appearing well below the surface. Tevake explained that te lapa streaks dart out from directions in which islands lie. The phenomenon is best seen eighty to a hundred miles out and disappears by the time a low atoll is well in sight. He stressed that it was quite different from ordinary surface luminescence. Tevake told me it was customary to steer by it on overcast nights." (R1)

References

R1. Lewis, David; "Wind, Wave, Star, and Bird," National Geographic Magazine, 146:751, 1974. (X1)

# GLW12 Moving, V-Shaped Phosphorescent Displays

Description. V-shaped or herring-bone patterns of phosphorescent light moving across the ocean surface. This phenomenon, like the moving parallel bands (GLW2), seems linked generically to the phosphorescent wheels (GLW4).

Background. V-shaped displays may be simply ramifications of moving, parallel band displays.

Data Evaluation. A single, carefully described observation. Rating: 3.

Anomaly Evaluation. Essentially the same as that for the moving, parallel band display. See GLW2. Rating: 2.

Possible Explanations. See discussion in GLW2.

Similar and Related Phenomena. All marine phosphorescent displays, especially the moving parallel bands (GLW2) and the wheel displays (GLW4).

Examples of V-Shaped Phosphorescent Displays

X1. February 14, 1977. Malacca Strait. A wheel-type display appeared to the starboard. "Soon afterwards a change in the pattern was observed. The spokes had formed an inverted V-shape with the apex pointing away from the ship and were rushing towards the ship's starboard side, reappearing on the port side and moving away. The pattern then changed twice more in quick succession; first to an anti-clockwise revolving wheel and then to a clockwise wheel, both times with the centre to starboard. As the vessel steamed past the latter wheel, a further wheel was seen ahead revolving anti-clockwise---the spokes of this wheel overlapping those of the other. The pattern changed twice more; first, back to the V-shape but moving in the opposite direction and finally to an anti-clockwise revolving wheel with the centre to starboard. By this time, the phenomenon was rapidly fading and soon disappeared altogether. The display lasted 10 minutes." (R1)

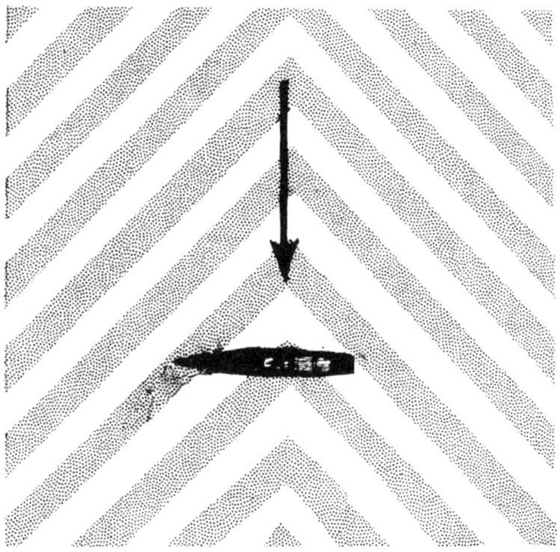

*Phosphorescent Vs sweep past a ship in the Malacca Strait, 1977*

References

R1. Banna, J.; "Bioluminescence," Marine Observer, 48:16, 1978. (X1)

# GLW13 Colored Rays Emanating from Ships

Description. A very rare, multicolored phosphorescent display consisting of rays of light emanating from the observing ship. The rays appear in the water like a horizontal fan and possess various colors.

Data Evaluation. A single observation. Rating: 3.

Anomaly Evaluation. Any multicolored phosphorescent display is anomalous, for almost all other displays are white or whitish-green. The beams radiating from the ship form a structure that is likewise difficult-to-explain. Rating: 2.

Possible Explanations. None.

Similar and Related Phenomena. None.

Examples of Colored Rays Emanating from Ships

X1. June 1, 1959. Arabian Sea. Elliptical flashes of light were observed in the sea. "At the same time the lookout on the forecastle reported coloured rays of light radiating from the stem of the ship, over the surface of the water to a distance of about 20 ft. Some of the colours noted and remembered by the lookout were red, purple, blue and green. The whole phenomenon lasted about 3/4 hour." (R1)

References

R1. Rattray, J.; "Phosphorescence," Marine Observer, 30:64, 1960. (X1)

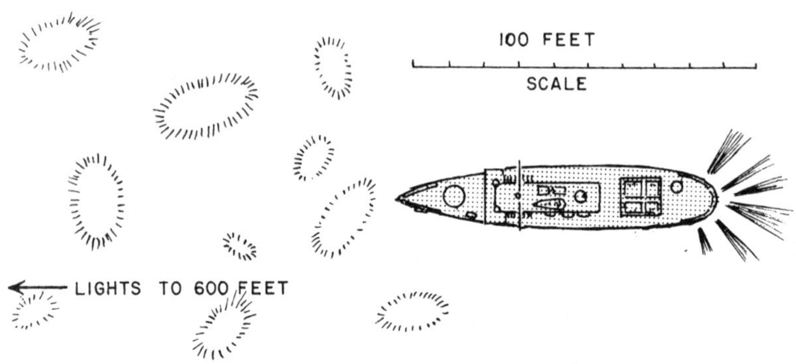


*Arabian Sea, 1959. Flashes of light ahead of the ship, diverging colored rays astern.*

# GLW14 Radar Detection of Phosphorescence

Description. Radar echoes apparently reflected by patches of marine phosphorescence.

Data Evaluation. The modicum of evidence relies upon the geographical coincidence of radar echos and phosphorescence. Since no other obvious targets are evident, marine phosphorescence is blamed. Rating: 2.

Anomaly Evaluation. Marine phosphorescence is generally considered to be a surface or subsurface phenomenon, providing no radar target whatsoever. Rating: 2.

Possible Explanations. Unseen mist or airborne particulate matter may exist in the air above the phosphorescent displays---perhaps caused by the bioluminescent organisms themselves. Bioluminescent displays might involve electromagnetic activity; perhaps with the production of ionization in the air above the phosphorescence. The ionized air might produce a radar echo. Some observers of phosphorescent displays have noted their similarity to some auroral displays.

Similar and Related Phenomena. Radar-stimulated bioluminescence (GLW10), low-level auroras (GLA4), luminous fogs and mists (GLA21), spurious radar echos (GER).

## Examples of Radar Detection of Phosphorescence

X1. September 5, 1954. Equatorial Pacific waters. "We were fixing the ship's position by cross bearings as well as by radar, and I noticed that the radar screen showed what appeared to be a rain squall in the area of phosphorescence. To the eye the light did have rather the appearance of a white squall of fierce wind and rain, except for the illumination. As we got closer the light showed up in horizontal streaks, like sandbanks or submerged reefs, and as we were going to meet the area before we could alter course it was quite alarming. The masses of phosphorescence and the echo on the radar both covered an area of about 2 miles square. The set was Marconi Radiolocator Mark IV, 3-mile range, sweep speed 21 r.p.m. At 0220 the ship met the area, and the nearest patch passed under the ship. Our breaking bow wave showed up dark against the light, which seemed to be at least 2 fathoms down. The long patches seemed to be round in section and had rounded ends, and while most of them were parallel to the horizon, some curved sharply away..... Light rain was falling from an overcast sky, with a lot of fractonimbus about. The ship reached the radar echo at the same time as the phosphorescence, but there was no noticeable increase in precipitation or wind force. The illumination was very brilliant but did not vary or pulsate, nor did it have any proper motion except relative motion caused by the ship's speed of 10 kt. The wind was light, about SE, force 2, with slight sea and swell. Air temp. 78°F, wet bulb 76°, sea temp. 80°. In the circumstances I came to the conclusion that the phosphorescence was actually shown on the radar screen." (R1, R2)

X2. March 18, 1977. North Atlantic Ocean. "Throughout the day spurious echoes had periodically appeared on the radar screen. The echoes, which resembled those of small clusters of fishing vessels, rarely closed the vessel to less than a range of 8 n. mile before disappearing from the screen. At 2200 another such cluster appeared on the radar screen directly ahead at a range of 10 n. mile. This cluster began to spread out on either side of the bow and a number of stronger echoes appeared in it until it gave an impression similar to than expected from a coastline. The echoes behaved as stationary objects and closed to a range of 5.5 n. mile when they again began to spread out until the ship was at the centre of a complete circle of echoes, none of which closed to within less than 5 n. mile of the ship. The radar appeared to be functioning normally throughout. At about this time, the sea took on a milky-white colour and the beam from the Aldis lamp revealed a mass of luminescent organisms each about the size of a flashlight bulb---yet each clearly distinguishable from the other. This phenomenon continued for about 45 minutes after which time both the luminescence and the various spurious radar echoes disappeared." (R3)

X3. June 20, 1977. South Atlantic Ocean. "At 0600 GMT, and for about 40 minutes, a more-pronounced area of bioluminescence was observed, during which time white caps from the wind waves were visible up to about 5-6 n. mile. During this period unusual echoes were observed on the radar screen. What was thought to be a patch of rain was observed moving towards the vessel from the west, against the wind. The patch had a rather distinct edge to it unlike those edges associated with rain areas. When the vessel was observed on the radar screen to be in the centre of the patch, there was no precipitation." (R4)

## References

R1. Hilder, Brett; "Phosphorescence and Radar," Marine Observer, 25:119, 1955. (X1)

R2. Hilder, Brett; "Radar and Phosphorescence at Sea," Nature, 176:174, 1955. (X1)

R3. Richards, A.W.; "Radar Echos and Bioluminescence," Marine Observer, 48:20, 1978. (X2)

R4. Turney, R.J.; "Bioluminescence," Marine Observer, 48:69, 1978. (X3)

# TIME-OF-EVENT INDEX

# PLACE-OF-EVENT INDEX

# FIRST-AUTHOR INDEX

# SOURCE INDEX

# SUBJECT INDEX